Integrated Pest Management on Rangeland

Westview Studies in Insect Biology
Michael D. Breed, Series Editor

About the Book and Editor

Grasslands comprise the largest and most diverse set of ecosystems in the United States and are among the most extensive in the world. Characterized by scanty rainfall, these western grasslands are too dry for crop production and are used almost exclusively for grazing livestock. The grasslands on the western edge of the Great Plains, known as the shortgrass region, support some unique species of plants and animals as well as some common to other rangeland ecosystems. Invertebrate plant pests, native vertebrate foragers, and weeds can adversely affect the efficient management of this economically and ecologically important region. The contributors—ecologists, entomologists, range scientists, vertebrate biologists, and economists—synthesize and review the available information on shortgrass rangelands, offering the first comprehensive treatment of pest management and ecology within this vital ecosystem.

John L. Capinera is professor and chair of the Department of Entomology and Nematology, University of Florida, Gainesville. He is the former head, Department of Entomology, Colorado State University.

Integrated Pest Management on Rangeland

A Shortgrass Prairie Perspective

edited by
John L. Capinera

Routledge
Taylor & Francis Group

LONDON AND NEW YORK

First published 1987 by Westview Press, Inc.

Published 2018 by Routledge
52 Vanderbilt Avenue, New York, NY 10017
2 Park Square, Milton Park, Abingdon, Oxon OX14 4RN

Routledge is an imprint of the Taylor & Francis Group, an informa business

Library of Congress Cataloging-in-Publication Data
Integrated pest management on rangeland.
  (Studies in insect biology)
  Includes index.
  1. Grasses—Diseases and pests—Integrated control—Great Plains. 2. Grasses—Diseases and pests—Great Plains. 3. Range plants—Diseases and pests—Integrated control—Great Plains. 4. Range plants—Diseases and pests—Great Plains. 5. Range management—Great Plains. 6. Range ecology—Great Plains. I. Capinera, John L. II. Series.
SB608.G8I58   1987   633.2   87-14711

ISBN 13: 978-0-367-01381-3 (hbk)
ISBN 13: 978-0-367-16368-6 (pbk)

DEDICATION

It is a pleasure to dedicate this book to a genuine pioneer in the field of rangeland pest management, *Dr. Robert Edward Pfadt.* Begining with employment in 1939 as a field supervisor with USDA in Montana, and with the University of Wyoming in 1940 as a field assistant in entomology, Bob Pfadt has had an unwavering interest in grasshoppers. He has dedicated much of his career to the study of the biology and management of these important rangeland pests. Few facets of grasshopper ecology have not been thoroughly considered by Bob, with the most significant contributions in the areas of insect-plant relationships and efficacious control.

Bob Pfadt has spent most of his career at the University of Wyoming, where he moved through the academic ranks and served as head of the Department of Entomology and Parasitology. His research interests, in addition to grasshoppers, also included livestock pests. Although perhaps most widely known for his universally used textbook, *Fundamentals of Applied Entomology*, Bob has made many important scholarly contributions; one of his early studies, published in 1949, on food plants as a factor in the ecology of migratory grasshoppers, remains a classic in the grasshopper literature. This volume contains a characteristically valuable contribution by Bob, Chapter 12. Now retired, Bob remains active at the University of Wyoming, and continues to pursue his interest in grasshoppers.

CONTENTS

Rangeland Pests and Their Management

Biological and Economic Models for Rangeland Pest Management

Future Developments

PREFACE

In concept, *integrated pest management* (IPM) is the application of ecological principles to the management of pest organisms. In practice, IPM is the application of a pest control method, or methods, within the context of enhanced awareness of economic considerations and ecological impact.

IPM is not a new idea, but a rediscovered concept. Prior to the increased availability and effectiveness of chemical pesticides of the last 40 years, the IPM concept was embraced as common sense. Now that the problems associated with over-reliance on pesticides (e.g., environmental pollution, increased production costs, pesticide resistance) have become so apparent, common sense once again dictates that alternatives be considered, when possible.

Rangeland, much like any other commodity, has a number of pest problems where the IPM approach is appropriate. Unlike many other commodities, however, rangeland is a multiple-use environment. In addition to extensive use for livestock production, rangeland is important in control of soil erosion, water conservation, wildlife production, recreation, and as a source of energy (e.g., natural gas, oil). The per-unit value of rangeland usually is quite low, precluding expensive management decisions. Also, much of the rangeland in the western United States is not privately owned, and so especially susceptible to government policy.

This book attempts to bring together some of the diverse problems and potentials in the rangeland ecosystem, focusing particularly on the western edge of the Great Plains, an area commonly known as the *shortgrass prairie*. It reflects the editor's perspective of IPM -- that there be a sound ecological base, that producer and consumer biology be understood, that a number of management tactics be available, that the economic implications of management decisions be known, and that computer models be important tools in integrating management decisions and assessing impact.

John L. Capinera

xiii

ACKNOWLEDGMENTS

This book is an outgrowth of a symposium, *Integrated Pest Management on Rangeland: State-of-the-art in the Shortgrass Prairie Ecosystem*, held on May 20 and 21, 1986 at the Colorado State University campus, Fort Collins, Colorado. The symposium was sponsored by USDA Regional Project W-161, Integrated Pest Management. This publication is the second in a series of publications sponsored by the Rangeland Subcommittee of W-161; the first was USDA, ARS-50, Integrated Pest Management on Rangeland: State of the Art in the Sagebrush Ecosystem, published in January 1987. I gratefully acknowledge the financial support of W-161, and the cooperation of the many participants of the symposium, some of whom contributed chapters to this book. In particular, W-161 coordinator Gary A. McIntyre was most helpful.

Completion of this book would not have been possible without the tireless efforts of David C. Thompson. Dave's mastery of computer technology, undaunted enthusiasm, and selfless dedication once again carried the day.

J.L.C.

THE RANGELAND RESOURCE

1. AN OVERVIEW OF THE WESTERN GRASSLANDS

John L. Capinera
Department of Entomology
Colorado State University, Fort Collins, Colorado 80523

Various terms have been used to describe grasslands. In North America, "prairie" has been applied to several grass-dominated communities, especially those of the Great Plains, but also including areas containing substantial amounts of shrubs. In the tropics such areas are termed "savanna", in South Africa "veld" is used, and in South America "pampa" has been applied.

GRASSLAND CLASSIFICATION

The Great Plains grassland region of North America is best referred to as a steppe region, and contains several physiographic regions or provinces (Daubenmire 1978, Sims et al. 1978, Risser et al. 1981). There is not complete agreement on a system of classification; three major systems are shown in Figures 1.1, 1.2, and 1.3.

The North American steppe has a variety of grasses present, but all die back to the ground each year. The grasses form groundcover ranging from continuous sod to very interrupted bunchgrass. Forbs vary in abundance, and tend to be more conspicuous in wetter areas. Low rainfall coupled with high evapotranspiration rates usually result in desiccation sufficient to kill all germinating tree seedlings. In cooler regions C_3 species comprise most of the green-shoot biomass, with a gradual transition to C_4 species in warmer areas (Sims and Coupland 1979). Shrubs sometimes are common. Decreasing moisture from east to west is the dominant factor which affects the occurrence of plant species comprising the steppe flora. Solar radiation is intense, often averaging 500-700 langleys/day. In general, there is an inverse relationship between precipitation and solar radiation (French 1979).

1

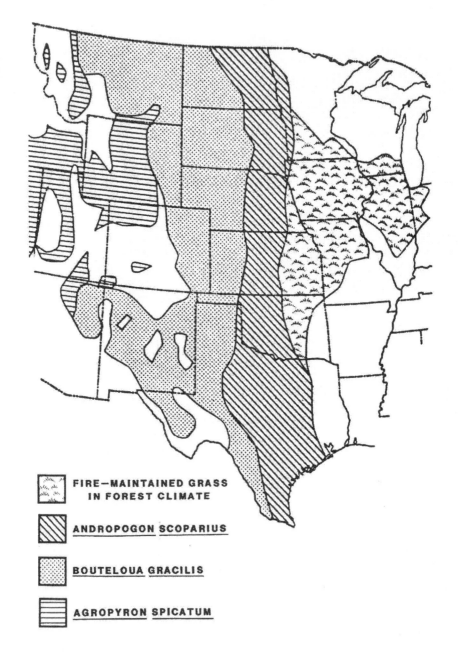

FIRE—MAINTAINED GRASS
IN FOREST CLIMATE

ANDROPOGON SCOPARIUS

BOUTELOUA GRACILIS

AGROPYRON SPICATUM

FIG. 1.1   Approximate distribution of major units of steppe in the
Great Plains region of the United States (from
Daubenmire 1978).

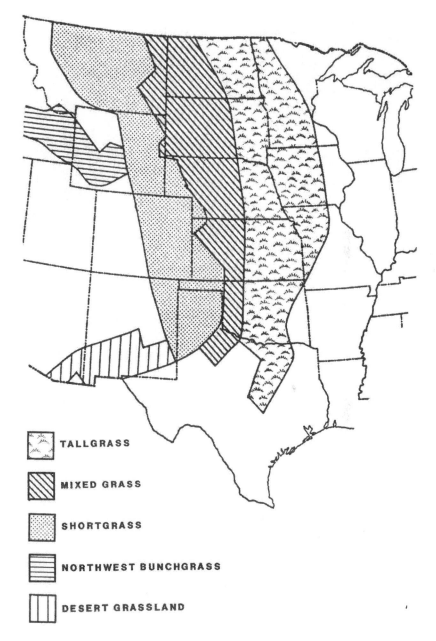

FIG. 1.2 Approximate distribution of major units of steppe in the Great Plains region of the United States (from Sims et al. 1978).

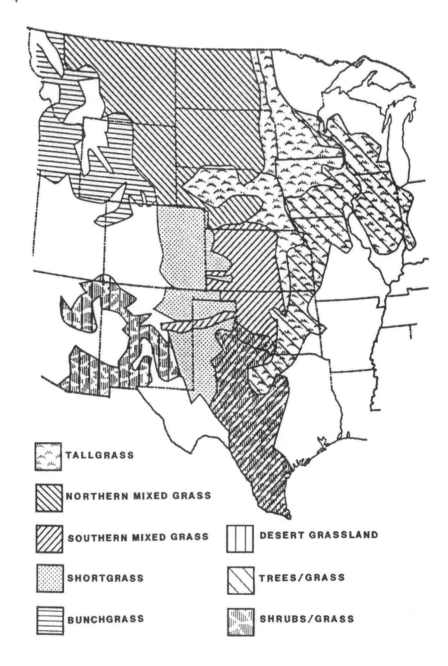

FIG. 1.3   Approximate distribution of major units of steppe in the
Great Plains region of the United States (from Risser et
al. 1981).

Tallgrass

The eastern portion, commonly called the tallgrass or true prairie, is characterized by taller grasses, principally *Agropyron*, *Andropogon*, and *Stipa* in the Dakotas; *Andropogon*, *Calamovilfa*, and *Stipa* in the Nebraska sandhills; and *Andropogon*, *Panicum*, and *Sorgastrum* throughout. There is considerable north-south diversity, and this region reportedly possesses the greatest number of major species within any North American grassland formation. In some areas tallgrass forms an understory for *Quercus* and *Juniperus*.

Mixedgrass

The central steppe region is occupied by a mixture of tall- and shortgrasses; consequently it is usually referred to as the mixed- or midgrass region. Various *Stipa* and *Bouteloua* spp. occur throughout the mixedgrass region, although the northern part contains substantial amounts of *Agropyron smithii* and *Koleria cristata*, and southern regions often contain *Schizachyrium scoparius*. At any specific location, precipitation and edaphic conditions may favor dominance of tall- or shortgrasses, and this relative dominance may shift over time.

Shortgrass

The western portion of the steppe region traditionally has been referred to as the shortgrass region, or sometimes the *Bouteloua gracilis* province (Figures 1.1, 1.2). Recently the northern portion (Wyoming and north) has been separated from the southern portion (Colorado and south). In this classification, the northern shortgrass steppe is referred to as the northern mixedgrass prairie, with the shortgrass region restricted to eastern Colorado and New Mexico, and western Kansas, Oklahoma, and Texas (Figure 1.3). The western steppe contains grasses adapted to xeric conditions, especially *Bouteloua gracilis*. The northern portions (northern mixedgrass) also contain abundant *Stipa comata* and *Agropyron smithii*, while in the southern portion these are replaced by *Buchloe dactyloides* and sometimes other species. In this book, the shortgrass region or shortgrass prairie is used in the traditional sense.

Other Grassland Types

Other grasslands occur, to a limited degree, within the area usually considered the Great Plains. In northwestern shrub steppe areas, bunchgrasses (commonly *Agropyron spicatum*) occur, sometimes in conjunction with *Bromus* spp. and *Artemesia tridentata*. The southwestern desert grasslands are dominated by such shortgrasses as *Bouteloua eriopoda*, *B. rothrockii*, and *Aristida*, and by mesquite *Hilaria mutica* and *H. belangeri*.

PRODUCTIVITY

Grasslands exist over a wide range of temperature and precipitation conditions, but there is a general trend of increased production with increasing precipitation and temperature. Precipitation is especially important in determining grassland characteristics (Lauenroth 1979). These relationships can also be seen from the data presented in Table 1.1.

Within the North American steppe region, estimates of aboveground net production ($g/m^2/yr$) range from 400 to 500 for the tallgrass, 200 to 300 for the mixedgrass, and about 200 for the shortgrass regions. The less significant components, southwestern desert and northwestern shrub steppe, exhibit aboveground net production levels of about 100 $g/m^2/yr$.

TABLE 1.1
Average annual precipitation (mm), temperature (^{o}C), and aboveground net primary production ($g/m^2/yr$) for several sites from the Great Plains region of North America (from Lauenroth 1979).

Site	Precipitation	Temperature	Production
Albion, Montana	329	5.9	42
Sundance, Wyoming	380	6.8	76
Ardmore, South Dakota	360	8.5	86
Sidney, Montana	313	5.3	122
New Town, North Dakota	350	4.8	144
Nunn, Colorado	311	8.2	145
Bismark, North Dakota	385	5.7	203
Dickinson, North Dakota	380	5.1	224
St. Lanatius, Montana	324	7.3	230
Long Valley, South Dakota	410	9.2	250
Cottonwood, South Dakota	387	8.2	282
Boulder, Colorado	470	11.8	298

Belowground production often is significant. Belowground biomass is frequently underestimated due to errors in sampling and root secretion. Ratios of belowground to green shoot biomass ranged from about 2, to 6, to 13, in desert, mixedgrass, and shortgrass regions, respectively (Sims and Coupland 1979).

FORAGE VALUE

The value of forage ($/ha) varies within the Great Plains region largely as a function of productivity (Hewitt and Onsager 1983). In 1977 dollars, average forage value per hectare generally was $4 - 8. Nebraska was exceptional in that average value exceeded $15 per hectare. More arid areas west of the Great Plains such as Arizona, Nevada, and Utah had forage values as low as $2 per hectare. Montana, Wyoming, Colorado, and New Mexico - states comprising the shortgrass region, had average forage values of $5.80, 4.42, 4.97, and 3.90 per hectare, respectively.

The value of forage affects what rangeland improvements, including integrated pest management (IPM) options, can be attempted. The low value of the rangeland resource obviously precludes regular, high-cost expenditures. Hewitt and Onsager (1983) estimated that control of grasshoppers alone cost $2.97 per hectare in 1979. Thus, it is difficult to justify a $3 per hectare expenditure to protect the $2 per hectare resource, such as is found in arid areas of the western United States. From the perspective of the rancher, who may be subsidized by state and local governments, the costs may be acceptable in the western Great Plains where forage values are considerably higher than control costs. Cost benefit analysis does not always prevail in pest management decision making, however. Hewitt and Onsager (1983) reported an inverse relationship between forage value and frequency of grasshopper control. Economic considerations and decision making processes in pest management are treated more extensively in Chapters 25-27.

REFERENCES

Daubenmire, R. 1978. Plant geography with special reference to North America. Academic Press, New York.
French, N.R. 1979. Introduction, p. 41-48. *In:* R.T. Coupland (ed.) Grassland ecosystems of the world: analysis of grasslands and their uses. Cambridge Univ. Press, Cambridge.

8

Hewitt, G.B., and J.A. Onsager. 1983. Control of grasshoppers on rangeland in the United States - a perspective. J. Range Manage. 36:202-207.

Lauenroth, W.K. 1979. Grassland primary production: North American grasslands in perspective, p. 3-24. *In:* N.R. French (ed.) Perspectives in grassland ecology. Springer-Verlag, New York.

Risser, P.G., E.C. Birney, H.D. Blocker, S.W May, W.J. Parton, and J.A. Wiens. 1981. The true prairie ecosystem. Hutchinson Ross Publishing Co., Stroudsburg, PA.

Sims, P.L., and R.T. Coupland. 1979. Producers, p. 49-72. *In:* R.T. Coupland (ed.) Grassland ecosystems of the world: analysis of grasslands and their uses. Cambridge Univ. Press, Cambridge.

Sims, P.L., J.S. Singh, and W.K. Lauenroth. 1978. The structure and function of ten western North American grasslands. I. Abiotic and vegetational characteristics. J. Ecol. 66:251-285.

2. GRAZING MANAGEMENT SYSTEMS FOR THE SHORTGRASS PRAIRIE

R. H. Hart, M. J. Samuel
USDA, ARS, High Plains Grasslands Research Station
Cheyenne, Wyoming 82009

J. W. Waggoner, Jr., C. C. Kaltenbach
Department of Animal Science, University of Wyoming
Laramie, Wyoming 82071

M. A. Smith
Department of Range Management, University of Wyoming
Laramie, Wyoming 82071

The connection between integrated pest management and grazing systems may not be readily apparent, but grazing management systems affect pest management in three ways. First, the plant species (particularly introduced and improved plant species) used in a system determine the pests with which we must be concerned. Second, management determines the rate of return we can expect from grazing and the amount of resources we will have available for pest management. Finally, a properly managed grazing system will produce a healthy, productive, stable plant community which will have greater resistance to pest attacks and which will recover more quickly after such attacks.

CLASSIFYING GRAZING SYSTEMS

Grazing systems may be divided into four "families" (Kothmann 1984). Continuous or season-long grazing provides a fifth category, but many range workers do not consider it a "system."

Rotationally deferred grazing systems provide deferment (a period without grazing, usually early in the season) for each pasture on a rotating basis. At the end of the rest period all pastures or only the pasture previously rested may be grazed for the remainder of the season. These systems are often called deferred-rotation systems, but this is grammatically ridiculous; it is deferment that is rotated, not rotation that is deferred.

9

Rest-rotation grazing systems were designed to benefit the plants without consideration of the nutritional needs of the animals. They require that each pasture in turn be closed to grazing until seed ripens, then be closed the following year to allow seedling establishment.

High-intensity low-frequency (HILF) grazing systems concentrate the grazing animals on a small portion of the total area for a short period, then provide a long period of rest before grazing the area again. The objective is to eliminate selective grazing by forcing the animals to graze all plants and all species, eliminating any competitive advantage of unpalatable plants.

Short-duration rotation grazing evolved from HILF grazing as an attempt to improve animal production. Like HILF, it uses high stocking densities, but for shorter periods so grazing is not so intensive. Also, more selectivity is permitted, the degree of defoliation is reduced, and shorter rest periods are required. This is a very flexible system with potential benefits for improving uniformity of grazing by livestock and improving the general level of herd management.

Any of the five systems can be used on a combination of range and improved or introduced pastures. These "complementary pasture systems" may improve range condition, carrying capacity, and animal performance more than any system applied to range alone.

RANGE GRAZING SYSTEMS

Published Research

Hickey (1969) reviewed 121 papers and concluded that rotations with four or more pastures produced some increase in livestock gain, carrying capacity, and/or forage production over continuous grazing, as did deferred-rotation or rest-rotation grazing. Herbel (1974) stated that no grazing system was definitely superior to continuous grazing on any range type where grazing was not normally season-long.

Rogler (1951) found in North Dakota that two-year-old steers on a 3-pasture rotation system gained more than steers on continuous grazing, but yearling steers gained more on continuous. McIlvain and Savage (1951) found no difference in cattle gains between 2-pasture rotation, 3-pasture rotation, and continuous grazing at comparable stocking rates in Oklahoma, but noted small advantages in density and vigor of the grasses in the rotated pastures. Sampson (1951) concluded that there was little benefit of rotation grazing on sod-forming grasses.

Increased knowledge of reproduction of perennial range plants in the shortgrass prairie indicates that reproduction from seed is of little importance (Hyder et al. 1975). Most studies purporting to show range improvement following implementation of rest-rotation grazing systems (Hormay and Talbot 1961, Ratliff et al. 1972) were poorly designed and interpreted.

Lewis et al. (1982) found that ewe gains were similar on continuous season-long and short-duration rotation grazing in South Dakota. Kirby et al. (1982) found that average daily gain (ADG) of cows and calves was similar on season-long and short-duration rotation systems in North Dakota. Marlow and Whitman (1983) found that steer ADG's were the same on continuous and HILF grazing in Montana. The short-duration systems were stocked more heavily, so produced higher gain/ha in these studies. This confounding of stocking rate and grazing system plagues many studies. Effects of stocking rates on a single system were determined by Bement (1974) and in studies reviewed by Hart (1978).

Grazing Systems Research At Cheyenne

Methods. From 1982 through 1985 we compared continuous season-long grazing, rotationally deferred grazing with one-fourth of the pasture deferred until 1 September each year, and short-duration rotation grazing with each pasture divided into eight paddocks. The range was typical mixed-grass prairie; the forage produced was approximately 50% blue grama, *Bouteloua gracilis*, 15% western wheatgrass, *Agropyron smithii*, 20% other graminoids (mostly needleandthread, *Stipa comata*, and needleleaf sedge, *Carex eleocharis*), and 15% forbs and half-shrubs (dominated by scarlet globemallow, *Sphaeralcea coccinea*; Drummond milkvetch, *Astragalus drummondi*; and fringed sagewort, *Artemisia frigida*).

Range was grazed from late May or early June until mid-October. All three systems were stocked at a heavy (4 steers/9 ha 1982-1984, 5 steers/9 ha 1985) and a moderate (4 steers/12 ha 1982-1984, 5 steers/12 ha 1985) rate, with each system-stocking rate combination in duplicate pastures. Only one pasture of a lightly-stocked (14-18 steers/84 ha) continuous treatment was used.

Peak standing crop of forage was estimated by clipping and/or electronic herbage meter in four exclosures per pasture in August of each year. Basal cover of each species and cover of litter and bare ground were estimated in July of each year on two transects in each heavily stocked pasture and on four transects in the lightly stocked continuous-grazed pasture. Cover was recorded at 10 points at each of 50 locations along each transect. Site and experimental procedures are described in detail by Hart et al. (1988a).

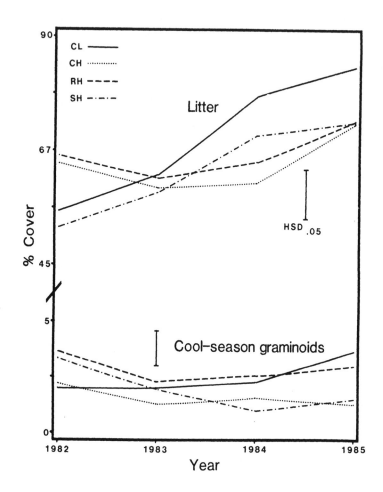

FIG. 2.1 Response of litter cover and basal cover of cool-season graminoids to grazing systems (C = continuous, R = rotationally deferred, S = short-duration rotation) and stocking rates (L = light, H = heavy), 1982-1985.

Results. Total forage production and production of blue grama, western wheatgrass, other graminoids, or forbs and half-shrubs did not differ among systems or stocking rates in any year. Utilization averaged 31% and 40% under moderate and heavy grazing, respectively, but did not differ among systems.

Grazing systems and stocking rates did not significantly affect basal cover of blue grama, western wheatgrass, lichens, or forbs and half-shrubs, or cover of bare ground. Litter cover increased from 1982 to 1985 (Figure 2.1) on the continuous light and short-duration heavy treatments, but on the continuous light treatment litter increased more and attained higher levels than on all other treatments by 1985. Basal cover of total cool-season graminoids (western wheatgrass, needleandthread, sedges, and some minor grasses) increased on the continuous light treatment from 1982 to 1985. These species had decreased or remained the same under heavy stocking over the 4 years. By 1985, only the rotationally deferred heavy treatment had as high a basal cover of cool-season graminoids as the continuous light treatment.

In 1982 and 1983, steers on the short-duration rotation system were rotated on a regular schedule; some paddocks were under or over-utilized and steer gains were depressed below gains on the other systems. In 1984 and 1985, steers were rotated according to the amount of forage in each paddock and the assumed growth rate of the forage. Less rest was needed when forage was growing rapidly so grazing and rest periods were shorter. Gains were the same on all systems at equivalent stocking rates. Average daily gain of steers remained high and constant at low stocking rates, then declined at high stocking rates. Because forage production varied among years, stocking rate was expressed as grazing pressure or steer days per tonne (1000 kg) of forage dry matter produced (Figure 2.2). Gains remained constant at 0.95 kg per day until grazing pressure exceeded 35 steer days per tonne of forage; then gains declined according to the equation ADG (kg) = 1.16 - 0.00656 (steer-days/tonne); r^2 = 0.62. In calculating the stocking rate response, gains under short-duration rotation grazing in 1982 and 1983 were omitted because management was less than optimum.

Average selling price of steers in the fall of 1985 was $1.37/kg. Carrying costs include the loss in value of the original weight of the steer; if a 250-kg steer was bought in the spring for $1.59/kg, this amounts to $55/steer. Interest cost for 150 days at 12.25% equals $20.01. Salt, implants, vaccination, trucking, other veterinary and supplement costs, and death loss might amount to $30 per steer. Total carrying cost (C) for 150 days equals $105.01 per steer or $0.70 per steer per day. Using the method outlined in Hart (1978), we calculated an optimum stocking rate of 53.5 steer-days or 0.356 steer/ha, equivalent to 2.81 ha/steer. Predicted ADG equals 0.84 kg, gain per ha equals 44.7 kg, and net profit equals $23.77 per ha.

The stocking rate recommended by the Soil Conservation Service (SCS 1982) is 29 steer-days/ha at a forage production of 1080 kg/ha, and net profit equals $17.44 per ha. Increasing to the optimum stocking rate (an 84% increase) produces a 36% increase in net profit; doubling stocking rate produces a 35% increase to $23.60 per ha. Such a large increase might not be sustained without serious damage to the range, but a 40% increase in

14

stocking rate above SCS recommendations would produce a 28% increase in profit, to $22.39 per ha. This is possible with all three systems, not just short-duration rotation grazing.

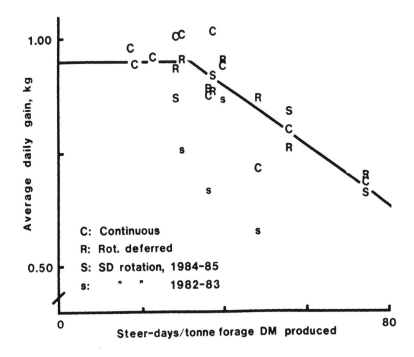

FIG. 2.2   Response of average daily gain of steers to grazing pressure under three grazing systems, 1982-1985.

COMPLEMENTARY PASTURE SYSTEMS

Published Research

McIlvain and Shoop (1973) found carrying capacity of range plus weeping lovegrass, *Eragrostis curvula*, and range plus wheat, *Triticum aestivum*, plus sudangrass, *Sorghum sudanense*, in Oklahoma was 82% and 89% more, respectively, than that of native range alone; beef production per acre was increased 73% and 100%. Sims (1984) reported a 90 to 110 kg increase in calf weaning weights and 30% increase in stocking rate when rye, *Secale cereale*, wheat and pearl millet, *Pennisetum americanum*, were used as complementary pastures with native range in Oklahoma.

Houston and Urick (1972) in Montana found higher cow and calf gains, calf crop, and pregnancy rates among cattle grazing crested wheatgrass, *Agropyron desertorum*,-alfalfa, *Medicago sativa*, in the spring than among cattle grazing native range at the same season. All cattle grazed range in summer and fall. Hart et al. (1983) doubled carrying capacity on a range plus spring-grazed crested wheatgrass system in Wyoming as compared to that on range alone, with no reduction in cow or calf gains, or conception rate. Smoliak and Slen (1974) reported native range, crested wheatgrass, and Russian wildrye, *Elymus junceus*, grazed rotationally in Alberta carried 87% more steers and produced 55 to 66% more beef per hectare than native range grazed continuously.

Kearl (1984) calculated that seeding 385 ha of a 5844-ha ranch to crested wheatgrass would, within 5 years, increase net ranch income 26% and the return to capital 35%. Spielman and Shane (1985) concluded it was profitable to seed crested wheatgrass if hayland and range forage resources were not limiting on the ranch in question.

None of these grazing and economic studies considered the impact of stocking rate on livestock performance and profitability. Stocking rate has an over-riding impact on profitability of grazing systems (Hart et al. 1988a, Quigley et al. 1984). Comparisions of systems are valid only at the optimum stocking rate or grazing pressure on each.

Complementary Pasture Systems Research At Cheyenne

Methods. From 1978 to 1985 we determined gains and reproduction of cattle on complementary pasture for 4 to 6 weeks in May and June and on range from June to October or November at a range of stocking rates on each. Complementary pastures were irrigated meadow bromegrass, *Bromus biebersteinii*, plus alfalfa or dryland crested wheatgrass. Forage production was estimated by clipping in exclosures when cattle were removed from the complementary pastures and at peak standing crop on range. Pastures were stocked with mixed herds of cow-calf pairs, yearling heifers, bulls during breeding season, and in some years on some pastures, dry cows and esophageally fistulated steers. Cow-calf pairs, dry cows, and bulls were considered as one animal unit (AU); steers and heifers as 0.75 AU. Details of sites and experimental procedure are given by Hart et al. (1987b).

Results. Gains of lactating cows, yearling heifers, and calves all declined linearly with increasing grazing pressure on irrigated bromegrass-alfalfa pastures (Table 2.1, Figure 2.3). Critical grazing pressure could not be determined because there was no indication that any of the grazing pressures imposed were below the critical grazing pressure. On dryland crested wheatgrass

16

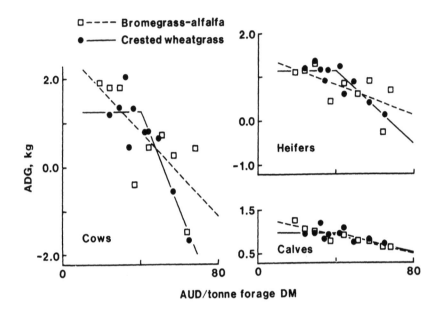

FIG. 2.3   Grazing pressure and average daily gains (ADG) of lactating cows, yearling heifers, and calves on irrigated bromegrass alfalfa or dryland crested wheatgrass pastures, 1978-1982.

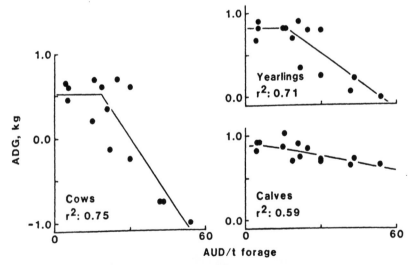

FIG. 2.4   Grazing pressure and average daily gains (ADG) of lactating cows, yearling heifers, and calves on native range, Cheyenne, WY, 1982-1985.

TABLE 2.1
Relationship of average daily gains of lactating cows, yearling heifers and calves, and conception rate of cows to grazing pressure (GP, animal-unit-days/tonne of herbage dry matter produced) on irrigated bromegrass alfalfa, dryland crested wheatgrass, and mixed-grass native range pastures.

| | | Estimates of Parameter | | | | |
Pasture & Parameter	Crit. GP	Below Crit. GP	Above	Critical	GP	r^2
Bromegrass-alfalfa						
Cow gain, kg	--	--	2.69 -	0.0477	GP	0.52
Heifer gain, kg	--	--	1.50 -	0.0170	GP	0.40
Calf gain, kg	--	--	1.33 -	0.0104	GP	0.84
Conception, %	41.1	100	129 -	0.708	GP	0.86
Crested wheatgrass						
Cow gain, kg	40.2	1.27	5.77 -	0.112	GP	0.84
Heifer gain, kg	40.1	1.16	2.85 -	0.0421	GP	0.84
Calf gain, kg	42.0	0.97	1.47 -	0.0119	GP	0.44
1979 conception, %	28.1	100	117 -	0.608	GP	0.83
1980 "	6.6	100	104 -	0.608	GP	0.91
1981-82 "	47.0	100	129 -	0.608	GP	0.96
1979,81,82 "	41.0	100	125 -	0.608	GP	--
Native range						
Cow gain, kg	18.9	0.52	1.39 -	0.0459	GP	0.75
Heifer gain, kg	15.6	0.81	1.16 -	0.0222	GP	0.71
Calf gain, kg	9.3	0.94	1.00 -	0.00607	GP	0.59

pastures and on native range (Figure 2.4), gains remained constant below the critical grazing pressure but declined linearly with increasing grazing pressure above the critical pressure. Calf gains decreased very slowly with increasing grazing pressure, because milk provided most of their nutritional needs. Because of the nutritional demands on the cow for milk production, cow gains declined precipitously with increasing grazing pressure.

The response of conception rate to grazing pressure followed the same model as did liveweight gain. On brome-alfalfa, conception rate was 100% below the critical grazing pressure of 41.1 AUD/tonne of herbage dry matter produced, then declined linearly with further increases in grazing pressure (Table 2.1, Figure 2.5).

The response on crested wheatgrass was more complex. The rate of decline in conception rate with increasing grazing pressure seemed to be similar in all years, but the critical grazing pressure varied widely, from 6.6 AUD/tonne in 1980 to 47.0 AUD/tonne in 1981 and 1982. June 1980 was the driest in 104 years; crested wheatgrass matured early and quality declined rapidly. Crude protein concentration of crested wheatgrass was 9.8% on 12 June 1980 vs. 12.5% on 11 June 1981 (Hart et al. 1988b).

Optimum Stocking Rate. Calculating optimum stocking rates for a cow-calf herd is more complex than for a steer herd. When cows lose weight and condition, conception rate declines (Herd and Sprott 1986). Thus a simplified objective might be to end the grazing season with cows at the same weight as at the start. If cows lose weight, a cost is incurred for the supplement needed to restore the loss; if cows gain, the value of the supplement saved can be considered a return. A cost also is incurred if a cow does not conceive, equal to the cost of a replacement heifer times 1.09 (109 pregnant heifers produce as many live calves as 100 pregnant cows; Hart et al. 1988b) minus the price received for the open (not pregnant) cow.

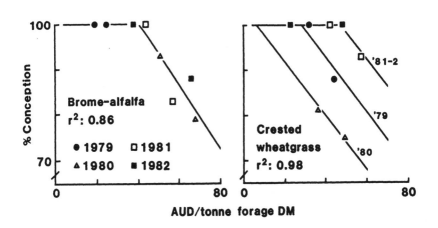

FIG. 2.5 Grazing pressure and conception rate of cows on irrigated bromegrass-alfalfa or dryland crested wheatgrass spring pastures, 1979-1982.

The most profitable stocking rate can be identified after conception rate, cow and calf gains, values, cost, and net return per hectare are computed over a range of stocking rates. The information in Table 2.2 was so generated. A 42-day grazing season from 1 June to 12 July was assumed on the improved pastures, with forage production of 2,500 and 1,600 kg/ha of dry matter on brome-alfalfa and crested wheatgrass, respectively. Cattle would be moved to native range (forage production 1,200 kg/ha) for 112 days from 13 July to 1 November. Calves would be weaned 20 September so they would be on range only 70 days.

TABLE 2.2
Net returns to land, labor and management from cow-calf pairs grazing crested wheatgrass-native range complementary pasture systems or native range alone at different stocking rates; "a" indicates combinations producing >90% of maximum return; "b" indicates combinations producing greater net return than maximum return on range alone.

AUD/ha (ha/AU) on crested wheat-grass	AUD/ha (ha/AU) on native range						
	10 (11.2)	20 (5.6)	30 (3.7)	40 (2.8)	50 (2.2)	60 (1.9)	70 (1.6)
	- - - - - - - - Net return, $/ha - - - - - - -						
20(2.10)	12.15	20.94b	24.79b	25.09b	23.32	20.04b	15.61
30(1.40)	12.82	23.04b	28.18b	29.27b	27.80b	24.33b	19.25
40(1.05)	13.19	24.25b	30.24b	31.93b	30.76b	27.25b	21.80b
50(0.84)	13.42	25.04b	31.63b	33.77a	32.86a	29.37b	23.67b
60(0.70)	13.58	25.59b	32.62a	35.12a	34.42a	30.97b	25.11b
70(0.60)	12.91	24.52b	31.25b	33.44a	32.38a	28.48b	22.06b
80(0.52)	11.37	21.72b	27.34b	28.48b	26.39b	21.43b	13.90
90(0.47)	9.80	18.80	23.20b	23.18b	19.95b	13.83	5.11
None	8.04	16.03	19.60	18.38	13.28	4.30	-8.55

Livestock prices of $1.50 per kg for weaned calves, $438 for a 500-kg dry cow in September, and $478 for a pregnant 360-kg heifer at the same date were taken from the 1980-84 averages given in Kearl (1985). Thus, the total cost of replacing an open cow is $40 times 1.09 or $43.60. Herd and Sprott (1985) indicate 4.1 kg of shelled corn or its energy equivalent are required for 1 kg of gain by a mature cow. The average price of corn was $0.105/kg in 1980-84 (USDA 1984); 1 kg of cow gain was valued at 4.1 times $0.105 or $0.43. Daily fixed costs per AU were estimated at $0.30 from data in Jose et al. (1985) and Kearl (1984); these include interest costs of $0.25 per day during a 154-day grazing season.

On a crested wheatgrass-range complementary pasture system, optimum stocking rates were 64 AUD/ha on crested wheat grass and 43 AUD/ha on range, equal to 0.66 and 2.60 ha/AU or 3.26 ha/AU on the entire system (one AUD/ha equals approximately 0.0089 and 0.024 AU/ha on range and crested wheatgrass, respectively, reflecting corresponding grazing seasons of 112 and 42 days). Net return was $35.70/ha. The optimum stocking rate on bromegrass-alfalfa plus range returned $39.08/ha on 3.30 ha/AU. Although this is more return than on the crested wheatgrass-range system, the difference is not enough to pay the additional costs of establishing and maintaining irrigated pasture.

The combination of 0.66 ha of crested wheatgrass and 2.60 ha of range per AU is equal to 1 ha of crested wheatgrass per 3.94 ha of range. This is a much higher ratio of wheatgrass to range than is usually used or recommended. Cordingly and Kearl (1975), Houston and Urick (1972), and Hart et al. (1983) used one ha of crested wheatgrass per 12.1, 8.6, and 11.3 ha of range, respectively. Our calculations indicate production at such a wide ratio of wheatgrass to range would be little different from that on optimally stocked native range alone. For example, we calculated a return of $21.72/ha from a system with 5.6 ha of range and 0.52 ha of crested wheatgrass per AU (Table 2.2) which is 1 ha of wheatgrass per 10.8 of range. This return is only 10% more than the $19.78/ha calculated for optimally-stocked range (3.4 ha/AU) alone.

In earlier studies, net returns from range alone may have been underestimated because the range was understocked. Houston and Urick (1972) provided 10.4 ha of range per AU and Hart et al. (1983) provided 10 ha/AU, but our calculations predict maximum return at 4.53 ha/AU. Predicted net return at a stocking rate of 10 AU/ha for a 154-day grazing season or 15.4 AUD/ha was $14.62/ha. The combination of 5.6 ha of range and .52 ha of wheatgrass (a ratio of 10.8 to 1) returned $21.72 or 49% more.

CONCLUSIONS

None of the rotation or deferred systems offer any immediate advantages in animal performance on range. However, rotationally deferred systems may improve range condition and eventually increase forage production enough to improve animal performance. Stocking rate or grazing pressure is the dominant factor determining animal performance on any grazing system.

The same is true of complementary pasture systems, but these offer real potential for improving carrying capacity. For maximum profitability, higher stocking rates and a higher ratio of improved pasture to range is needed than is usually recommended. Dryland pastures are likely to be more profitable components than are irrigated pastures.

Even with optimal stocking rate, ratio of improved pasture to range, and management, returns from grazing systems are small. At the prices used in our analyses, net returns were $23 to $35 per hectare or $10 to 15 per acre. And these are returns to land, labor and management, which must cover all costs of these three components, including pest management systems. Thus, only the most inexpensive pest management systems are feasible on range-based production systems, and it must be shown that these are cost-effective.

For example, suppose a grasshopper infestation reduces forage production on range by 10%, to 972 kg/ha. If the rancher maintains the stocking rate of 53.5 steer days/ha, which was optimum at full forage production, his returns will be reduced from $23.97 to $23.30/ha. If he anticipates the grasshopper damage and reduces his stocking rate to 50.3 steer days/ha, the optimal rate at the reduced level of forage production, returns will be $23.40/ha. Profit is reduced so little when grasshoppers take 10% of the forage production that no control measures can be justified.

But suppose we have a really severe infestation, one which reduces forage production by 50% to 540 kg/ha. If the rancher does not reduce his stocking rate, his steers will return only $2.13/ha. But if he reduces his stocking rate to the new optimum of 27.9 steer days/ha, his returns will be $13.00/ha and he will have lost only $10.97/ha. Unless the grasshopper population can be reduced to normal levels for less than $10.97/ha, the most effective way to handle an infestation as severe as this is by reduction of stocking rate.

Similar examples could be calculated for black grass bug, *Labops hesperius* (Uhler), infestations of crested wheatgrass or spotted alfalfa aphid, *Therioaphis maculata* (Buckton), infestations of bromegrass-alfalfa pastures in a complementary pasture-range system. The conclusions likely would be the same; unless infestations cause severe damage or control measures are very inexpensive the best strategy may be to reduce stocking rate and await natural decline in pest abundance.

22

REFERENCES

Bement, R.E. 1974. Strategies used in managing blue-grama range on the Central Great Plains, p. 160-166. *In:* K.W. Kreitlow and R.H. Hart (eds.) Plant morphogenesis as the basis for scientific management of range resources. USDA Misc. Publ. 1271.

Hart, R.H. 1978. Stocking rate theory and its application to grazing on rangelands, p. 547-550. *In:* D.N. Hyder (ed.) Proc. 1st Intl. Rangel. Congr. Society for Range Management, Denver, CO.

Hart, R.H., P.S. Test, M.J. Samuel, and M.A. Smith. 1988a. Cattle and vegetation responses to grazing systems and stocking rates on the Wyoming High Plains. J. Range Manage. (in press).

Hart, R.H., J.W. Waggoner Jr., D.H. Clark, C.C. Kaltenbach, J.A. Hager and M.B. Marshall. 1983. Beef cattle performance on crested wheatgrass plus native range vs. native range alone. J. Range Manage. 36:38-40.

Hart, R.H., J.W. Waggoner Jr., and C.C. Kaltenbach, and L.D. Adams. 1988b. Stocking rate effects on cattle gains and reproduction on complementary pasture-native range systems. J. Range Manage. (in press).

Herbel, C.H. 1974. A review of research related to development of grazing systems on native ranges of the western United States, p. 138-149. *In:* K.W. Kreitlow and R.H. Hart (eds.) Plant morphogenesis as the basis for scientific management of range resources. USDA Misc. Publ. 1271.

Herd, D.B., and L.R. Sprott. 1985. Body condition, nutrition and reproduction of beef cows. Texas Agr. Ext. Serv. Bull. B-1526.

Hickey, W.C. Jr. 1969. A discussion of grazing management systems and some pertinent literature (Abstract & excerpts) 1895-1966. Forest Service, USDA, Albuquerque, NM & Lakewood, CO.

Hormay, A.L., and M.W. Talbot. 1961. Rest-rotation grazing--a new management system for perennial bunchgrass ranges. USDA Prod. Res. Rep. 51.

Houston, W.R., and J.J. Urick. 1972. Improved spring pastures, cow-calf production, and stocking rate carry-over in the northern Great Plains. USDA Tech. Bull. 1451.

Hyder, D.N., R.E. Bement, E.E. Remmenga, and D.F. Hervey. 1975. Ecological responses of native plants and guidelines for management of shortgrass range. USDA Tech. Bull. 1503.

Jose, D., L. Bitney, D. Duey, P. Miller, J. Robb, and L. Sheffield. 1985. Estimated crop and livestock production costs. Nebraska Coop. Ext. Serv. Ext. Circ. 84-872.

Kearl, W.G. 1984. Economics of range reseeding. Wyoming Agr. Exp. Sta. AE-79-01-2R.

Kearl, W.G. 1985. Average prices of cattle and calves, eastern Wyoming and western Nebraska: 1968-1984. Wyoming Agr. Exp. Sta. B-730.

Kirby, D.R., M.D. Parman, T.J. Conlon, and P.E. Nyren. 1982. Short duration grazing in the mixed grass prairie of North Dakota, p. 186-189. In: D.D. Briske and M.M. Kothmann (eds.) Proc. Nat. Conf. on Grazing Technology. Dep. Range Sci., Texas A and M University.

Kothmann, M.M. 1984. Concepts and principles underlying grazing systems: a discussant paper, p. 903-916. In: National Research Council/National Academy of Sciences. Developing strategies for rangeland management. Westview Press, Boulder, CO.

Lewis, J.K., L.S. Bilger, D.M. Engle, T.P. Weber, and L. Blome. 1982. Comparison of high performance short duration and repeated seasonal grazing in the northern mixed prairie, p. 193-196. In: D.D. Briske and M.M. Kothmann. (eds.) Proc. Nat. Conf. on Grazing Technology. Dep. Range Sci., Texas A and M University.

Marlow, C.B., and R.W. Whitman. 1983. High intensity-low frequency grazing vs. continuous grazing on introduced/native pasture, p. 4-7. In: C. Flaherty (ed.) 1983 Anim. and Range Res. Highlights (5th Annual) Range and Animal Sci. Dep., Montana State Univ. Agr. Exp. Sta. Res. Rep. No. 213.

McIlvain, E.H., and D.A. Savage. 1951. Eight-year comparisons of continuous and rotational grazing on the Southern Plains Experimental Range. J. Range Manage. 4:42-47.

McIlvain, E.H., and M.C. Shoop. 1973. Use of farmed forage and tame pasture to complement native range, p. 1-1 to 1-19. In: F. H. Baker (ed.) Great Plains Agr. Council Publ. 63.

Quigley, T.M., J.M. Skovlin, and J.P. Workman. 1984. An economic analysis of two systems and three levels of grazing on ponderosa pine-bunchgrass range. J. Range Manage. 37:309-312.

Ratliff, R.D., J.N. Reppert, and R.J. McKonnen. 1972. Rest-rotation grazing at Harvey Valley...range health, cattle gains, cost. USDA For. Serv. Res. Paper PSW-77.

Rogler, G.A. 1951. A twenty-five year comparison of continuous and rotation grazing in the Northern Plains. J. Range Manage. 4:35-41.

Sampson, A.W. 1951. A symposium on rotation grazing in North America. J. Range Manage. 4:35-41.

24

Sims, P. L.   1984.   Use of forage systems to improve the productivity of stocker and cow-calf operations.   Proc. O-K Beef Cattle Conf., p. H-1 to H-6.   Arkansas City, KS, 15-16 Nov. 1984.

Smoliak, S., and S.B. Slen.   1974.   Beef production on native range, crested wheatgrass, and Russian wildrye pastures.   J. Range Manage. 27:433-436.

Soil Conservation Service.   1982.   Technical guide, section 11 E. Major land resource area (67) 15-17" precipitation zone. Southern Plains. USDA Soil Cons. Serv. Casper, WY.

Spielman, K.A., and R.L. Shane.   1985.   Ranch resource differences affecting profitability of crested wheatgrass as a spring forage source. J. Range Manage. 38:365-369.

US Dept. of Agriculture.   1984.   Agricultural statistics.   U.S. Govt. Printing Off., Washington, DC.

3. FORAGE QUALITY: PRIMARY CHEMISTRY OF GRASSES

Larry R. Rittenhouse and L. Roy Roath
Department of Range Science
Colorado State University, Fort Collins, Colorado 80523

Forage quality is a general term that characterizes the ability of an animal to extract essential nutrients from plants. As such, quality becomes a function of both the animal and the plant. The ultimate evaluation of forage quality is its ability to sustain animal life processes. This paper is concerned with the primary chemistry of grasses and how structural and functional processes in the plant influence nutrient availability; emphasis is on vertebrate herbivores.

Plants provide different amounts of nutrients to different organisms. Nutrient availability to an animal is dependent on the ability of the animal to locate, ingest, and utilize the dry matter in the plant. Variables of importance are: (1) mobility (including bipedalism), (2) anatomy of mouth parts, and (3) the digestive system. Less mobile animals must complete their life cycles within a relatively limited spatial arena; therefore, the concentration and availability of nutrients must be adequate to meet the nutritive requirements of the organism during all life phases. More mobile animals can select among both temporal and spatial availability of nutrients to meet their needs. Animals differ in anatomical features that potentially constrain the physical ability to prehend plants or penetrate epidermal tissue. The anatomy of the plant often presents barriers to extraction of nutrients. Animals whose gut configuration allows microbial fermentation of plant material are not constrained to enzymatic digestion, but sacrifice efficiency of utilization of nutrients.

Forage quality is a function of four primary factors, (1) phylogenetic origin, anatomy, and photosynthetic pathway, (2) proportion of various plant parts, (3) phenological stage of plant development, and (4) conditions of growth (Figure 3.1). Energy from the sun is captured in the process of photosynthesis and stored in numerous kinds of chemical bonds. The compounds of greatest interest are those considered essential to sustaining life processes in animals. Utilization of these compounds may be modified by other compounds not considered essential to the animal.

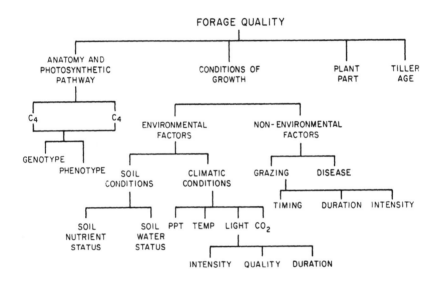

FIG. 3.1 Plant and environmental factors influencing forage quality.

Most work in cell biology focuses on processes in the cell and individual organelles located within the cell. Little comprehensive literature is available on the average chemistry of the cell or the plant in general. Reasons for this are that compounds found in the cell are frequently transitory and continuously changing. Chemical reactions occur at different rates. Rates and kinds of reactions are subject to many plant and environmental variables. The level of metabolites in the aboveground parts of the plant is dependent on the size of the mobile nutrient pool, the rate of incorporation into permanent structural parts, and allocation below and above ground (Figure 3.2). The size of the metabolic pool and the rate of uptake and synthesis of materials is highly dependent on conditions of growth. Lechtenberg et al. (1972) documented in tall fescue plants changes in leaf sugar content of more than 20% in a 12-hr period, with sucrose accounting for most of the change in concentration. Concurrently, fructosans demonstrated little change. Sudangrass also showed great changes in sugar content, but showed marked changes in starch at the same time (Lechtenberg et al. 1973). Over a 24-hr period the carbohydrate content in a cell and among plant parts differs dramatically, and the balance of individual compounds within the cell changes even more.

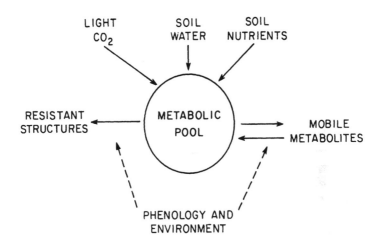

FIG. 3.2   A schematic showing the partitioning of the metabolic pool in plants.

The intracellular space (lumen) is occupied by chloroplasts, mitochondria, cytoplasm, and the cell nucleus, each with its own membrane tissue.   For purposes of this paper, however, the intracellular space will be considered as a unit, recognizing that each may contain varying amounts of each cell constituent (Brady 1976).   The cell contents of grasses are composed of several thousand individual compounds, many of which are extremely transitory.   The compounds can be grouped chemically into carbohydrates (including simple and complex sugars and starch), lipids, proteins, nucleic acids, and other non-protein nitrogen. Changes in cell contents are a response to both short-term metabolic processes (modified by environmental and nonenvironmental factors that influence growing conditions) and longer-term phenologic changes fixed by the genetics of the plant.
Cell contents are generally considered to be completely available to all animals with a digestibility of 98% (Van Soest 1982).   Analytically, cell content percentage is approximated by

difference (1 - neutral detergent fiber). The actual utilization of components of cell contents is not reported. However, indirect evidence of the high biological value of cell contents comes from the high partial efficiencies for animal maintenance and growth from plants with high cell contents.

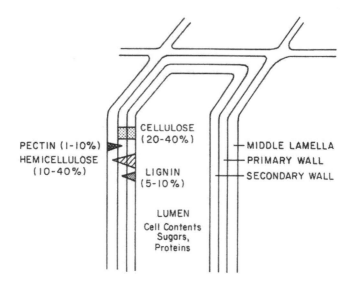

FIG. 3.3 Schematic diagram of a plant cell.

The intercellular space (cell wall) is composed of the middle lamella, and the primary and secondary wall (Figure 3.3). The wall of various kinds of cells differs in carbohydrate composition. Cell walls without secondary thickening are mostly α-cellulose, hemicelluloses, pectin, and glycoproteins with possibly a small amount of polyphenolic compounds. Some cells, such as bundle sheath cells, may produce secondary cell wall by thickening. During the growth of secondary cell wall, α-cellulose, hemicelluloses, and lignin are deposited. The thickening results more from the α-cellulose than the hemicellulose deposition. Lignin is associated with the microfibrillar structure of the wall and the lignin-polysaccharide complex becomes a rigid structure. Opalate silica may also be found in the cell wall matrix of some plants. The degree of incrustation tends to be species specific.

Nutrients are extracted from the cell by the animal through hydrolysis, enzymes secreted by the animal, or extracellular microbial enzymatic activity. As a result, forage quality to animals with simple digestive systems is quite different from animals that utilize end products of microbial breakdown and synthesis. Simple-gutted animals have specific dietary requirements to meet their carbohydrate, protein, vitamin, and fat needs. Microbial fermenters have much less specific dietary requirements, because the microbes synthesize most of the proteins and vitamins needed by the animal and can use volatile end-products of fermentation as sources of energy.

PRECHEMISTRY OF CELL CONTENTS OF GRASSES

Glucose, fructose, sucrose, starches, and fructosans are major 'soluble' carbohydrates in plant cells. The water-soluble sugars, i.e., glucose, fructose, sucrose, and fructosan are readily hydrolyzed and absorbed as sources of energy for the organism. Microbes rapidly ferment these compounds to produce volatile fatty acids (VFAs). Sugar and starch are formed mainly within the chloroplasts of tropical plant cells; whereas, in temperate species the starch is formed within the mesophyll cells (Brown 1977). Starch accumulated in the leaves of most grasses is insoluble in water, depending on the amylopectin content. Insoluble starch is broken down enzymatically and is less digested in the rumen than soluble starch. Tropical grasses store high concentrations of starch in their leaves. Temperate grasses accumulate sucrose and fructosans rather than starch, mainly in their stems (Norton 1982).

Nitrogen containing compounds are found in extremely diverse protein and non-protein forms. As much as 25% of the nitrogen in the plant cell may be in non-protein form. Non-protein nitrogen includes amino acids, amides, ureides, nitrates, and ammonia. Protein may be soluble or insoluble. Most of the protein in the leaves of temperate grasses is found in the chloroplasts with smaller amounts in the mitochondria, cytoplasm and nucleus. Distribution of protein in tropical grasses is less known (Norton 1982). Up to 50% of the soluble protein of temperate grasses is RuBP (Ribulose bisphosphate) carboxylase. The level of the same enzyme in tropical grasses represents only about 20 percent of the nitrogen in the cell and that is restricted to the bundle sheath. Ribulose bisphosphate carboxylase of the Calvin-Benson cycle incorporates CO_2 (in the bundle sheath cell of C_4 plants) into 3-phosphoglyceric acid. This becomes important because cell wall structure of maturing C_4 grasses renders cell contents relatively unavailable for either hydrolysis or microbial activity (Owensby, pers. comm.).

The proportion of soluble to insoluble protein seems to remain fairly constant in grasses as they mature. The tannin content of cell contents interacts with protein and may render some of the protein insoluble, especially in the rumen. Therefore, insoluble protein is higher in diets high in tannins than in diets that contain low levels of tannins. Tropical grass cells contain much lower levels of nitrogen than temperate grass cells, and the rate of decline in nitrogen with plant maturity is much greater in tropical than temperate plants. Norton (1982) reviewed the literature and found that 53 percent of all tropical grasses contained less than 9 percent crude protein compared with 32 percent of the temperate grasses. This fact has important implications in regard to the nutritive requirements of the organism and its ability to complete life-cycle processes.

Simple-gutted animals utilize products of enzymatic hydolysis, i.e., short-chain peptides and amino acids. Therefore, the biological value of N in the plant depends on the balance of amino acids available. Non-protein nitrogen is not utilized to a significant extent by these organisms. Microorganisms found in the digestive system of ruminant and non-ruminant herbivores deamminate plant protein and resynthesize it into high biological value protein. These animals can, of course, utilize both endogenous and exogenous sources of non-protein nitrogen to meet their needs. Not all protein is soluble in the rumen or cecum. Bypass nitrogenous compounds may be hydrolyzed in the abomasum of the ruminant and absorbed directly from the upper gut. Insoluble nitrogenous compounds that reach the cecum are excreted in the feces.

Lipid content of grass cells is generally considered low and of little consequence to forage quality. It is true that glycerol-based lipids are generally low in grasses, especially as the plant matures. Many of the lipids in grasses are glycolipids. It is also true that the content of non-glycerol based lipids is usually low, but there is growing evidence of the influence of these compounds on forage quality and ingestion. Waxes may serve as important barriers to cell attack by microbes. Terpene-based compounds, often referred to as 'essential oils' may reduce palatability of plants as well as interfere with digestive processes. Other terpene-based compounds are parts of important molecules, such as phytols and carotenoids. Literature on differences in lipid content of temperate and tropical grasses was not located.

PRIMARY CHEMISTRY OF THE CELL WALL OF GRASSES

Variations in the anatomy of the cell wall of grasses has a major impact on forage quality. These differences are grouped according to the phylogenetic origin of the plants and are referred to as either tropical with a C_4 metabolic pathway, or temperate with a C_3 metabolic pathway (Brown 1977 as he referred to

Haberlandt 1884). The Kranz anatomy of tropical plants is distinguished from temperate plants by a radial arrangement of chlorenchyma cells (bundle sheath) around the vascular bundles and other structural anomalies. The bundle sheath cells vary in size and depth, depending on species. These cells have no intercellular air spaces and have thick, suberized outer walls (Norton 1982). The mesophyll cells of tropical plants are more densely packed than in temperate grasses.

The primary and secondary structure of the cell wall were introduced earlier in the manuscript. The reader is referred to Selvendran (1983) for a more detailed review of the chemistry of plant cell walls.

The cell wall of grasses is composed mainly of cellulose (an oligosaccharide) and the non-sugar heteropolysaccharide, hemicellulose. Tropical grasses have higher contents of hemicellulose than temperate grasses. They have similar cellulose contents.

Cellulose (Figure 3.4) is found in both the primary and secondary walls of cells. The basic structure is ß-1,4 linked D-glucan chains arranged in an ordered manner within the microfibrils (Figure 3.5). Microfibrils are interconnected by non-glucan polymers which in the case of grasses are primarily xyloglucans.

Hemicellulose is an integral part of the cell wall matrix and is composed of ß-1,4 xylose units with branched methylglucuronic acid, glucose, arabinose, and/or galactose. In ruminants, hemicellulose and cellulose appear to be digested to the same extent and at approximately the same rate. The rate of digestion of cell wall material is faster in ruminant than non-ruminant herbivores, primarily because of the difference in levels of available substrates (Van Soest 1986). Wedig et al. (1986) found that orchardgrass had a higher concentration of xylose and a lower concentration of galactose than did alfalfa. Lignin content was negatively related to hemicellulose digestibility. Lignin appeared to inhibit digestion primarily by bonding with hemicellulose.

Lignin is a heterogeneous, non-carbohydrate compound that is not digested either by ruminal microorganisms or intestinal enzymes. It is a phenylpropanoid polymer. This compound is an end-product(s) of the Shikimic acid pathway as are several of the aromatic substances found in plants. The lignin in grasses arises from one of three alcohols: p-coumaryl alcohol, coniferyl alcohol, and sinapyl alcohol. Of these, p-coumaryl alcohol is the most prominent. Lignin contains considerable quantities of nitrogen in the form of aromatic amino acids, such as, phenylalanine, tyrosine and tryptophan.

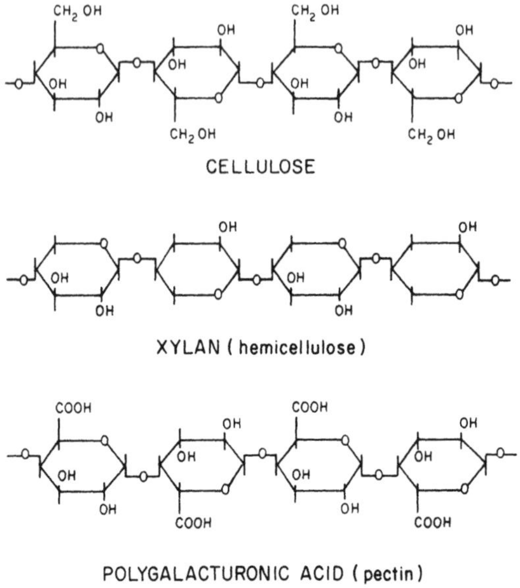

FIG. 3.4 The chemical structure of cellulose, hemicellulose, and pectin.

Pectic substances are a complex mixture of colloidal polysaccharides that can be partially extracted from the cell wall with water. Pectin content of the cell wall tends to be inversely related to the amount of lignification. Therefore, the lignin content of grasses seldom exceeds 15 percent. The 'insoluble' pectic material is always found in close association with other cell wall constituents, particularly the α-cellulose fraction. The currently held view is that pectines are partially esterified rhamnogalacturonans in which the α-1,4 linked D-galacturonan chains are interspersed with L-rhamnopyranosyl residues and the other neutral sugars are present in the side-chains only. Rhamnogalacturonans are thought to be the backbone chains of pectins.

Van Soest and Jones (1968) found that digestibility of temperate grasses was inversely related to silica content. The influence of silica on digestibility can be a drop of 2 to 3 units dry matter digestibility per unit of silica. The ultimate effect of silica on forage quality depends on the form of the silica, i.e., if it is intracellular in soluble form, or intercellular in opalate form, or extracellular (see Van Soest 1982 for a more complete discussion).

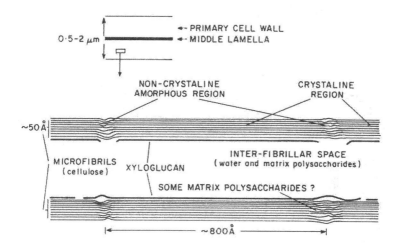

FIG. 3.5 Diagrammatic representation of the arrangement of cellulose in cell wall.

ENVIRONMENT AND FORAGE QUALITY

Growing conditions have a large influence on plant metabolic processes, and the consequent rate of growth and phenological development. Nutrients in the plant are a part of the mobile metabolic pool and are incorporated into structural parts of the plant either above or below ground. Both environmental and non-environmental factors influence the rates of above processes.

Light

Solar radiation is the source of energy for photosynthesis. Under reduced light, both tropical and temperate grasses show an increase in cell wall percentage and lignification, and a decrease in the levels of water soluble carbohydrates (WSC) (Table 3.1). Nitrate and other NPN levels in shaded plants are usually higher than in unshaded plants. Crude protein levels show increases in concentration, as might be expected from higher NPN and nitrate levels. This is related to larger cell volume and to younger phenological stages of growth. Silica content increases appear to be associated with cell wall constituents. Light quality is seldom limiting; however, changes in the red/far red ratio of light have been shown to influence tillering in some plants (Derigibus and Trlica 1985). The effect of daylength, per se, is inconsistent. Whatever effect increasing or decreasing day length has on initiation of flowering and other processes, however, will affect forage quality. Increased day length in far northern latitudes can compensate for reduced temperatures to allow growth processes.

TABLE 3.1
Influence of environmental factors on composition of forages.

	Temp	Light	N	Water	Grazing
Yield	+	+	+	+	-
WSC	-	+	-	-	+
Nitrate	-	-	+	?	?
Cell Wall	+	-	+/-	+	-
Lignin	+	-	+	+	-

Soil Water

Most literature indicates that limited soil water either has no effect or a positive effect on forage quality, unless the limitation is so severe as to cause death of the tissue. Under moisture stress conditions plants tend to produce fewer flowering tillers; therefore, the leaf to stem ratio is larger and the overall quality of the sward is higher. The rate of cell wall thickening is reduced and the level of cell solubles tends to increase. Aging of tissue is not necessarily hastened by drought. Further, the rate of decline in nitrogen content and digestibility is slower for stressed leaves than well-watered (Wilson 1982). As available water is increased, total yield of dry matter increases and the rate and extent of cell wall thickening and lignification increases. Fisher (1980) reported that phosphorus content is often low in water-stressed plants.

Temperature

The effect of temperature on forage quality of tropical vs temperate grasses must be considered relative to their optimal temperatures for growth. Further, there is an important interacting effect of available soil water and temperature on plant growth. With that in mind, the literature is consistent in that high temperature has a strong detrimental effect on the forage quality (dry matter digestibility) of both tropical and temperate grasses. High temperatures reduce soluble carbohydrates, nitrates, and protein, while cell wall constituents and lignification increase. It is not uncommon to observe lower rates of gain of animals on summer regrowth than fall regrowth (depending on latitude). Johnson et al. (1973) show that when water is not limiting, the rate of decline in dry matter digestibility of herbage with age is slower during the cooler autumn and winter than the hotter months of spring and summer. Most grasses are well adapted to gradually decreasing temperatures. As temperatures decline and daylength changes, soluble metabolites, especially carbohydrates and nitrogen, may be translocated to belowground organs, thus reducing quality of aboveground plant parts. In early spring as plants initiate growth, temperatures may be too low to sustain optimal levels of photosynthesis, and soluble C-and N-based metabolites may accumulate. The build-up of organic acids in the presence of excess N, and possibly K, and low levels of Mg can result in grass tetany.

Soil Nutrient Status

Ambient temperature regimes will regulate the rate of mineralization of important nutrients such as nitrogen, phosphorus, magnesium, zinc, and sulphur. Precipitation, temperature, and morphogenesis determine the status and rate of organic matter turnover and hence available nutrients in the soil.

Disease and other Non-environmental Stress

Although there is dearth of literature to document the effect of disease-stress on forage quality, one would expect it would be similar to the effect of water stress. One would also expect a strong interaction with leaf age; however, in all probability pathogens would be most likely to damage tissue during the early growth stages causing reduction in biomass and quality.

Grazing

When an actively growing above-ground part is injured, there is a shift in energy allocation, provided growing conditions are adequate. Hormonal control of plant growth dictates that vegetative growth will continue, creating a plant which is younger phenologically. This plant has a greater cell contents and thinner cell walls than a plant which was not defoliated. Cell wall thickening is delayed and soluble carbohydrates and nitrogen level are increased. Grazing animals typically choose plant parts which are highest in nutrient content, contingent on availability. Maturity decreases the overall selectivity displayed by animals but does not eliminate preference for chosen plants or plant parts.

REFERENCES

Beaty, E.R., and J.L. Engel. 1980. Forage quality measurements and forage research - a review, critique and interpretation. J. Range Manage. 33:49-54.

Brady, N.C. Advances in agronomy. Vol. 26. Academic Press, New York.

Brown, W.F. 1977. The Kranz syndrome and its subtypes in grass systematics, p. 1-97. *In:* R.W. Kiger (ed.) Memoirs of the Torrey Botanical Club. Vol. 23. Fisher-Harrison Corp., Durham, NC.

Derigibus V.A., and M.J. Trlica. 1985. Tillering response to enrichment of red light beneath the canopy in a humid natural grassland. J. Appl. Ecol. 22:199-206.

Fisher, M.J. 1980. The influence of water stress on nitrogen and phosphorus uptake and concentrations in Townsville style (*Stylosanthes humilis*). Aust. J. Exp. Agr. and Anim. Husb. 20:175-180.

Johnson, W.L., J. Guerrero, and D. Pezo. 1973. Cell-wall constituents and *in vitro* digestibility of Napier grass (*Pennisetum purpureum*). J. Anim. Sci. 37:1255-1261.

Laredo, M.A., and D.J. Minson. 1975. The voluntary intake and digestibility by sheep of leaf and stem fractions of Lolium perenne. J. Br. Grassld. Soc. 30:73-77.

Lechtenberg, V.L., D.A. Holt, and H.W. Youngberg. 1972. Diurnal variation in nonstructural carbohydrates of *Festuca arundinacea* (Schreb.) with and without N fertilizer. Agron. J. 64:302-305.

Lechtenberg, V.L., D.A. Holt, and H.W. Youngberg. 1973. Diurnal variation in non-structural carbohydrates of *Sorghum sudanense* (Stapf) as influenced by environment. Agron. J. 65:579-583.

Norton, B.W. 1982. Differences between species in forage quality, p. 90-110. *In:* J.B. Hacker (ed.) Nutritional limits to animal production from pastures. Commonwealth Agr. Bur., United Kingdom.

Selvendran, R.R. 1983. The chemistry of plant cell walls, p. 95-147. *In:* G.G. Birch and K.J. Parker (eds.) Dietary fibre. Applied Science Publishers, London.

Struik, P.C., B. Dienum, and J.M.P. Hoefsloot. 1985. Effects of temperature during different stages of development on growth and digestibility of forage maize (*Zea mays* L.). Netherlands J. Agr. Sci. 33:405-420.

Van Soest, P.J. 1982. Nutritional ecology of the ruminant. O & B Books, Inc., Corvallis, OR.

Van Soest, P.J. 1986. Comparative fiber requirements for ruminant and non-ruminants, p. 52-60. *In:* Proc. Cornell Nutrition Conference. Cornell Univ., Ithaca, NY.

Wedig, C.L., E.H. Jaster, and K.J. Moore. 1986. Composition and digestibility of alfalfa and orchardgrass hemicellulose monosaccharides by Holstein steers. J. Dairy Sci. 69:1309-1316.

Wilson, J.R. 1982. Environmental and nutritional factors affecting herbage quality, p. 111-131. *In:* J.B. Hacker (ed.) Nutritional limits to animal production from pastures. Commonwealth Agr. Bur., United Kingdom.

4. FORAGE QUALITY: SECONDARY CHEMISTRY OF GRASSES

Richard A. Redak
Department of Entomology
Colorado State University, Ft. Collins, Colorado 80523

Traditionally, the secondary chemistry of the Poaceae has been viewed by ecologists as poorly developed and relatively ineffective in deterring herbivory (Owen and Wiegert 1981, McNaughton 1983, Coughenour 1985). It has been thought that, unlike most plants, grasses depend on silicification, lignification, trichomes, and a basal meristem instead of relying on secondary metabolites as defenses against herbivory (Owen and Wiegert 1981, McNaughton 1983). Compared to the diversity of secondary compounds present in the Angiospermae (McNaughton 1983), it is true that the chemistry of the Poaceae is relatively undeveloped. This comparison is an invalid one, however. It is unreasonable to assume that the diversity of secondary compounds within a single family, the Poaceae (8000 species) is comparable to that of an entire Division, the Angiospermae (230,000 species) (Cronquist 1981). Although grasses may be somewhat adapted to herbivory (McNaughton 1983, Coughenour 1985), as evidenced by a basal meristem, the Poaceae should not be considered to be chemically undefended against herbivory. This chapter documents that members of the Poaceae are not chemically "depauperate" (McNaughton 1983) and do in fact contain a wide variety of secondary compounds which can and do deter herbivory. This chapter is not a comprehensive review of grass secondary chemistry, for which the reader is referred to Gibbs (1974).

ALKALOIDS

Numerous members of the Poaceae have been shown to contain alkaloids. Species of *Alopercus, Aristida, Arundo, Avena, Bambusa, Bromus, Chloris, Dactylis, Echinochloa, Eleusine, Festuca, Holcus, Hordeum, Imperata, Lolium, Muehlenbergia, Oryza, Oxythenanthera, Panicum, Phalaris, Phragmites, Poa, Pogonatherum,*

Setaria, Sorghum, Trichachne, Thelepogon, and *Zea* have all been shown to contain various alkaloids (Willamen and Schubert 1961, Kingsbury 1964, Raffaut 1970, Willamen and Li 1970, Gibbs 1974). Although many of these compounds are unknown, several have been identified and are shown in Figure 4.1.

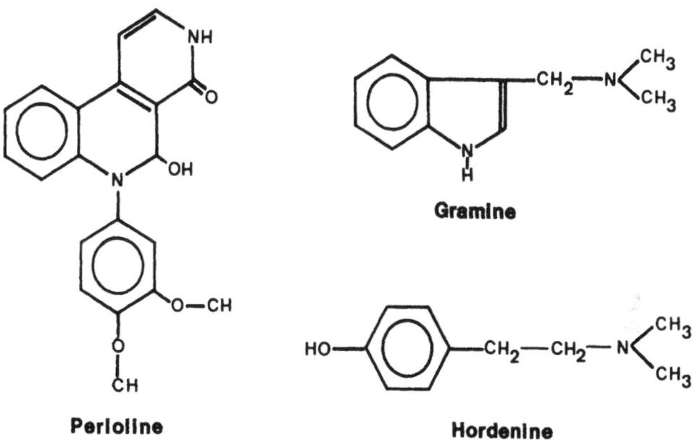

Gramine

Perloline

Hordenine

FIG. 4.1 Representative alkaloids found in grasses.

Most of the known grass alkaloids are derived from the amino acids tyrosine and tryptophan (Smith 1975, Goodwin and Mercer 1983). That alkaloids are present in grasses at concentrations that can affect herbivory is well documented. Alkaloids found in the genera *Phalaris* and *Festuca* (e.g., gramine and hordenine) have been shown to be toxic to cattle and sheep. Such alkaloids are associated with heart failure, general convulsions, neurological disorders, overall weight loss, and decreased rumen performance (Gallagher et al. 1964, Gallagher et al. 1966, Bush et al. 1970). Additionally, alkaloids found in *Phalaris* have been associated with decreased palatability and digestibility to cattle and sheep (Simmons and Marten 1971, Marten et al. 1976, Fairbourn 1982). When administered in artificial diets at concentrations resembling those of *Phalaris* and *Hordeum*, alkaloids such as gramine and other similar tryptamines have also been shown to decrease survival of the corn leaf aphid, *Rhopalosiphum maidis* (Fitch), the oat bird cherry aphid, *Rhopalosiphum padi* (Linn.) (Zuniga and Corcuera 1986), and the greenbug aphid, *Schizaphis graminum* (Rhodani) (Corcuera 1984, Zuniga et al. 1985). Additionally, Corcuera (1984) showed that gramine acted as a feeding deterrent to both *Rhopalosiphum* and *Schizaphis*. When incorporated within agar diets, both gramine and hordenine performed as feeding deterrents for the generalist grasshopper, *Melanoplus bivittatus* (Say) (Harley and Thorsteinson 1967). Additionally, gramine resulted in reduced adult dry weights when this grasshopper species was fed artificial diet containing gramine (Harley and Thorsteinson 1967).

CYANOGENIC GLYCOSIDES

Certain members of the Poaceae have been shown to be cyanogenic. Species of *Agropyron, Agrostis, Andropogon, Avena, Bambusa, Bothriochloa, Bouteloua, Briza, Cortaderia, Dactylotaenium, Eleusine, Festuca, Glyceria, Hordeum, Holcus, Leptochloa, Lolium, Oryza, Panicum, Poa, Saccharum, Sorghum, Tridens, Triticum,* and *Zea* have all been shown to contain cyanogenic glycosides (Gibbs 1974, Seigler 1976a, b). Members of the genus *Sorghum* are perhaps the most studied species with regard to cyanogenesis (Wolf and Washko 1967, Barnett and Caviness 1968, Gillingham et al. 1969, Loyd and Gray 1970, Reay and Conn 1970). The compound of interest is dhurrin (Figure 4.2).

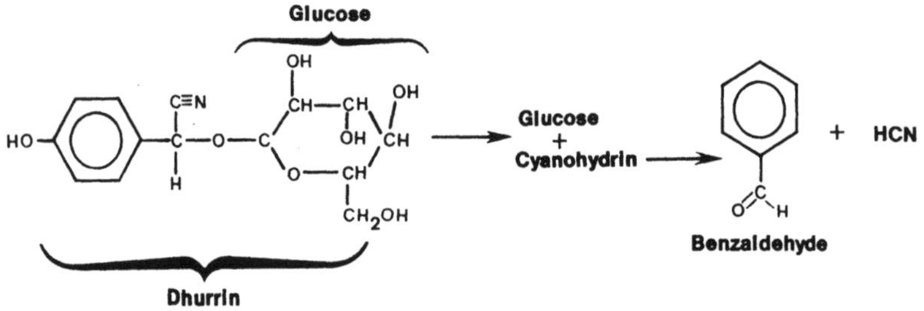

FIG. 4.2 Synthesis of hydrogen cyanide from dhurrin.

Upon tissue damage, through herbivore feeding, dhurrin is enzymatically converted to glucose and cyanohydrin. The latter compound is then enzymatically converted to one molecule of benzaldehyde and one molecule of hydrogen cyanide (Figure 4.2). Hydrogen cyanide is the compound which will ultimately affect the herbivore. Dhurrin, through the release of cyanide, has been shown to be quite toxic to mammals (Kingsbury 1964). Presumably, the cyanide released from dhurrin during feeding by cows and sheep disrupts the rumen fauna (Barnett and Caviness 1968, Conn 1979) and eventually irreversibly binds with a variety of mammalian and microfaunal cytochrome enzymes (Conn 1979). Cyanide is also toxic to many insect species (Rodriguez and Levin 1976). It presumably also binds irreversibly to insect cytochrome enzymes. Cyanide in sorghum has been implicated in deterring feeding by greenbug aphid, *Schizaphis graminum* (Rhodani) (Jones 1983), the locust, *Locusta migratoria* L. (Woodhead and Bernays 1978), and western corn rootworm, *Diabrotica virgifera virgifera* LeConte (Branson et al. 1969).

BENZOXAZINONES

Benzoxazinones such as 2,4-dihydroxy-7-methoxy-1,4-benzoxazin-3-one (DIMBOA) also occur within the Poaceae (Klun et al. 1967, Baker and Smith 1977, Long et al. 1977, Robinson et al. 1978, Sanders et al. 1981). Most of the studies have been conducted using corn, *Zea mays*. DIMBOA is found in the plant usually as a glucoside at concentrations as high as 1% dry weight (Klun and Robinson 1969). The glucoside typically undergoes enzymatic hydrolysis to form the aglycon (Figure 4.3).

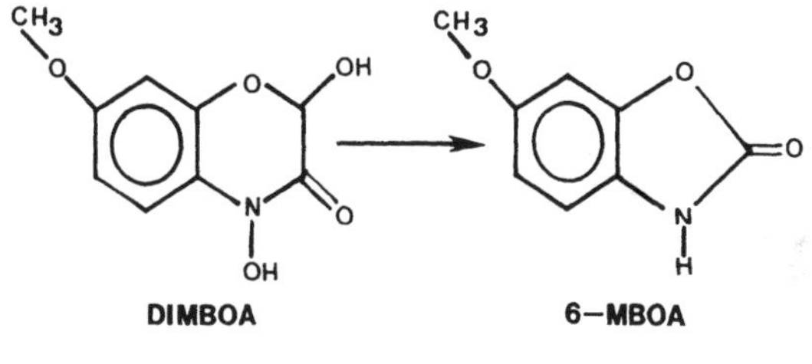

DIMBOA   6—MBOA

FIG. 4.3 Aglycons of DIMBOA AND 6-MBOA.

Upon tissue damage, the aglycon may be converted to 6-methoxybenzoxazolinone (6-MBOA) (Figure 4.3) (Kluhn and Robinson 1969). The aglycon form of DIMBOA has been shown to inhibit larval development of the European corn borer, *Ostrinia nubilalis* (Hubner), on corn (Kluhn et al. 1967, Kluhn and Robinson 1969, Scriber et al. 1975, Robinson et al. 1978, 1982). DIMBOA has been shown to increase the mortality of the greenbug and corn leaf aphids when raised on artificial diets (Long et al. 1977, Argandona et al. 1982, 1983). Additionally, DIMBOA has been shown to be a major resistance characteristic in corn to bacterial soft rot, *Erwinia* spp. (Corcuera et al. 1978, Woodward et al. 1978, Lacy et al. 1979) and to the plant pathogenic fungus, *Helminthosporium turcicum* Pass. (Couture 1971). 6-MBOA has been shown to positively influence mammalian reproductive cycles (Berger et al. 1981, Sanders et al. 1981). At present, DIMBOA and 6-MBOA have only been found in a few genera of the Poaceae. However, this may reflect that only a few grass species have been investigated for these compounds.

PHENOLICS

Like almost all plants, grasses contain a wide variety of monomeric phenolic compounds (Gibbs 1974, Buritt et al. 1984). Included in this category are simple phenolic acids, alcohols, aldehydes, coumarins, flavonoids, etc. Little is known about the ecological roles that these compounds may play, if any. Many occur as intermediate products in the synthesis of tannins, lignins, and plant pigments (Goodwin and Mercer 1983). Some studies have suggested that a variety of simple phenolics may affect herbivory (Todd et al. 1971, Dreyer and Jones 1981, Dreyer et al. 1981, Buchsbaum et al. 1984, Lindroth and Batzli 1984). Dreyer and Jones (1981) and Todd et al. (1971) have shown that several phenolic compounds, some of which have been reported in grasses (e.g., quercitin, p-hydroxybenzoic acid, luteolin, p-coumaric and caffeic acids), deter feeding by the greenbug aphid and green peach aphid, *Myzus persicae* (Sulz.). Dreyer et al. (1981) showed that the phenolic fraction of polar solvent extracts of *Sorghum* exhibited feeding deterrent activity towards the greenbug aphid. For vertebrate herbivores, Lindroth and Batzli (1984) have shown that the flavonoid quercitin is toxic to the prairie vole, *Microtus ochrogaster* (Wagner), and Buchsbaum et al. (1984) have shown that soluble phenolics in grasses influence palatability to geese. Certain grass phenolics have also been found to be phytotoxic. Species of *Sorghum, Zea, Avena, Bromus, Phleum, Sporobolus* have all been found to contain phenolic compounds which play a role in plant-plant, plant-fungal, or plant-bacterial interactions (Rice 1984 and references therein). Most of these compounds are allelopathic agents which inhibit seed germination, seedling growth, bacterial nitrification, and fungal growth (Rice 1984). Examples of phenolic compounds that may play an ecological role in grasses are shown in Figure 4.4. Quercitin is toxic towards aphids and mammals (Todd et al. 1971, Dreyer et al. 1981, Lindroth and Batzli 1984). Vanillic acid, found in corn, wheat, sorghum and oats, functions as a plant inhibitor (Rice 1984), and p-coumaric acid, found in *Bromus, Sorghum,* and *Sporobolus* was found to be toxic to aphids (Todd et al. 1971) and functions as an plant allelopathic agent.

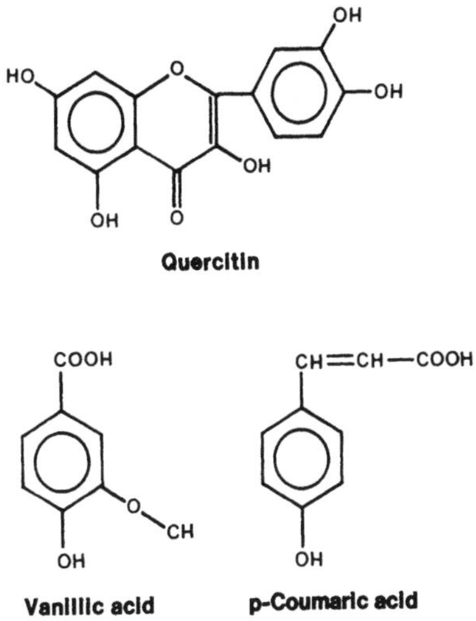

FIG. 4.4 Representative phenolics found in grasses.

POLYPHENOLICS (TANNINS)

In addition to the relatively simple phenolics described above, members of the Poaceae also possess condensed tannins or procyanidins (Burns 1971, Rice and Pancholy 1973, Capinera et al. 1983, Roehrig and Capinera 1983, Buchsbaum et al. 1984, Butler et al. 1984, Rice 1984). Apparently, grasses do not contain hydrolyzable tannins (Swain 1979). The basic structure of condensed tannins is shown in Figure 4.5. For the most part, they consist of repeating units of hydroxyflavanols (Goodwin and Mercer 1983). Typically, condensed tannins occur within grasses at concentrations ranging from 1.0 to 5.0% dry weight (based on astringency estimates, Buchsbaum et al. 1984, Roehrig and Capinera 1983). Presumably, upon tissue damage, tannins bind to plant proteins and carbohydrates, and herbivore digestive enzymes, thus reducing the digestibility of the tissue (Feeny 1976, Rhoades and Cates 1976; but see Bernays 1981, Zucker 1983, and Martin et al. 1985 for alternative mechanisms by which tannins may function in deterring herbivory).

Grasses contain complex polyphenolics that function in a number of ecological processes. Capinera et al. (1983) suggested that range caterpillar, *Hemileuca olivae* Cockerell, preferred C_4 plants over C_3 plants due to a lower protein-precipitating capacity (estimated by astringency) associated with C_4 plants. Roehrig and Capinera (1983) showed that when incorporated within artificial diets, condensed tannins acted as feeding deterrents to the range caterpillar. Additionally, Capinera and Epsky (unpublished) have shown that grasshopper development time is increased and biomass is decreased when individuals are fed diets containing quebracho tannin (a condensed tannin) (Table 4.1).

Procyanidin
(N=1−10)

FIG. 4.5 Procyanidin (=tannin) molecule.

Finally, grass tannins have also been shown to affect ecosystem processes. Low concentrations of condensed tannins (2-200 ppm), presumably originating from species of *Andropogon*, *Sorgastrum*, *Panicum*, and *Aristida*, were able to completely inhibit bacterial nitrification (Rice and Pancholy 1973).

PROTEINASE INHIBITORS

Species of *Aegilops*, *Andropogon*, *Bouteloua*, *Hordeum*, and *Triticum* have all been found to contain appreciable levels (μg/g) of proteinase inhibitors (Applebaum and Konijn 1966, Weiel and Hapner 1976, Ross and Detling 1983). Proteinase inhibitors in

TABLE 4.1
Grasshopper development time and biomass for the migratory
grasshopper, *Melanoplus sanguinipes* (F.), when fed quebracho
tannin-containing diet or tannin-free (control) diet (Capinera and
Epsky, unpublished).

DEVELOPMENT TIME (days)

Tannin Level

Instar	Control	5%[a]	Control	1%[b]
2	6.4	7.1*	6.4	6.4
3	7.0	7.8*	6.7	7.6*
4	7.9	9.2*	7.9	8.8*
5	9.2	10.9*	11.0	11.4
6	14.0	14.9	13.8	15.5*
adult male	---	---	---	---
adult female	---	---	---	---

GRASSHOPPER BIOMASS (mg)

Tannin Level

Instar	Control	5%[a]	Control	1%[b]
2	---	---	---	---
3	---	---	---	---
4	29.1	28.9	26.6	26.3
5	63.9	55.1*	53.2	52.7
6	108.2	101.2*	110.4	100.4*
adult male	194.8	177.8*	205.4	174.9*
adult female	220.5	187.7	227.3	206.2

*Significant differences (p < 0.05; t-test) between treatment and
control.
[a]n = 28-50 for nymph; 15-30 for adult.
[b]n = 10-13 for nymph; 5-8 for adult.

tomato and potato apparently are an inducible herbivore defense
(Green and Ryan 1972, 1973, Ryan 1981). Unlike responses found
in most other plant species, the proteinase inhibitor activity found
in grasses apparently is not inducible (Ross and Detling 1983).
Within the Poaceae, the role of proteinase inhibitors in unclear.
They may provide some degree of resistance to the flour beetle,
Tribolium castaneum (Herbst.) (Applebaum and Konijn 1966) and

are hypothesized to provide resistance in barley to grasshoppers (Weiel and Hapner 1976). Unfortunately, these are the only studies concerning the role of grass proteinase inhibitors, and further research is needed to determine if their existence is widespread in the Poaceae.

TERPENES

Terpenes and terpenoids occur within the Poaceae (Gibbs 1974). Most of the research concerning grass terpenoids has been confined to identifying and quantifying these compounds (Kami 1977, Gibbs 1974, Mody et al. 1974, 1975, Saeed et al. 1978). Four genera of grasses have been reported to contain monoterpenes, 2 genera contain sesquiterpenes, 36 genera were found to contain triterpenoids and only one genus (*Zea*) was found to contain tetraterpenoids. With the exception of plant hormones, diterpenoids seem to be exceptionally rare in grasses (Gibbs 1974). Additionally several steroidal saponin compounds (which like terpenes, are derived from the mevalonic acid-isoprene pathway) are also found in grasses. At least 25 genera have been shown to contain a variety of saponins, and at least 6 genera contain several types of steroids (Gibbs 1974). The ecological roles, if any, that grass terpenes may play are poorly known. Since terpenes have been implicated in deterring herbivory in several systems (Beck and Reese 1976, Bryant 1981, Chapman et al. 1981, Krischik and Denno 1983, Redak and Cates 1984), their potential ecological roles in grasses should be investigated.

PLANT HORMONES

Although phytohormones are not usually considered secondary plant products, they will be covered in this chapter due to the roles they may play influencing graminivores. Most available studies concern the effects that grass plant hormones have on insect herbivores. Abscissic acid, a growth inhibiting hormone, and giberellin A_3, a growth promoting hormone (Figure 4.6), were associated with decreased fecundity, egg viability, and reproductive rate in the grasshopper *Aulocara ellioti* (Thomas) (Visscher 1980). Indole acetic acid, also a growth promoting hormone (Figure 4.6), was associated with increased grasshopper longevity, fecundity, and egg viability (Visscher 1982).

Visscher (1982) also found that lower levels of giberellin A_3 (application levels of 18 mg/l vs. 60 mg/l or 600 mg/l) increased adult longevity, fecundity, and egg viability. Depending on exposure time, the growth hormone ethylene was associated with either increased or decreased nymphal development time in the grasshopper *Melanoplus sanguinipes* (Fabricius) (Chrominski et al. 1982). Long exposure time (24 h) decreased development times, while shorter exposure times (6 h) increased nymphal development time. The longer periods of ethylene exposure (12 and 24 h) decreased adult longevity (Chrominski et al. 1982).

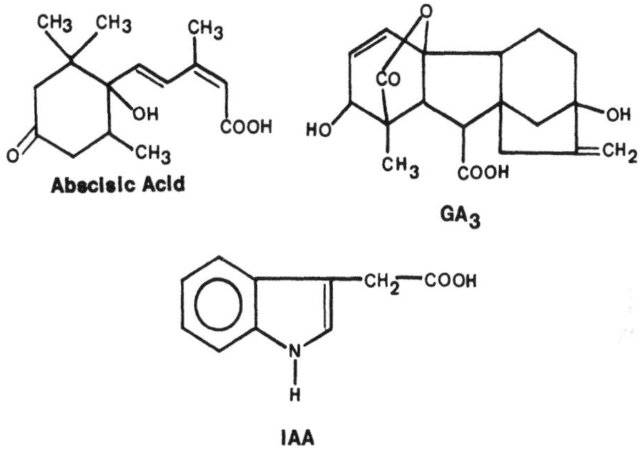

FIG. 4.6 The plant hormones abscisic acid, giberellin A$_3$ (GA$_3$), and indole acetic acid (IAA).

The effects of phytohormones, especially giberellins and ethylene, may be concentration specific. Unfortunately, in the aforementioned studies the concentrations of the phytohormones were not directly measured within the grass being fed upon. Additionally, it was not clear that the observed responses were directly related to the plant hormones *per se* or due to hormone mediated changes which may have occurred within the plant.

SILICA

Silica (SiO$_2$) is often found in the Poaceae at levels that are of ecological importance. Although usually not considered to be a plant secondary compound, silica will be discussed in this chapter because it is generally considered to be a quantitative, and perhaps inducible, plant defense (Rhoades 1983). Silica appears to be a potent defense in grasses against stem-boring insects. Rice plants high in silica content are somewhat resistant to the rice stem borer, *Chilo suppressalis* (Walker) (Djamin and Pathak 1967). Plants high in silica exhibited lower stem borer damage. Stem borers attempting to feed on resistant rice strains exhibited severely worn mandibles and were relatively unsuccessful in boring into the plant stem. Presumably, insects with worn mandibles are less able to feed and damage the plant than relatively normal insects. Additionally, silica levels were inversely correlated with larval density of stem-boring *Oscinella* species on Italian ryegrass, *Lolium multiflorum* L. (Moore 1984). Additional references concerning silica and its effects on insect herbivores may be found in Moore (1984). Mammalian herbivores are also adversely affected by silica in grass plants. Silica has been linked to teeth wear in sheep (Baker et al. 1959), silica urolithiasis (Swingle 1952, Whiting et al. 1958), esophageal cancer (Parry and Hodson 1982, Sangster et al. 1983), and reduced digestibility of forage (Van Soest and Jones 1968). Any of these pathological effects of ingesting silica may reduce herbivory. It appears that silica may be an inducible plant defense. Grasses under long-term mammalian herbivore pressure tend to have higher silica levels than grasses under low herbivore pressure (McNaughton and Tarrants 1983, McNaughton et al. 1985, Brizuela et al. 1986). Whether or not increased silica levels can be induced by short-term artificial herbivory is unclear. For African grass species, McNaughton and Tarrants (1983) and McNaughton et al. (1985) showed that silica levels increased in plants which were artificially defoliated. However, for North American grass species, Brizuela et al. (1986) failed to find increased silica levels following artificial defoliation.

CONCLUSIONS

The Poaceae have long been assumed to be lacking in significant levels of secondary metabolites. One reason for this assumption is the paucity of published studies that document whether grasses contain significant levels of various secondary compounds, and whether such compounds play active roles in ecological processes. Evidence in the chemotaxonomic and poisonous plant literature suggests, however, that grasses contain a plethora of secondary compounds (Kingsbury 1964, Gibbs 1974). The ecological roles, if any, that these compounds may play is less well documented. Further research elucidating the roles that secondary chemistry plays in grasses is sorely needed. This is especially true for rangeland grass species. Knowledge of the chemical basis for herbivore and pathogen resistance could be immensely important, as these traits possibly could be incorporated into commercial varieties, thereby providing low-cost, effective, long-term protection against excessive forage losses by a variety of rangeland pests.

ACKNOWLEDGMENTS

John L. Capinera, Rex G. Cates, Louis B. Bjostad, Anthony Joern, and Charles M. MacVean kindly commented on the manuscript. This work was partially supported by the Western Regional Integrated Pest Management Program and Colorado Agricultural Experiment Station.

REFERENCES

Applebaum, S.W., and A.M. Konijn. 1966. The presence of a *Tribolium* protease inhibitor in wheat. J. Insect Physiol. 12:665-669.

Argandona, V.H., L.J. Corcuera, H.M. Niemeyer, and B.C. Campbell. 1983. Toxicity and feeding deterrency of hydroxamic acids from Graminae in synthetic diets against the green bug, *Schizaphis graminum*. Entomol. Exp. Appl. 34:134-138.

Argandona, V.H., G.F. Pena, H.M. Niemeyer, and L.J. Corcuera. 1982. Effect of cysteine on stability and toxicity to aphids of a cyclic hydroxamic acid from Gramineae. Phytochemistry 21:1573-1574.

Baker, E.A., and I.M. Smith. 1977. Antifungal compounds in winter wheat resistant and susceptible to *Septoria nodorum*. Ann. Appl. Biol. 87:67-73.

Baker, G., L.H.P. Jones, and I.D. Wardrop. 1959. Cause of wear in sheeps' teeth. Nature 184:1583-1584.

Barnett, R.D., and C.E. Caviness. 1968. Inheritance of hydrocyanic acid production in two sorghum X sudangrass crosses. Crop Sci. 8:89-91.

Beck, S.D., and J.C Reese. 1976. Insect-plant interactions: nutrition and metabolism. Rec. Adv. Phytochem. 10:463-512.

Berger, P.J., N.C. Negus, E.H. Sanders, and P.D. Gardner. 1981. Chemical triggering of reproduction in *Microtus montanus*. Science 214:69-70.

Bernays, E.A. 1981. Plant tannins and insect herbivores: an appraisal. Ecol. Entomol. 6:353-360.

Branson, T.E., P.L. Guss and E.E. Ortman. 1969. Toxicity of sorghum roots to larvae of the western corn rootworm. J. Econ. Entomol. 62:1375-1378.

Brizuela, M.A., J.K. Detling, and M.S. Cid. 1986. Silicon concentration of grasses growing in sites with different grazing histories. Ecology 67:1098-1100.

Bryant, J.P. 1981. Phytochemical deterrence of snowshoe hare browsing by adventitious shoots of four Alaskan trees. Science 213:889-890.

Buchsbaum, R., I. Valiela, and T. Swain. 1984. The role of phenolic compounds and other plant constituents in feeding by Canada geese in a coastal marsh. Oecologia 63:343-349.

Burns, R.E. 1971. Method for estimation of tannin in grain sorghum. Agron. J. 63:511-512.

Burritt, E.A., A.S. Bittner, J.C. Street, and M.J. Anderson. 1984. Correlations of phenolic acids and xylose content of cell wall with in vitro dry matter digestibility of three maturing grasses. J. Dairy Sci. 67:1209-1213.

Bush, L.P., C. Streeter, and R.C. Buckner. 1970. Perloline inhibition of in vitro ruminal cellulose digestion. Crop Sci. 10:108-109.

Butler, L.G., D.J. Riedl, D.G. Lebryk, and H.J. Blytt. 1984. Interaction of proteins with sorghum tannin: mechanism, specificity and significance. J. Amer. Oil Chem. Soc. 61:916-920.

Capinera, J.L., A.R. Renaud, and N.E. Roehrig. 1983. Chemical basis for host selection by *Hemileuca olivae*: role of tannins in preference of C_4 grasses. J. Chem. Ecol. 9:1425-1437.

Chapman, R.F., E.A. Bernays, and S.J. Simpson. 1981. Attraction and repulsion of the aphid, *Cavariella aegopodii*, by plant odors. J. Chem. Ecol. 7:881-888.

Chrominski, A., S.N. Visscher, and R. Jurenka. 1982. Exposure to ethylene changes nymphal growth rate and female longevity in the grasshopper *Melanoplus sanguinipes*. Naturwissenschaften 69:45-46.

Conn, E.E. 1979. Cyanide and cyanogenic glycosides, p. 387-412. *In:* G.A. Rosenthal and D.H. Janzen (eds.) Herbivores: their interaction with secondary plant metabolites. Academic Press, New York.

Corcuera, L.J. 1984. Effects of indole alkaloids from gramineae on aphids. Phytochemistry 23:539-541.

Corcuera, L.J., M.D. Woodward, J.P. Helgeson, A.Kelman, and C.D. Upper. 1978. 2,4-Dihydroxy-7-methoxy-2h-1,4-benzoxazin-3(4h)-one, an inhibitor from *Zea mays* with differential activity against soft rotting *Erwinia* species. Plant Physiol. 61:791-795.

Coughenour, M.B. 1985. Graminoid responses to grazing by large herbivores: Adaptations, exaptations, and interacting processes. Ann. Missouri Bot. Gard. 72:852-863.

Couture, R.M., D.G. Routley, and G.M. Dunn. 1971. Role of cyclic hydroxamic acids in monogenic resistance of maize to *Helminthosporium turcicum*. Physiol. Plant Path. 1:515-521.

Cronquist, A. 1981. An integrated system of classification of flowering plants. Columbia Univ. Press, New York.

Djamin, A., and M.D. Pathak. 1967. Role of silica in resistance to asiatic rice borer, *Chilo suppressalis* (Walker), in rice varieties. J. Econ. Entomol. 60:347-351.

Dreyer, D.L., and K.C. Jones. 1981. Feeding deterrency of flavonoids and related phenolics towards *Schizaphis graminum* and *Myzus persicae*: aphid feeding deterrents in wheat. Phytochemistry 20:2489-2493.

Dreyer, D.L., J.C. Reese, and K.C. Jones. 1981. Aphid feeding deterrents in sorghum. Bioassay, isolation, and characterization. J. Chem. Ecol. 7:273-284.

Fairbourn, M.L. 1982. Alkaloid affects in vitro dry matter digestibility of *Festuca* and *Bromus* species. J. Range Manage. 35:503-504.

Feeny, P. 1976. Plant apparency and chemical defense. Rec. Adv. Phytochem. 10:1-40.

Gallagher, C.D., J.H. Koch, and H. Hoffman. 1966. Poisoning by grass. New Sci. 31:412-414.

Gallagher, C.D., J.H. Koch, R.M Moore, and J.D. Steel. 1964. Toxicity of *Phalaris tuberosa* for sheep. Nature 204:542-545.

Gibbs, R.D. 1974. Chemotaxonomy of flowering plants. v. I-IV. McGill-Queen's Univ. Press, Montreal.

Gillingham, J.T., M.M. Shirer, J.J. Starnes, N.R. Page, and E.F. McClain. 1969. Relative occurrence of toxic concentrations of cyanide and nitrate in varieties of sudangrass and sorghum-sudangrass hybrids. Agron. J. 61:727-730.

Goodwin, T.W., and E.I. Mercer. 1983. Introduction to plant biochemistry. Pergamon Press, Elmsford, NY.

Green, T.R., and C.A. Ryan. 1972. Wound-induced proteinase inhibitor in plant leaves: a possible defense mechanism against insects. Science 175:776- 777.

Green, T.R., and C.A. Ryan. 1973. Wound-induced proteinase inhibitor in tomato leaves. Some effects of light and temperature on the wound response. Plant Physiol. 51:19-21.

Harley, K.L.S., and A.J. Thorsteinson. 1967. The influence of plant chemicals on the feeding behavior, development, and survival of the two striped grasshopper, *Melanoplus bivittatus* (Say), Acrididae: Orthoptera. Can. J. Zool. 45:305-319.

Jones, C.G. 1983. Phytochemical variation, colonization, and insect communities: the case of bracken fern, p. 513-558. *In:* R. F. Denno and M. S. McClure (eds.) Variable plants and herbivores in natural and managed systems. Academic Press, New York.

Kami, T. 1977. Composition of the essential oils of sudangrass and hybridsorgo, forage sorghums. J. Agric. Food Chem. 25:1295-1299.

Kingsbury, J.M. 1964. Poisonous plants of the United States and Canada. Prentice-Hall, Englewood Cliffs, NJ.

Klun, J.A., and J.F. Robinson. 1969. Concentration of two 1,4-benzoxazinones in dent corn at various stages of development of the plant and its relation to resistance of the host plant to the European corn borer. J. Econ. Entomol. 62:214-220.

Klun, J.A., C.L. Tipton, and T.A. Brindley. 1967. 2,4-dihydroxy-7-methoxy-1,4-benzoxazin-3-one (DIMBOA), an active agent in the resistance of maize to the European corn borer. J. Econ. Entomol. 60:1529-1533.

Krischik, V.A., and R.F. Denno. 1983. Individual, population, and geographic patterns in plant defense, p. 463-512. *In:* R.F. Denno and M.S. McClure (eds.) Variable plants and herbivores in natural and managed systems. Academic Press, New York.

Lacy, G.H., S.S. Hirano, J.I. Victoria, A. Kelman, C.D. Upper. 1979. Inhibition of soft-rotting *Erwinia* spp. strains by 2,4-dihydroxy-7-methoxy-2H-1,4,benzoxazin-3(4H)-one in relation to their pathogenicity on *Zea mays*. Phytopathology 69:757-763.

Lindroth, R.L., and G.O. Batzli. 1984. Plant phenolics as chemical defenses: effects of natural phenolics on survival and growth of prairie voles (*Microtus orchrogaster*). J. Chem. Ecol. 10:229-244.

Long, B.J., G.M. Dunn, J.S. Bowman, and D.G. Routley. 1977. Relationship of hydroxamic content in corn and resistance to the corn leaf aphid. Crop Sci. 17:55-58.

Loyd, R.C., and E. Gray. 1970. Amount and distribution of hydrocyanic acid potential during the life cycle of plants of three sorghum cultivars. Agron. J. 62:394-397.

Marten, G.C., R.M. Jordan, and A.W. Hovin. 1976. Biochemical significance of reed canary grass alkaloids and associated palatability variation to grazing sheep and cattle. Agron. J. 68:909-914.

Martin, M.M., D.C. Rockholm, and J.S. Martin. 1985. Effects of surfactants, pH, and certain cations on precipitation of proteins by tannins. J. Chem. Ecol. 11:485-494.

McNaughton, S.J. 1983. Physiological and ecological implications of herbivory, p. 657-677. *In:* O.L. Lange, P.S. Nobel, C.B. Osmond, and H. Ziegler (eds.) Physiological plant ecology. vol. 3. Springer-Verlag, Berlin.

McNaughton, S.J., and J.L. Tarrants. 1983. Grass leaf silicification: natural selection for an inducible defense against herbivores. Proc. Natl. Acad. Sci. 80:790-791.

McNaughton, S.J., J.L. Tarrants, M.M. McNaughton, and R.H. Davis. 1985. Silica as a defense against herbivory and a growth promotor in African grasses. Ecology 66:528-535.

Mody, N.V., J. Bhattacharyya, D.H. Miles, and P.A. Hedin. 1974. Survey of the essential oil in *Spartina cynosuroides*. Phytochemistry 13:1175-1178.

Mody, N.V., A.A. de la Cruz, D.H. Miles, and P.A. Hedin. 1975. The essential oil of *Distichlis spicata*. Phytochemistry 14:599-601.

Moore, D. 1984. The role of silica in protecting Italian ryegrass (*Lolium multiflorum*) from attack by dipterous stem-boring larvae (*Oscinella frit* and other related species). Ann. Appl. Biol. 104:161-166.

Owen, D.F., and R.G. Wiegert. 1981. Mutualism between grass and grazers: an evolutionary hypothesis. Oikos 36:376-378.

Parry, D.W., and M.J. Hodson. 1982. Silica distribution in the caryopsis and inflorescence bracts of foxtail millet (*Setaria italica* (L.) Beauv.) and its possible significance in carcinogenesis. Ann. Bot. 49:531-540.

Raffaut, R.F. 1970. A handbook of alkaloids and alkaloid containing plants. John Wiley and Sons, New York.

Reay, P.F. and E.E. Conn. 1970. Dhurrin synthesis in excised shoots and roots of sorghum seedlings. Phytochemistry 9:1825-1827.

Redak, R.A. and R.G. Cates. 1984. Douglas-fir (*Pseudotsuga menziesii*)-spruce budworm (*Choristoneura occidentalis*) interactions: the effect of nutrition, chemical defenses, tissue phenology, and tree physical parameters on budworm success. Oecologia 62:61-67.

Rhoades, D.F. 1983. Herbivore population dynamics and plant chemistry, p. 155-220. *In:* R.F. Denno and M.S. McClure (eds.) Variable plants and herbivores in natural and managed systems. Academic Press, New York.

Rhoades, D.F., and R.G. Cates. 1976. A general theory of plant anti-herbivore chemistry. Rec. Adv. Phytochem. 10:168-213.

Rice, E.L. 1984. Allelopathy. Academic Press, New York.

Rice, E.L., and S.K. Pancholy. 1973. Inhibition of nitrification by climax ecosystems. II. Additional evidence and possible role of tannins. Amer. J. Bot. 60:691-702.

Robinson, J.F., J.A. Klun, and T.A. Brindley. 1978. European corn borer: a nonpreference mechanism of leaf feeding resistance and its relationship to 1,4- benzoxazin-3-one concentration in dent corn tissue. J. Econ. Entomol. 71:461-465.

Robinson, J.F., J.A. Klun, W.D. Guthrie, and T.A. Brindley. 1982. European corn borer (Lepidoptera: Pyralidae) leaf feeding resistance: DIMBOA bioassays. J. Kansas Entomol. Soc. 55:357-364.

54

Rodriguez, E., and D.A. Levin. 1976. Biochemical parallelisms of repellents and attractants in higher plants and arthropods, p. 214-270. *In:* J.W. Wallace and R.L. Mansell (eds.) Biochemical interactions between plants and insects. Rec. Adv. Phytochem. 10:214-270.

Roehrig, N.E., and J.L. Capinera. 1983. Behavioural and developmental responses of range caterpillar larvae, *Hemileuca olivae*, to condensed tannin. J. Insect Physiol. 29:901-906.

Ross, C.W. and J.K. Detling. 1983. Investigations of trypsin inhibitors in leaves of four North American prairie grasses. J. Chem. Ecol. 9:247-257.

Ryan, C.A. 1981. Proteinase inhibitors, p. 351-370. *In:* A. Marcus (ed.) The biochemistry of plants. vol. 6. Academic Press, New York.

Saeed, T., P.J. Sandra, and M.J.E. Verxele. 1978. Constituents of the essential oil of *Cymbopogon jawarancusa*. Phytochemistry 17:1433-1434.

Sanders, E.H., P.D. Gardner, P.J. Berger, N.C. Negus. 1981. 6-methoxybenzoxazolinone: a plant derivative that stimulates reproduction in *Microtus montanus*. Science 214:67-69.

Sangster, A.G., M.J. Hodson, and D.W. Parry. 1983. Silicon deposition and anatomical studies in the inflorescence bracts of four *Phalaris* species with their possible relevance to carcinogenesis. New Phytol. 93:105-122.

Scriber, J.M., W.M. Tingey, V.E. Gracen and J.L. Sullivan. 1975. Leaf-feeding resistance to the European corn borer in genotypes of tropical (low-DIMBOA) and U. S. Inbred (high-DIMBOA) maize. J. Econ. Entomol. 68:823-826.

Seigler, D.S. 1976a. Plants of the northeastern United States that produce cyanogenic compounds. Econ. Bot. 30:395-407.

Seigler, D.S. 1976b. Plants of Oklahoma and Texas capable of producing cyanogenic compounds. Proc. Oklahoma Acad. Sci. 56:95-100.

Simmons, A.B., G.C. Marten. 1971. Relationship of indole alkaloids to palatability of *Phalaris arundinacea* L. Agron. J. 63:915-919.

Smith, T.A. 1975. Recent advances in the biochemistry of plant amines. Phytochemistry 14:865-890.

Swain, T. 1979. Tannins and Lignins, p. 657-682. *In:* G.A. Rosenthal and D.H. Janzen (eds.) Herbivores: their interactions with secondary plant metabolites. Academic Press, New York.

Swingle, K.F. 1952. The chemical composition of urinary calculi from range steers. Amer. J. Vet. Res. 14:493-498.

Todd, G.W., A. Getahun, and D.C. Cress. 1971. Resistance in barley to the greenbug, *Schizaphis graminum*. 1. Toxicity of phenolic and flavonoid compounds and related substances. Ann. Entomol. Soc. Amer. 64:718-722.

VanSoest, P.J., and L.H.P. Jones. 1968. Effect of silica in forages upon digestibility. J. Dairy Sci. 51:1644-1648.

Visscher, S.N. 1980. Regulation of grasshopper fecundity, longevity and egg viability by plant growth hormones. Experientia 36:130-131.

Visscher, S.N. 1982. Plant growth hormones affect grasshopper growth and reproduction, p. 57-62. *In:* J.H. Visser and A.K. Minks (eds.) Proceedings of 5th international symposium on insect-plant relationships. Wageningen, Netherlands.

Weiel, J., and K.D. Hapner. 1976. Barley proteinase inhibitors: a possible role in grasshopper control. Phytochemistry 15:1885-1887.

Whiting, F., R. Connell, and S.A. Forman. 1958. Silica urolithiasis in beef cattle. Can. J. Comp. Med. Vet. Sci. 22:332-337.

Willamen, J.J., and H.L. Li. 1970. Alkaloid-bearing plants and their contained alkaloids. 1957-1968. Lloydia 33(3A), September supplement.

Willamen, J.J., and B.G. Schubert. 1961. Alkaloid-bearing plants and their contained alkaloids. USDA Tech. Bull. 1234.

Wolf, D.D., and W.W. Washko. 1967. Distribution and concentration of HCN in a sorghum-sudan-grass hybrid. Agron. J. 59:381-382.

Woodhead, S. and E.A. Bernays. 1978. The chemical basis of resistance of *Sorghum bicolor* to attack by *Locusta migratoria*. Entomol. Exp. Appl. 24:123-124.

Woodward, M.D., L.J. Corcuera, J.P. Helgeson, A. Kelman, and C.D. Upper. 1978. Factors that influence the activity of 2,4-dihydroxy-7-methoxy-2H-1,4-benzoxazin-3(4H)-one on *Erwinia* species in growth assays. Plant Physiol. 61:803-805.

Zucker, W.V. 1983. Tannins: does structure determine function? An ecological perspective. Am. Nat. 121:335-365.

Zuniga, G.E., and L.J. Corcuera. 1986. Effect of gramine in the resistance of barley seedlings to the aphid *Rhopalosiphum padi*. Entomol. Exp. Appl. 40:259-262.

Zuniga, G.E., M.S. Salgado, and L.J. Corcuera. 1985. Role of and indole alkaloid in the resistance of barley seedlings to aphids. Phytochemistry 24:945-947.

5. GRASS RESPONSE TO HERBIVORY

James K. Detling
Natural Resource Ecology Laboratory and
Department of Range Science
Colorado State University, Fort Collins, Colorado 80523

Grasslands are unique among terrestrial ecosystems in that they support relatively high herbivore loads compared with other ecosystems. Thus, while herbivores typically consume less than 5-10% of the annual net primary production in most terrestrial ecosystems (Chew 1974, Wiegert and Evans 1967), they commonly consume half or more of the annual aboveground net primary production (ANPP) in grasslands. Data reviewed by Detling (in press) indicated that native mammalian herbivores removed on the average about 1/3 of the ANPP while invertebrates consumed or wasted about 5-15% of the ANPP of the world's grasslands and savannas. By comparison, cattle on managed grasslands around the world consume from about 20-80% of ANPP (Detling in press). Lacey and Van Poollen (1981) considered that moderate grazing of western U.S. rangelands constituted 40-60% removal of current year's growth by livestock.

In spite of the relatively heavy grazing by native or domesticated herbivores on grasslands, many grass species, particularly those native to the shortgrass and mixedgrass prairies of the western Great Plains in North America, are quite resistant to defoliation at all but the highest grazing intensities or frequencies. Because plant response to herbivory is mediated by a variety of interactive processes in grassland ecosystems (McNaughton et al. 1981), it is necessary to examine not only individual plants, but plants growing in mixtures in the field, to appreciate the causal mechanisms involved in plant responses to herbivory.

In this paper, I review a number of individual plant responses to grazing, and ways in which these responses may be modified by other environmental variables in the field. Although this paper focuses on important species native to the western Great Plains, results of experiments on plants from other rangelands are discussed when necessary to illustrate a point.

RESPONSES OF INDIVIDUAL PLANTS TO GRAZING

Because of the importance of photosynthetic carbon fixation as a carbon and energy source utilized in grass regrowth (Richards 1986), a number of investigators have recently focused attention on the photosynthetic response of range grasses to defoliation. A variety of such responses have been reported, and the nature and magnitude of the response appears to depend upon the type and severity of damage done, the specific plant parts on which photosynthesis is being measured, and the time frame over which photosynthesis is measured following defoliation.

Available evidence indicates that when an individual leaf blade is damaged, the net photosynthetic rate (P_N) per unit of remaining leaf area of the damaged leaf is decreased. Detling et al. (1979a) damaged individual leaves of western wheatgrass, *Agropyron smithii* Rydb., at several places along their length to simulate observed grasshopper feeding patterns. In all cases the damage decreased the P_N rate in the remaining damaged leaf blades. For example, when approximately 25% of the leaf area was removed by cutting 4 cm long notches from the base and tip of the blade, P_N per unit of remaining tissue was reduced by one third almost immediately and remained at this level over the next week (line C of Figure 5.1). Thus, in this example, the photosynthetic capacity of the individual damaged leaf was reduced by approximately half when the lost leaf area and reduced photosynthetic rate of remaining leaf tissue were both considered.

In contrast to results such as these, many investigators have reported increased P_N rates of leaves in response to defoliation. This type of response, which is illustrated in line B of Figure 5.1, often has been observed in remaining undamaged leaves of partially defoliated plants. Thus, for example, when 50-75% of the tillers of western wheatgrass were removed by clipping, P_N rates of remaining undamaged leaves averaged 25-35% greater than leaves of the same age on undamaged control plants (Painter and Detling 1981, Detling and Painter 1983). As a result of such compensatory photosynthesis, photosynthetic capacity of grass canopies may not be reduced as much by defoliation as would be expected solely on the basis of the amount of leaf area removed by grazers.

58

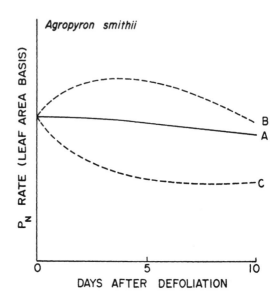

FIG. 5.1   Net photosynthetic rates (P_N) of individual leaf blades of western wheatgrass, *Agropyron smithii*. Line A, undamaged leaf on an undamaged plant; line B, undamaged leaf on a plant from which 50-75% of the connected tillers had been removed; line C, damaged leaf from which 25% of the blade had been removed to simulate grasshopper damage. Redrawn from data of Detling et al. (1979a), Painter and Detling (1981), Detling and Painter (1983).

In spite of the apparent potential which compensatory photosynthesis has for enhancing plant regrowth following defoliation, the extent to which it actually contributes to plant regrowth has not been entirely resolved. Nowak and Caldwell (1984) studied compensatory photosynthesis of two other *Agropyron* species in the field. In these species, *A. desertorum* (Fisch.) Schult. and *A. spicatum* (Pursh) Scribn. & Smith, increases in P_N rates were usually lower than those reported above, and the magnitude of compensatory photosynthesis was largest in the two oldest leaves that were present when the plants were defoliated. Because these leaves represented only a small proportion of total photosynthetic leaf area, they considered that their contribution to total carbon assimilation would be rather small. However, in another study of the same two species, Caldwell et al. (1981) observed that leaf blades on regrowing tillers of defoliated plants had significantly higher photosynthetic capacities than leaf blades on non-defoliated plants. This may be partially attributable to the greater average age of leaves on non-defoliated plants.

The aforementioned photosynthetic responses to defoliation were obtained by measuring P_N rates of single leaves. The defoliated plant, however, typically consists of a mixture of partially damaged leaves, undamaged leaves, and young leaves which expanded following grazing. To understand the effects of defoliation on subsequent primary production, then, it is important to consider the photosynthetic rate of the entire regrowing plant. In another laboratory experiment, Detling et al. (1979b) measured the P_N rates of entire shoot systems of defoliated and non-defoliated *Bouteloua gracilis* (H.B.K.) Griffiths plants. As indicated in Figure 5.2a, mean P_N rate per unit leaf area of defoliated plants (line B) was reduced by half or more immediately following defoliation. However, by three days after defoliation, mean P_N rate on the regrowing defoliated plants had exceeded that of the non-defoliated plants (line A) and remained higher for at least the next week. The immediate decline in photosynthesis following defoliation probably resulted from a combination of reduced photosynthetic rates in remaining damaged leaf blades, and the resulting high proportion of photosynthetically less efficient leaf sheaths in the defoliated plants. The higher mean P_N rate of shoots in the defoliated plants after several days likely resulted from a rapidly increasing proportion of newly produced leaf blades which have a high photosynthetic capacity, and from an increase in photosynthetic rate of any remaining undamaged leaf blades.

Another frequently observed response of grasses to grazing is an alteration in how the plant distributes and utilizes photosynthates. Generally, grazed plants utilize proportionately more of their current photosynthetic production for the synthesis of new shoots at the expense of roots (Ingham and Detling 1984, Richards 1984). For example, in another experiment with *B. gracilis*, Detling et al. (1980) observed that while almost 40% of the new growth of non-defoliated plants was in roots and crowns, only about 20% of the total new growth of defoliated plants occurred there (Figure 5.3). By contrast, less than 40% of the new growth of non-defoliated plants was allocated to production of leaf blades, while new leaf blade growth accounted for approximately 60% of the total new growth of defoliated plants. The allocation of a high proportion of current photosynthates to synthesis of new leaf blades at the expense of other plant parts in defoliated plants appears to be an adaptation which permits rapid restoration of plant photosynthetic capacity following grazing. In fact, more grazing tolerant ecotypes of *A. smithii* apparently allocate a proportionately greater amount of their new growth to synthesis of new leaf blades than do less grazing tolerant ecotypes following defoliation (Detling et al. 1986).

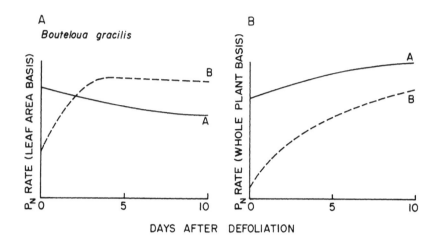

FIG. 5.2 Net photosynthetic rates (P_N) of entire blue grama *Bouteloua gracilis*, plants after defoliation to a height of 4 cm. (A) Mean P_N rate per unit leaf area of non-defoliated (line A) and defoliated (line B) plants. (B) Mean P_N rate per entire plant for non-defoliated (line A) and defoliated (line B) plants. Redrawn from data of Detling et al. (1979b).

The net effects of the defoliation-induced changes in P_N rates and carbon allocation patterns discussed above on total plant carbon balance of *B. gracilis* are shown in Figure 5.2b (Detling et al. 1979b). Immediately following clipping to 4 cm to simulate grazing by large herbivores, total plant net CO_2 exchange remained positive but was less than 10% as great as that of similar non-defoliated plants. Over the next 10 days, however, net whole plant CO_2 uptake increased more rapidly in defoliated plants so that by 10 days after defoliation, photosynthetic production of defoliated plants was already 80% of that of the non-defoliated plants.

Although there have been a few reports that grazing either increases or has no effect on root standing crops or production (Sims and Singh 1978a,b) root biomass is usually reduced by grazing by aboveground herbivores (Weaver 1950, Schuster 1964). For example, Ingham and Detling (1984) evaluated seasonal changes in root biomass beneath *A. smithii* and *Schizachyrium scoparium* (Michx.) Nash on heavily grazed prairie dog colonies and adjacent lightly grazed uncolonized areas. Root biomass was consistently lower on the heavily grazed site and annual net root production (Figure 5.4) was 1/3 to 1/2 lower on more heavily grazed prairie dog colonies. These reductions in root biomass and production were apparently caused by a combination of reduced photosynthetic capacity of the shoot systems and reductions in the proportion of photosynthates which were used in root growth.

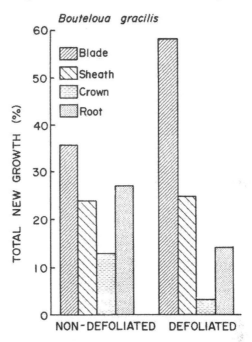

FIG. 5.3 Effect of defoliation on allocation of biomass produced in the initial 10 days following defoliation of blue grama. Redrawn from data of Detling et al. (1980).

A traditional view in range ecology has held that carbohydrate reserves are a primary source of carbon for regrowth following defoliation (Trlica 1977). Consequently, range scientists have attempted to relate plant regrowth potential to levels of soluble carbohydrate reserves in roots and crowns. As indicated in literature reviewed by Richards and Caldwell (1985) and Richards (1986), such correlations have frequently not been successful. In a carefully conducted series of field experiments, Richards and Caldwell (1985) compared the amount of carbon supplied to regrowth from storage and photosynthesis in two bunchgrass species of *Agropyron*. They found that the more grazing tolerant species, *A. desertorum*, produced more regrowth than the less tolerant species, *A. spicatum*, but that differences

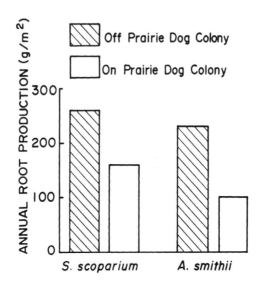

FIG. 5.4   Comparison of annual net root production in the upper 10 cm of soil beneath shoots of *Schizachyrium scoparium* and *Agropyron smithii* on a heavily grazed prairie dog colony and an adjacent, uncolonized lightly grazed area. Redrawn from data of Ingham and Detling (1984).

within and between species were not correlated with nonstructural carbohydrate concentrations, total carbohydrate pools, or amounts of carbohydrates utilized during regrowth. They further concluded that current photosynthesis was the principal source of carbon for regrowth, and that utilization of carbon reserves exceeded utilization of photosynthetically fixed carbon for only 2-5 days following defoliation in plants exhibiting maximal regrowth rates. When regrowth potential was reduced even more by removal of apical meristems, utilization of current photosynthates during regrowth immediately exceeded utilization of stored reserves as a carbon source. Thus, carbohydrate reserve levels appeared rarely to limit regrowth of these grazed grasses. Instead, Richards and Caldwell (1985) concluded that meristematic limitations were the dominant control on shoot regrowth.

PLANT RESPONSES TO GRAZING IN AN ECOSYSTEM CONTEXT

Most of the grazing effects described above resulted directly from removal of leaves or other photosynthetic parts of grasses.

In addition to such direct effects, however, grazing may cause habitat and microclimatic changes in the system. For example, grazing might be expected to alter the soil water balance through a variety of potentially offsetting effects. A reduction in the total live leaf area per unit of ground area might result in less total transpirational water loss from the system. Similarly, reductions in leaf standing crop would ultimately lead to a diminution of litter production (Coppock et al. 1983a), and this in turn might result in a decrease in litter interception of precipitation, especially from small rainfall events. At the same time, litter reduction would possibly lead to an increase in bare soil evaporation losses. Because of potentially offsetting effects such as these, Archer and Detling (1986) and Svejcar and Christiansen (1987) evaluated grazing effects on water relations of native and seeded grasslands, respectively. Results of the native grassland study in western South Dakota (Archer and Detling 1986) suggested that there were relatively few differences in leaf water potential of grasses in heavily grazed prairie dog colonies and adjacent lightly grazed areas. The study of Svejcar and Christiansen (1987), however, indicated that both leaf water potential and leaf conductance were significantly higher throughout the growing season in a heavily grazed Caucasian bluestem, *Bothriochloa caucasica* (Trin.) C.E. Hubb., stand than a nearby lightly grazed stand in Oklahoma (Figure 5.5). These results suggest that compensatory growth of grasses in heavily grazed sites may result, at least partially, from conservation of soil water and improved plant water relations.

Another indirect effect of grazing concerns alterations in competitive interactions among plants of the same or other species, or even among interconnected tillers of the same individuals. In western South Dakota, Archer and Detling (1984) found that leaf growth of *Andropogon gerardii* Vitman and *Carex filifolia* Nutt. which were defoliated biweekly was comparable to that of non-defoliated plants if competition from nearby plants was also reduced by defoliation. However, there was little or no new leaf growth on defoliated tillers, and survival of those tillers was reduced, when simulated grazing events did not simultaneously remove leaves of surrounding plants. Therefore, a selective herbivore which defoliates only one or a few plants in a community may have a more deleterious effect on survival of those plants than an herbivore which also removes the surrounding vegetation.

From the standpoint of forage production for livestock, it is important to consider how grazing affects subsequent ANPP. Clearly, the nature of the response is complicated and will vary with the type of plant grazed; the frequency, intensity and time of grazing; rainfall and other climatic conditions; range condition;

64

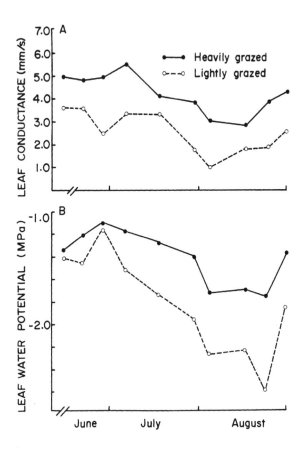

FIG. 5.5 Intraseasonal changes in (A) leaf conductance and (B) leaf water potential of Caucasian bluestem plants from heavily and lightly grazed sites. Redrawn from data of Svejcar and Christiansen (1987).

and many other factors. In some quarters, at least, there has developed the widespread but controversial (Belsky 1986, McNaughton 1986) belief that moderate grazing results in an increase in ANPP (McNaughton 1979). Lacey and Van Poollen (1981) summarized results from 12 studies in western North American rangelands which had reported moderate levels (40-60%) of utilization of the current year's forage production. While the investigators who conducted the studies used a variety of different techniques and approaches to the problem, results of the analyses of Lacey and Von Poollen (1981) (Figure 5.6) indicated that ANPP was either reduced or unaffected by the levels of grazing reported in the reviewed studies. Although this suggests that primary production is not apparently maximized by moderate grazing, the quality of forage on grazed areas is often higher than that on similar lightly grazed or ungrazed areas (Coppock et al. 1983a,b). Specifically, the nitrogen concentration and digestibility of leaves on grazed plants is higher than in ungrazed plants. This probably occurs partly from the fact that the average leaf age of regrowing grasses is less than that of the leaves of ungrazed grasses. In part, however, it may be a result of increased nitrogen uptake or increased allocation of nitrogen to the shoots of regrowing grazed grasses (Jaramillo 1986). Thus, it is possible that even though grazing may reduce subsequent ANPP, incorporation of nitrogen into aboveground plant tissues may be enhanced, or at least not substantially reduced, by moderate levels of grazing.

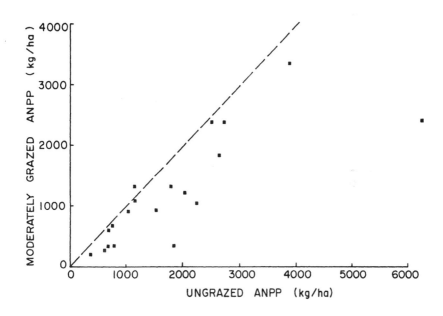

FIG. 5.6 Comparison of aboveground net primary production (ANPP) on moderately grazed and adjacent ungrazed rangelands in the western United States. Dashed diagonal line represents 1:1 correspondence. Drawn from data summarized in the review of Lacey and Van Poollen (1981).

66

REFERENCES

Archer, S., and J.K. Detling. 1984. The effects of defoliation and competition on regrowth of tillers of two North American mixed-grass prairie graminoids. Oikos 43:351-357.

Archer, S., and J.K. Detling. 1986. Evaluation of potential herbivore mediation of plant water status in a North American mixed-grass prairie. Oikos 47:287-291.

Belsky, A.J. 1986. Does herbivory benefit plants? A review of the evidence. Am. Nat. 127:870-892.

Caldwell, M.M., J.H. Richards, D.A. Johnson, R.S. Nowak, and R.S. Dzurec. 1981. Coping with herbivory: photosynthetic capacity and resource allocation in two semiarid *Agropyron* bunchgrasses. Oecologia 50:14-24.

Chew, R.M. 1974. Consumers as regulators of ecosystems: an alternative to energetics. Ohio J. Sci. 74:359-370.

Coppock, D.L., J.K. Detling, J.E. Ellis, and M.I. Dyer. 1983a. Plant-herbivore interactions in a North American mixed-grass prairie. I. Effects of black-tailed prairie dogs on intraseasonal aboveground plant biomass and nutrient dynamics and plant species diversity. Oecologia 56:1-9.

Coppock, D.L., J.E. Ellis, J.K. Detling, and M.I. Dyer. 1983b. Plant-herbivore interactions in a North American mixed-grass prairie. II. Responses of bison to modification of vegetation by prairie dogs. Oecologia 56:10-15.

Detling, J.K. Grasslands and savannas: regulation of energy flow and nutrient cycling by herbivores. *In:* L. R. Pomeroy and J. A. Alberts (eds.) Ecosystems: analysis and synthesis. Springer Verlag, New York. (in press).

Detling, J.K., M.I. Dyer, and D.T. Winn. 1979a. Effect of simulated grasshopper grazing on CO_2 exchange rates of western wheatgrass leaves. J. Econ. Entomol. 72:403-406.

Detling, J.K., M.I. Dyer, and D.T. Winn. 1979b. Net photosynthesis, root respiration, and regrowth of *Bouteloua gracilis* following simulated grazing. Oecologia 41:127-134.

Detling, J.K., M.I. Dyer, C. Procter-Gregg, and D.T. Winn. 1980. Plant-herbivore interactions: examination of potential effects of bison saliva on regrowth of *Bouteloua gracilis* (H.B.K.) Lag. Oecologia 45:26-31.

Detling, J.K., and E.L. Painter. 1983. Defoliation responses of western wheatgrass populations with diverse histories of prairie dog grazing. Oecologia 57:65-71.

Detling, J.K., E.L. Painter, and D.L. Coppock. 1986. Ecotypic differentiation resulting from grazing pressure: evidence for a likely phenomenon, p. 431-433. *In:* P.J. Joss, P.W. Lynch, and O.B. Williams (eds.) Rangelands: A resource under siege. Proceedings of the Second International Rangeland Congress, Australian Academy of Sciences, Canberra, Australia.

Ingham, R.E., and J.K. Detling. 1984. Plant herbivore interactions in a North American mixed-grass prairie. III. Soil nematode populations and root biomass on *Cynomys ludovicianus* colonies and adjacent uncolonized areas. Oecologia 63:307-313.

Jaramillo, V. 1986. The effect of grazing history and competition on the response of *Bouteloua gracilis* to defoliation. Unpublished M.S. Thesis, Colorado State Univ.

Lacey, J.R., and H.W. Van Poollen. 1981. Comparison of herbage production on moderately grazed and ungrazed western ranges. J. Range Manage. 34:210-212. .

McNaughton, S.J. 1979. Grazing as an optimization process: grass-ungulate relationships in the Serengeti. Am. Nat. 113:691-703.

McNaughton, S.J. 1986. On plants and herbivores. Am. Nat. 128:765-770.

McNaughton, S.J., M.B. Coughenour, and L.L. Wallace. 1981. Interactive processes in grassland ecosystems, p. 167-193. *In:* J.R. Estes, R.J. Tyrl, and J.N. Brunken (eds.) Grasses and grasslands: systematics and ecology. Univ. Oklahoma Press, Norman.

Nowak, R.S., and M.M. Caldwell. 1984. A test of compensatory photosynthesis in the field: implications for herbivory tolerance. Oecologia 61:311-318.

Painter, E.L., and J.K. Detling. 1981. Effects of defoliation on net photosynthesis and regrowth of western wheatgrass. J. Range Manage. 34:68-71.

Richards, J.H. 1984. Root growth response to defoliation in two *Agropyron* bunchgrasses: field observations with an improved root periscope. Oecologia 64:21-25.

Richards, J.H., and M.M. Caldwell. 1985. Soluble carbohydrates, concurrent photosynthesis and efficiency in regrowth following defoliation: a field study with *Agropyron* species. J. Appl. Ecol. 22:907-920.

Richards, J.H. 1986. Plant response to grazing: the role of photosynthetic capacity and stored carbon reserves, p. 428-430. *In:* P.J. Joss, P.W. Lynch and O.B. Williams (eds.) Rangelands: A Resource Under Siege. Proceedings of the Second International Rangeland Congress. Australian Acad. Sci., Canberra.

68

Schuster, J.L. 1964. Root development of native plants under three grazing intensities. Ecology 45:63-70.

Sims, P.L., and J.S. Singh. 1978a. The structure and function of ten western North American grasslands. II. Intra-seasonal dynamics in primary producer compartments. J. Ecol. 66:547-572.

Sims, P.L., and J.S. Singh. 1978b. The structure and function of ten western North American grasslands. III. Net primary production, turnover, and efficiencies of energy capture and water use. J. Ecol. 66:573-597.

Svejcar, T., and S. Christiansen. 1987. Grazing effects on water relations of Caucasian bluestem. J. Range Manage. 40:15-18.

Trlica, M.J. 1977. Distribution and utilization of carbohydrate reserves in range plants, p. 73-96. In: R.E. Sosebee (ed.) Rangeland plant physiology. Society for Range Management, Denver, CO.

Weaver, J.E. 1950. Effects of different intensities of grazing on depth and quantity of roots of grasses. J. Range Manage. 3:100-113.

Wiegert, R.G., and F.C. Evans. 1967. Investigations of secondary productivity in grasslands, p. 499-518. In: K. Petrusewicz (ed.) Secondary productivity of terrestrial ecosystems: principles and methods. Vol. II. Warszawa: Pànstwowe Wydawnictwo Naukowe.

6. ESTIMATION OF FORAGE REMOVAL BY RANGELAND PESTS

Charles D. Bonham
Range Science Department
Colorado State University, Fort Collins, Colorado 80523

The magnitude of forage removal by rangeland pests has been of interest for centuries. Historical documents abound with references to vegetation removal and destruction by pests such as locusts.

A definition of "rangeland pests" is usually restricted to herbivores since they eat "forage". "Forage" is defined by livestock owners as vegetation which can be used as feed by their livestock. It follows, therefore, that the basic problem in estimation of forage removal by rangeland pests is to measure the forage available to feed livestock. The problem is really one of not enough forage to go around which leads to the focus being placed on consumers other than livestock, and these consumers are labeled as "pests". The magnitude of forage removal by rangeland pests often is significant; however, estimates of this removal may not be accurate. An individual or a small number of individuals usually have been used in an attempt to measure the amount of forage removed by insects and other pests on rangelands. Since it has been difficult to use a large number of individuals over a long time interval, estimates of real forage removal by rangeland pests have been made on the basis of extrapolation.

DAMAGE BY PESTS

Feeding behavior of rangeland pests has been studied extensively over the past 50 years. Much of the early work was concerned with only a few destructive species (Gangwere 1961). Most studies have emphasized diet selection or environmental factor effects on feeding habits of these pests (Parker 1930, Isely 1938, Oma and Hewitt 1984). The effect of an insect population on vegetation is the product of two opposing processes: the consumption rate of insects and the growth rate of the plant.

These factors interact with age, population size, temperature, and other factors of the insect and plant populations, as well as those of the environment.

Bullen (1966) used data from a number of sources and concluded that grasshoppers at a density of 20 per sq. yd. cause, on average, a vegetative loss of 7 lbs. per acre per day in North American rangelands. Infestation at this level, over a 3-month period during the summer, would consume 600 lbs. per acre and would be equivalent to the complete destruction of typical shortgrass prairie. On poor grazing land, grasshoppers at densities as low as 3 per sq. yd. destroy more than 50% of the vegetation, which is the maximum permitted in good range management practices. The amount of damage caused by grasshoppers is a function of number of variables including amount of vegetation eaten daily by a single insect, differences in food preferences between species, fluctuations in population size, and relative mobility of the insect stages.

Mulkern et al. (1964) and Kelly and Middlekauf (1961) pointed out that some grasshoppers cut blades and stems near the crown and eat only a part of them. The resulting damage is far more serious than just the loss of grass consumed. These authors also concluded that grasshoppers feed closer to the ground than do livestock and, therefore, retard growth, prevent reseeding, and even kill plants.

Leafhoppers of the genus *Athysanella* are typical inhabitants of the shortgrass regions of North America. They are sucking insects and the effects of their feeding are not obvious, but they do contribute to loss of forage on rangelands. It is possible that their total effect on grassland productivity is as great as some of the more obvious invertebrate consumers. It has been reported that their biomass is often greater than that of all other foliage feeders combined (Blocker 1969). Blocker also reported that harvester ants could be very serious in shortgrass prairie regions. Harvester ants have been exterminated in studies and as a result, *Agropyron smithii* and other grasses showed great increase in forage production.

METHODS OF STUDY

A number of methods have been employed in studies of feeding habits of rangeland pests. Some direct field observations have been made (Criddle 1933, Anderson and Wright 1952), but most studies emphasize crop analysis and differential feeding trials (Ueckert and Hansen 1971, Isely 1938). Mitchell (1975) made a detailed study of grasshopper food preferences on shortgrass prairie. His study involved feeding trials which included collection of food plant material from ungrazed and irrigated areas of a shortgrass prairie site. The study was conducted with the insects contained in cages. The amount of forage consumed was not measured directly but was estimated using an equation which incorporated initial available forage, unavailable forage, and destroyed forage.

Putnam (1962) studied foraging behavior of some grasshoppers on native grasslands of British Columbia. Like other studies previously conducted, this one also made use of cages to conduct feeding trials. Losses of forage caused by grasshopper feeding did not consider the nutritional value per unit weight of the unconsumed residues. Since grasshoppers eat the same plant parts that domestic grazing animals prefer to feed upon, Putnam (1962) concluded that his data may underestimate true forage losses.

Hewitt (1980) made a study of the tolerance of 10 species of *Agropyron* fed upon by black grass bug, *Labops hesperius* Uhler. He used two different sizes of cages for the study. A large cage was placed over several plants in two different treatments while a small cage was placed over individual plants in another treatment. Bugs were collected and placed in cages with *Agropyron* species. By taking plant measurements such as number of culms per plant, seed weight, forage production, and percentage leaf damage, he determined that some *Agropyron* species were tolerant to grass bug feeding at various densities.

Moroka et al. (1982) made a study of the effects of kangaroo rat on southern New Mexico desert rangelands. Emphasis was placed on rat mounds and the amount of damage done to forage by the presence of these mounds. Size of mounds was calculated, and aerial cover and composition of plants was obtained by a point method. The highest mound density occurred in black grama rangeland, while the lowest density occurred in a mesquite grassland. Cruz and Turpin (1983) made a study of larval infestations of the small armyworm. This study, which emphasized corn, revealed that the relationship of leaf damage ratings and yield was linear and inverse. Yield losses were directly related to reduction in kernel numbers on ears from infested plants. Few studies using quantitative information such as that used by these authors exist in the study of native plant species and insect infestation. However, Hewitt and Berdahl (1984) studied food preferences of grasshoppers upon alfalfa cultivars which were adapted for rangeland interseeding. Both laboratory and field tests were conducted in this study. Evaluation of the plants consisted of classifying defoliation stages of each plant within a plot. Categories included all leaves present, between 1 and 25% destroyed, between 26 and 50% destroyed, etc. These categories were assigned rating values of 0 to 5, and data analyses were conducted on the ratings.

Hingtgen and Clark (1985) studied the impact of small mammals on the vegetation of reclaimed land in the northern Great Plains. Vegetation sampling emphasized standing crop for each plant species present on the reclaimed site. These plots were caged to keep small mammals from grazing and then harvested after the growing season ended. Paired plots were then selected, and mammals allowed to graze on the vegetation. Assumptions were made concerning the rate of intake for the mammals, and this rate, in turn, was converted to an estimate of consumption rates for the small mammals. No direct measurements were made for the consumption of forage by individual species of small mammals.

Oma and Hewitt (1984) conducted a study of food consumption by various grasshopper species as affected by the presence of a microbial agent, *Nosema locustae* Canning. The study utilized food consumption tests to obtain dried leaf weights before and after feeding. Cages were used to contain the grasshoppers, and differential leaf weights were used to calculate losses of plant material.

Roundy et al. (1985) studied the effects of jackrabbit grazing on crested wheatgrass seedlings. Field experiments included the use of exclosures and paired plots inside and outside of each exclosure. Forage yield was determined by clipping the paired plots and taking differences inside and outside of exclosures. This study also considered the occurrence of induced drought; i.e., seedlings received different amounts of water over the study period. One-meter-square plots were used to obtain clippings.

FORAGE LOSS

Forage loss has been defined by Walker (1982) as:

$$W = \frac{M - Y}{M} \times 100$$

where
W = loss in yield due to pests (in %),
M = maximum, potential or attainable yield in absence of pests, and
Y = yield in presence of pests.

The amount of forage must be estimated in the absence of pests which feed upon it. Methods to obtain this estimate are well-known and are presented frequently in the literature. Therefore, determination of M is not a problem. The same can also be said of the yield of forage in the presence of pests, Y. Any number of harvesting techniques for forage production could be used to make this estimate; however, the determination of both

M and Y at an accuracy level which would then permit an accurate determination of W is difficult. This difficulty arises namely as a result of economic considerations. Economic considerations usually prohibit such a large-scale sampling effort. Therefore, W often is not estimated with any desirable degree of precision from field studies.

Difficulties in measuring forage loss result from causes other than direct removal of vegetation:

1. Photosynthetic damage done by sucking insects, for example, can result in a reduction in area of photosynthetic tissue (both leaves and stems). It is very difficult to measure loss accounted for through this damage.

2. Loss of forage by feeders on plant sap also results because uptake and translocation of water and nutrients are interrupted in their passage from roots or leaves to storage organs such as seeds. Likewise, this source of forage loss is difficult to measure.

3. Another major difficulty in measurement is caused by interactions which occur by different kinds of pests feeding on different plant parts. Then it becomes improbable that accurate measures, in the field, of forage losses due to a specific pest can be made with any accuracy.

4. Another difficulty encountered in field measurements is that of forage yield compensation. That is, plants that have not been fed upon by a pest may increase their forage production and may compensate for the reduced yield of forage by adjacent plants which have been fed upon by the pests. This measurement of yield compensation could be accounted for only by individual plant measurements. Thus, one would have to seek out plants which were utilized by the pests and those not utilized. Then differences in forage produced by these plants are measured to determine the net forage loss or gain.

    A complication also arises in some cases since an increase in forage yield does occur when low infestation levels of rangeland pests are present. Resources are often not available to obtain adequate measures of these losses and possible subsequent increases that do occur at various infestation levels of the pests.

5. Variation in forage yield is significant even in the absence of pests, and presents a measurement problem. Natural variation often accounts for a large degree of the total variation found in yields. The variation

introduced by pests which feed on plants further complicates the assessment of forage losses.

6. The difficulty in measurement of forage standing crop includes the separation of live and recent dead forage for each plant species. Therefore, some effort in the estimation process must be made to separate at least three categories of forage: live, recent dead, and old dead material. There are techniques available for such a separation, but economic considerations may prevent their use for estimation of forage losses over large areas.

7. Lastly, a determination of what rangeland pests would consider "forage" has not been fully worked out. Old dead forage that has been produced by plants and is left standing may or may not be utilized by some rangeland pests.

STATIC ESTIMATES OF FORAGE LOSSES

Static estimates of forage losses are estimates of loss on a unit area basis obtained at one time during the growing season. Most studies conducted on a static basis utilize the time period which occurs during the end of the growing season, resulting in an estimate for a particular vegetation type only. This static estimate gives no opportunity to estimate production for a species producing forage throughout the season. Since material dies in a continuous fashion, the static estimate of forage production and/or loss is not an accurate one.

There are at least three methods in current use for making static estimates of forage. One method is to make an educated guess in terms of weight (grams) per unit area. This guess is accomplished only after much practice with what constitutes a certain amount per given area (grams/unit area) for the species under consideration. These estimates may be well suited for large-scale studies where hundreds or even thousands of acres are under consideration for some type of pest management program.

Secondly, a more precise estimate may be obtained by weighing the amount of material produced by plant species per unit area. Agronomic methods, such as clipping a bounded unit area and weighing the amount of forage obtained from that unit area, have been more widely accepted than the estimation process. The agronomic harvesting method certainly requires much more time than estimating and, therefore, would be much more expensive than that of an estimate. Therefore, a third method has been suggested and used by many workers who are interested in measuring forage production. Both the estimate and the weight are combined into what is known as double sampling (Ahmed and Bonham 1982, Ahmed et al. 1983). In this procedure, some unit area is selected whereby estimates are made of the amount of forage per unit area, and then the same areas are clipped and weighed to obtain actual measurements. A correction is then made for these estimates by the use of double sampling equations. This process is discussed in detail in another section.

DYNAMIC ESTIMATES OF FORAGE

A dynamic estimate is an estimate of the amount of forage which is produced and utilized on a unit area over several periods of time. The following equation was modified from Germain (1982) and represents a dynamic estimate of standing crop (SC_t):

$$SC_t = SC_{t-1} + G_{t-1,t} + E$$
SC_t = standing crop biomass at time t.
$G_{t-1,t}$ = incremental growth from time t-1 to t.
E = amount of error, including natural variation.

In this equation, SC_t represents the initial level of forage available and then the sum of the increments of growth produced between the first time interval and the last time interval. E is the amount of error, which includes natural variation, that is made in the sampling process. Commonly, this method is used in association with cages, whereby plots are caged and the pest is introduced into the cage. Ingestion of forage by the pest is measured by fecal materials or other methods to indicate indirectly the amount of forage that has been removed. The uncaged plots and the caged plots containing the pest are clipped after a period of time; differences would indicate the amount of forage that has been removed by the pests from time t to t-1. However, the uncaged plots are also subject to being grazed by other herbivores or exposure to a different microenvironment.

The caged-uncaged plot scheme has been used to obtain dynamic estimates of the amount of forage removed by livestock as well as some rangeland pests. In this method, the caged plot is used to keep the herbivore from grazing forage, while the uncaged would represent forage subjected to grazing over a period of time. On the other hand, it might be useful to cage a pest and leave the other paired plot uncaged to measure how much has been removed by the pest and repeat the process over a number of time units. In this case, the equation (Germain 1982) for the estimate will be:

$$ANPP = U_1 + \sum_{t=2}^{k} (U_t - C_t)$$

ANPP = annual net primary production
U_t = standing crop biomass at time t in uncaged plots.
C_t = standing crop biomass at time t in caged plots.

U_t is the uncaged plot, while C_t is the caged plot. In this case, note the assumption that caging was done after growth has begun. Any number of time units can be used. Studies and subsequent analyses conducted with this method have indicated that a number of problems exist in the methodology. Argument still exists as to the validity of this approach to measure forage losses of herbivores. Nevertheless, modification of the approach might be feasible. Assumptions needed for the paired plot approach include that pairs of plots have the same initial forage amounts whether caged or uncaged, and that uncaged plot forage is not removed by herbivores, other than those being investigated, by death of plants, etc. Also, the cage is presumed to have no effect on primary production; if caging does, then the experiment should be conducted with cages containing herbivores and cages without herbivores.

A large sample size would allow most of the assumptions to be met in field studies. Usually, however, 10 plots or fewer for a given pasture are used and, therefore, these assumptions are probably not met.

In order to go from a small caged-plot study to pasture estimates of forage removal, a large sample size is required. The adequate sample size may be in excess of 100 observations per vegetation type or pasture. In order to accommodate large sample sizes, a double-sampling method should be considered.

DOUBLE-SAMPLING METHOD

The double-sampling method, as previously mentioned, includes both estimating and harvesting the forage in a known unit area. A small number of plots are harvested and estimated, while a much larger number of plots are estimated only (Ahmed and Bonham 1982). The weight estimate equation is as follows:

$$Y = A + BX$$

where
Y = actual forage weight/unit area, and

X = forage weight estimated by sight.

In this equation, A is a site or plant-specific constant, while B is a measure of the rate of change in the actual weight compared to the weight that has been estimated visually. This method is widely used in rangeland sampling to estimate amount of forage. Abundant literature exists which documents both pros and cons for using the method. While it has been widely accepted as a reliable method to determine forage production on rangelands, it has not been documented that it is a reliable method to determine forage losses by pests. In order to implement the double-sampling method to determine forage losses of pests, it would be necessary to obtain estimates of forage using the method before the forage has been grazed and/or removed by pests. Then, after grazing has occurred, the double-sampling method would be again implemented to determine the estimated forage remaining after grazing by pests.

If the caged-plot method was used, then double-sampling would include estimating and harvesting a number of plots outside of cages, and only estimating inside cages. The advantage of double-sampling in this case would be that a large number of samples could be taken because harvesting in cages is unnecessary. The amount of forage removed by pests could then be estimated and confidence levels could be placed on this estimated amount. To date, this approach has not been documented in literature.

SINGLE-PLANT OR SHOOT METHOD

A single-plant or shoot method has been used in some studies where a pair of plants is selected and one plant has been attacked by the pest while another has remained unaffected by the pest (Walker 1983). In this method, an estimate is made of the amount of forage that has been removed by the pest and the single plant or shoot that has not been attacked will also be estimated for its weight. The difference between the two estimates indicates the amount of forage removed by the pest. The proportion of plants in the pasture that is attacked is used to estimate the damage that has occurred.

In some studies, plant tillers, stems or even leaves, have been tagged, and periodic checks are made of these tagged plant parts to determine the amount of forage that has been removed from each part. This would provide a dynamic estimate of the amount of forage that is removed by a certain pest during the season. However, the method may not accurately indicate what kind of pest is removing the material, unless provision is made to actually document which pests are present. The advantages of the single-plant or shoot method include speed and simplicity. The method is very rapid and is simple to use compared to a caged-plot method wherein estimation, harvesting, and subsequent corrections by regression are needed.

ASSESSMENT METHODS TO DETERMINE FORAGE LOSSES

There are probably as many methods available as there are rangeland pests removing forage. No study has been reported in the literature which was conducted in order to determine optimum techniques to measure forage consumption by rangeland pests. In any case, an assessment method should include the following considerations:

1. It should be efficient; the method should be analyzed for costs, as well as time, since both are an indication of efficiency.
2. The size of the sampling unit should be considered. A small sampling unit (e.g., 0.5 m^2) may arrive at statistically sufficient estimators of forage amounts at a lower economic cost than larger plot sizes. Different vegetation types and rangeland conditions will dictate the particular plot size and shape that is most efficient for the vegetation under consideration. Therefore, studies should be conducted on optimum size and shape of sampling units to obtain forage estimates.
3. The number of observations needed to obtain a statistically adequate estimate in a particular vegetation type should be calculated from sample data before laying out the actual experiment.
4. Time of measurement of the forage certainly should be a consideration for any assessment method since vegetation phenology, as well as feeding habits of the pests, depend upon time of the season. In any case, a dynamic estimate would be much better than a one-time or static estimate
5. The growth stage of the pests has to be considered in any assessment since various growth stages may feed on different species of plants.
6. Precision and accuracy should be incorporated into a selection of a methodological assessment. Precision is the repeatability of the measurement process, while accuracy is the nearness to which the assessment method measures what is actually available in terms of forage. A method may be repeatable or precise but not be accurate.

CONCLUSION

In summary, it is obvious that an in-depth study of assessment methods has not been conducted for all conditions encountered in determining forage losses by rangeland pests. However, the documentation thus far on determination of forage using double-sampling methods enables one to use this method in various combinations of caged vs. uncaged plots in order to determine specifics relative to a given pest. While it may not be adequate for all rangeland pests, it would suffice to estimate forage removal for most.

REFERENCES

Ahmed, J., C.D. Bonham, and W.A. Laycock. 1983. Comparison of techniques used for adjusting biomass estimates by double sampling. J. Range Manage. 36:217-221.

Ahmed, J., and C.D. Bonham. 1982. Optimum allocation in multivariate double sampling for biomass estimation. J. Range Manage. 35:777-779.

Anderson, N.L., and J.C. Wright. 1952. Grasshopper investigations on Montana rangelands. Montana Agr. Exp. Sta. Bull. 486.

Blocker, H.D. 1969. The impact of insects as herbivores in grassland ecosystems, p. 290-299. In: R. Dix and R. Beidleman (eds.) The grassland ecosystem: a preliminary synthesis. Range Sci. Dept. Sci. Series No. 2, Colorado State Univ.

Bullen, F.T. 1966. Locust and grasshoppers as pests of crops and pasture--a preliminary economic approach. J. Applied Ecol. 3:147-168.

Criddle, N. 1933. Studies in the biology of North American Acrididae, development and habits, p. 474-494. In: World's grain exhibition and conference, Regina, Canada. Vol. 2. Can. Soc. Tech. Agr., Ottawa.

Cruz, I., and F.T. Turpin. 1983. Yield impact of larval infestations of the fall armyworm (Lepidoptera: Noctuidae) to midwhorl growth stage of corn. J. Econ. Entomol. 76:1052-1054.

Gangwere, S.K. 1961. A monograph on food selection in Orthoptera. Trans. Am. Entomol. Soc. 87:67-230.

Germain, L.P. 1982. Aboveground net primary production of rangeland under four stocking densities in southeastern Colorado. Unpublished M.S. Thesis,Colorado State Univ.

Hewitt, G.B. 1980. Tolerance of ten species of *Agropyron* to feeding by *Labops hesperius*. J. Econ. Entomol. 73:779-782.

Hewitt, G.B., and J.D. Berdahl. 1984. Grasshopper food preferences among alfalfa cultivars and experimental strains adapted for rangeland interseeding. Environ. Entomol. 13:828-831.

Hingtgen, T.M., and W.R. Clark. 1984. Impact of small mammals on the vegetation of reclaimed land in the Northern Great Plains. J. Range Manage. 37:438-441.

Isely, F.B. 1938. The relations of Texas Acrididae to plants and soils. Ecol. Monogr. 8:551-604.

Kelly, G.D., and W.W. Middlekauff. 1961. Biological studies of *Dissosteria spurcata* Saussure with distributional notes on related California species (Orthoptera: Acrididae). Hilgardia 30:395-424.

Mitchell, J.E. 1975. Variation in food preferences of three grasshopper species (Acrididae: Orthoptera) as a function of food availability. Am. Midl. Nat. 94:267-283.

Moroka, N, R.F. Beck, and R.D. Pieper. 1982. Impact of burrowing activity of the bannertail kangaroo rat on southern New Mexico desert rangelands. J. Range Manage. 35:707-710.

Mulkern, G.B., D.R. Toczek, and M.A. Brusven. 1964. Biology and ecology of North Dakota grasshoppers. II. Food habits and preferences of grasshoppers associated with sand hills prairie. Res. Rep. North Dakota Agr. Exp. Sta. No. 11.

Oma, E.A., and G.B. Hewitt. 1984. Effect of *Nosema locustae* (Microsporida: Nosematidae) on food consumption in the differential grasshopper (Orthoptera: Acrididae). J. Econ. Entomol. 77:500-501.

Parker, J.R. 1930. Some effects of temperature and moisture upon *Melanoplus mexicanus* Saussure and *Camnula pellucida* Scudder (Orthoptera). Montana Agr. Exp. Sta. Bull. 223.

Putnam, L.G. 1961. The damage potential of some grasshoppers (Orthoptera: Acrididae) of the native grasslands of British Columbia. Can. J. Plant Sci. 42:596-560.

Roundy, B.A. G.J. Cluff, J.K. McAdoo, and R.A. Evans. 1985. Effects of jackrabbit grazing, clipping, and drought on crested wheatgrass seedlings. J. Range Manage. 38:551-555.

Ueckert, D.N., and R.M. Hansen. 1971. Dietary overlap of grasshoppers on sandhill rangeland in northeastern Colorado. Oecologia 8:276-295.

Walker, P.T. 1983. Crop losses: the need to quantify the effects of pests, diseases and weeds on agricultural production. Agr. Ecosystems Environ. 9:119-158.

RANGELAND PESTS AND THEIR MANAGEMENT

7. ECOLOGY AND MANAGEMENT OF PRICKLYPEAR CACTUS ON THE GREAT PLAINS

W.A. Laycock and B.S. Mihlbachler
Department of Range Management
University of Wyoming, Laramie, Wyoming 82071

Many species of cactus occur on the Great Plains. The main emphasis of this paper will be on plains pricklypear, *Opuntia polyacantha* Haw., the most abundant and widespread species. According to Turner and Costello (1942), plains pricklypear ranges from British Columbia to Manitoba, east to Wisconsin, south to Missouri, and west to New Mexico, Arizona and Utah. It is most abundant in northern New Mexico, eastern Colorado, eastern Wyoming, western Kansas and western Nebraska.

The total area infested with *Opuntia* spp. in the United States was estimated by Platt (1959) to be 31.8 million ha. This included 26.3 million ha in the southern Great Plains, 2.6 million ha in the northern Great Plains, 2.0 million ha in the central Great Plains, 810,000 ha in the Pacific Southwest, and 40,000 ha in the Canadian Plains.

Costello (1941) stated that more than 2 million ha in eastern Colorado and eastern Wyoming are infested by plains pricklypear to a degree that presents serious problems in range management. Johnson et al. (in press) estimated that 520,000 ha of western South Dakota have moderate to heavy levels of infestation of plains pricklypear. Hoffman (1967) stated that pricklypear presents a range management problem on approximately 30 million ha of Texas rangeland and that it occurs with a cover of 10 percent or more on about 14 million ha and is a severe problem on about 8 million ha. However, Lundgren et al. (1981) estimated that pricklypear infests 10.3 million ha of rangeland in Texas (28 percent of the rangeland in the state). On the high plains and rolling plains in the panhandle and northern Texas a total of 2.7 million ha were reported to be infested with pricklypear, with 1.7 million ha considered light, 0.7 million ha moderate, and 0.2 million ha dense.

ECOLOGY

Much of the information in this section was taken from a study of the ecology of plains pricklypear on the central Great Plains by Turner and Costello (1942). Plains pricklypear is a common associate of blue grama, *Bouteloua gracilis* (H.B.K.) Lag., and buffalo grass, *Buchloe dactyloides* (Nutt.) Engelm., on the shortgrass steppe. It occurs on clay loam to sandy loam soils but is rarely found on pure sand. It usually does not occur in areas with more abundant soil moisture or heavier soil texture in the central Great Plains, but does occur on dense clay range sites in South Dakota in association with western wheatgrass, *Agropyron smithii* Rydb., and green needlegrass, *Stipa viridula* Trin., (Johnson et al. in press).

The aboveground parts of the plant are fleshy branched stems consisting of flat circular pads, joints, or cladodes with numerous spines. These prostrate stems occur in dense clumps. Flowers are green, yellow-green, bright yellow or pink in color. Fruits are dry and spiny when mature and numerous seeds are produced in favorable years. Seed is dispersed in late summer and germinates the following late spring or early summer. Rabbits eat the fruits and disperse viable seeds widely (Weaver and Albertson 1956) and rodents sometimes bury caches of seed (Timmons and Wenger 1940).

The root system of plains pricklypear consists of larger lateral roots somewhat woody in nature, extending 1.2 m or more. Smaller more succulent roots occur beneath the plants and at intervals along the lateral roots (Harvey 1936). A small cluster of pricklypear 0.3 m in diameter may have a root system 1.2-1.8 m in diameter. According to Turner and Costello (1942) the root system is largely confined to the top 7-10 cm of soil and has a network of rootlets that can take advantage of light rains (2.5-4.0 mm). They found the extensive lateral root to be shallower than the fibrous roots which penetrated somewhat deeper directly under pricklypear plants. Dougherty (1986) found 89% of the root length and 95% of the root biomass of plains pricklypear in northeastern Colorado occurred in the top 7.5 cm of soil.

Watering experiments conducted by Dougherty (1986) in northeastern Colorado indicated that plains pricklypear picked up water from artificial waterings of as little as 2.5 mm, but effects on plant water lasted only one day. Greater amounts of water (5, 10, and 20 mm) resulted in a significant water increase in pricklypear plants for at least three days. Dodd and Lauenroth (1975) found that keeping soil water at field capacity during the growing season for three years resulted in little change in biomass of plains pricklypear.

In addition to seeds, pricklypear reproduces by buds from roots and development of roots from detached pads. Weaver and Albertson (1956) found numerous plants originating from root sprouts from the extensive root system.

The method of reproduction which is most pertinent to attempts to mechanically control pricklypear is its ability to produce roots from detached pads or joints. This ability has been widely recognized in the literature but quantification of this phenomenon has been rare. It was used early in the century (Griffiths 1908, 1912) to propagate several species of pricklypear grown for forage. Each segment can produce roots and establish a new plant. In Arizona, Martin and Tschirley (1969) reported large increases in the number of jumping cholla, *O. fulgida* Engelm., plants in the first year following cabling as a result of the rooting of scattered joints and fruits.

In northeastern Colorado, Laycock (1983) reported rooting of large numbers of plains pricklypear pads broken from plants by a severe hailstorm in July 1978. By August 1979, 13-34% of the segments had rooted and the population of rooted segments was equivalent to 1,400-2,400 new plants/ha. After a dry late summer, fall, and winter most rooting took place in May and June in 1979, when more than 200 mm of precipitation fell, more than 9 months after the segments had been detached.

RELATIONSHIP BETWEEN GRAZING AND CACTUS ABUNDANCE

"Conventional Wisdom" has long indicated that pricklypear cactus increases on overgrazed ranges. One of the first statements to this effect was by Weaver (1920). The Weaver and Clements (1938) book "Plant Ecology" included a picture caption stating "Overgrazed range in western Nebraska indicated by abundance of cactus (*Opuntia*)". One of the major textbooks in range management (Stoddard and Smith 1943) contained a photograph with the caption "A shortgrass range which is badly invaded by cactus (*Opuntia* sp.) as a result of heavy grazing. In such large quantities this plant is a reliable indicator of misuse (Photograph by J.E. Weaver)".

Are there any earlier references that would give some idea of the amount of pricklypear present on the plains prior to European man's introduction of domestic livestock? In 1805, when Lewis and Clark were in the vicinity of the present city of Great Falls, Montana, they described the area as "mountainous country covered with sharp fragments of flint-rock; these bruised and cut their feet very much, but were scarcely less troublesome than the prickly pear of the open plains, which have now become so abundant that it is impossible to avoid them, and the thorns are so strong that they pierce a double sole of dressed deer skin" (Lewis 1903).

Dorn (1986) reviewed journals of travellers through Wyoming in the early and middle 1800's and some typical references to pricklypear follow. John C. Fremont, in 1842, described the area where the North Platte flows through a narrow cut in the hills (Fremont Canyon) and stated that cacti were the prevailing plants in this area. J. Robert Brown described the area about 6 km west of Fort Laramie along the Platte in 1856 "We find a scarcity of grass...Charley and I drove the oxen up on the high bluffs for grass...We had to drive the cattle down again, on account of the vast amount of prickly pears which almost covered the ground." Custer, on his expedition to the Black Hills in 1874, described some of the vegetation in northeastern Wyoming as--"cactus and prickly pear prevailing with very little grass". Dorn (1986) concluded that the prevalence today of cactus, sagebrush, and other shrubs in Wyoming was *not* caused by livestock overgrazing. That statement appears to be true on the entire Great Plains.

Weaver and Albertson (1956) stated that pricklypear increased greatly in abundance during the long cycle of drought of the 1930s, but decreased in remarkable degree during the four years with more than normal precipitation after the drought. This may be where the ideas originated that cactus increases with overgrazing because rather heavy grazing often was associated with the drought at least in the early years. Cook (1942) studied the reasons for increase of pricklypear with drought and decrease with above average precipitation. He concluded that during wet weather, insects help control cactus populations. *Dactylopius*, a mealybug which lives on the juices of the pricklypear plant, and larvae of the cactus moth, *Melitara dentata* Grote, which eats the center of fleshy stems of pricklypear, both cause death. Moist warm weather is conducive to population increases of both insect species and thus to the decline of cactus. During dry spells, conditions don't favor these insects, grasses are low in vigor, and the pricklypear can increase.

The studies documenting abundance of pricklypear under different grazing intensities have indicated that grazing has little effect on pricklypear. Klipple and Costello (1960) reported that frequency of plains pricklypear increased from 1940-1953 in northeastern Colorado under all grazing treatments. The increase was largest under no grazing, intermediate under moderate use, and least under heavy use. Hyder et al. (1966), at the same location, showed that the amount of cactus was not significantly affected by different intensities of grazing. Bement (1968) concluded that changes in grazing intensity cannot be depended upon as a management practice either to increase or decrease pricklypear.

Holscher (1944) observed that cactus increased rapidly near Miles City, Montana, during the drought of the 1930's but light and moderate grazing had similar effects. Subsequent studies in the same area indicated that grazing intensity had little effect on pricklypear abundance (Reed and Peterson 1961). Cactus was more affected by weather patterns; as precipitation increased and temperatures decreased, cactus numbers declined. Houston (1963) analyzed data collected from 1937 through 1961 at Miles City and found that cattle numbers had the greatest influence on the upland soils where more green pads were present under light grazing than under heavy grazing.

One of the reasons for the impression that pricklypear increases with overgrazing, at least on the shortgrass plains, is the illusion created by the presence or absence of herbaceous vegetation. In lightly grazed pastures the cactus is camouflaged by the ungrazed vegetation. In heavily grazed pastures, the grass is shorter which makes the pricklypear appear to be much more abundant (Bement 1968, Speirs 1978). In northeastern Colorado, Bement developed visibility of cactus as an indicator of the amount of ungrazed grass (Marvin Shoop, personal communication).

Thomas and Darrow (1956) found no relationship between pricklypear density and stocking rate or density in Texas. On a semi-desert grassland in southern Arizona, Glendening (1952) found that increases in cactus density for some species were strongly related to the grazing activities of cattle and rodents. Detached pads of jumping cholla and cane cholla were spread by grazing animals and became established. No consistent grazing effect was identified for Engelmann pricklypear, *O. engelmanni* Salm-Dyck.

PROBLEMS CAUSED BY CACTUS

Hindrance to Livestock

Thick stands of cactus can be a hindrance to livestock movement and distribution (Lundgren et al. 1981 and others). Stands thick enough to cause these problems occur mainly in the southwest but also occurred in the rest of the Great Plains during the drought of the 1930's.

Reductions of Space, Yield, and Availability

The ideas that pricklypear occupies space or reduces total and available forage yield are related. Costello (1941) implied that cactus reduced grass yield by occupying space where grasses could grow, as well as decreasing forage availability. Bement (1968) experimentally removed plains pricklypear from plots in northeastern Colorado. For five years, absence of pricklypear did not result in an increase in blue grama production, but it did make 110 kg more forage per ha available for grazing.

In the same area in northeastern Colorado, Laycock et al. (1986) found that removal by machine or by hand grubbing did not increase forage production of grasses or other herbaceous species. Grass production on the untreated plots averaged 560 kg/ha compared to 520-580 kg/ha on the machine harvested and grubbed plots.

In a sagebrush-grass community in east central Wyoming, available forage was only 170 kg/ha acre on ranges containing 4500 kg/ha of plains pricklypear (Smith et al. 1985). In the third year following a fall burn, cactus was nearly absent and available forage had increased to 450-780 kg/ha. Most of this increase was caused by sagebrush removal but availability also was increased because of the removal of the cactus. J.L. Dodd (unpublished data) found no significant change in total herbaceous biomass following fall burning in low-fuel (460 kg/ha) cactus-grassland sites in east-central Wyoming. Two years after fall burning, cactus density declined 54% and available forage increased 121%. In South Dakota, Johnson et al. (in press) reported utilization of grass as high as 40% in areas where pricklypear was controlled with picloram two years previously. Utilization on untreated areas averaged 18%.

In northeastern Wyoming, Hyde et al. (1965) found that treating plains pricklypear with a mechanical beater removed 90% of the pricklypear and resulted in an average forage yield of 910 kg/ha compared to 440 kg/ha in an untreated area. No data were given on the species composition, but a general species list for all study sites included big sagebrush, *Artemisia tridentata* Nutt. If any substantial amount of sagebrush were present in the treated area, the increase in forage yield may have been from sagebrush removal rather than from pricklypear removal. Hyde et al. (1965) reported that forage production was 810 kg/ha after blading compared to 740 kg/ha on the untreated area.

In southern Arizona, Reynolds and Bohning (1956) reported that perennial grass production was reduced 10% by the presence of jumping cholla, cane cholla, and Engelmann pricklypear.

Many of the studies involving herbicide treatment of cactus have not reported the effect of the removal of cactus on associated vegetation. In South Dakota, Johnson et al. (in press) reported that in areas where cactus had been reduced with picloram (Tordon 22K and 2K) non-cactus vegetation production increased from 700 kg/ha on untreated areas to 830-1110 kg/ha on treated areas in the second year following treatment.

Animal Health Problems

Many of the animal health problems caused by pricklypear have been reported from the southwest. According to Merrill et al. (1980), pricklypear causes heavier death losses in sheep and goats than any of the plants commonly considered poisonous. The spines and glochids (small spines) work into the sheep's lips and tongue and cause severe ulceration. The resulting decline in condition can reduce or stop lactation in mother animals which can lead to loss of young. Spines entering the fourth stomach can damage the pyloric area and produce ulcers. Eating large amounts of seeds can also cause trouble because they are not digested and may accumulate in the rumen. Before screwworm control, many deaths from screwworm infestations in the mouth were caused by cactus injuries.

In a survey in Texas, Lundgren et al. (1981) reported that 45 to 80% of ranchers responding indicated that pricklypear was a range management problem. Problems listed were reduced forage production (no data were given) and hindrance to livestock movement. Fifty to 90% reported pricklypear as a major livestock health problem. Hoffman et al. (1964) reported that some cattle that have been fed burned pricklypear may continue eating unburned pricklypear with spines. These "pear eaters" usually remain in poor condition. They also can tear off and scatter pricklypear pads, helping spread the pricklypear. However, in Colorado, Shoop et al. (1977) found no problems with heifers grazing unburned plains pricklypear after they had been fed singed pricklypear.

BENEFITS OF PRICKLYPEAR

Weaver and Albertson (1956) reported that clumps of pricklypear protected the native grasses growing in or near them after the drought of the 1930's. At the end of the drought these remnant grasses were a local source of seed to revegetate the ranges. Others indicated that pricklypear aids in holding the soil and slows runoff (Hoffman et al. 1964, Reed and Peterson 1961).

Some of the main benefits of pricklypear to wildlife are for food and, in the southwest where plants are bigger, for shelter (Lundgren et al. 1981, Keasey 1981). Martin et al. (1961) stated that pricklypear provides food for at least 44 species of birds and animals. Vaughan (1967) found that plains pricklypear was the most important food of the pocket gopher on the shortgrass plains. Keasey (1981) reported that the jackrabbit, packrat, and javelina feed on the joints, and that the Harris ground squirrel and other rodents feed upon the fruit. Sparks (1968) reported that cactus may constitute up to 30% of the summer diet of the black-tailed jack rabbit on sandhill ranges in Colorado.

Pricklypear is a part of the diet of some of the larger native ungulates. Peden (1976) reported plains pricklypear as a minor part of the diet of bison on shortgrass plains. Stelfox and Vriend (1977) reported frequent use of pricklypear by pronghorn antelope on burned areas on a military reserve in southern Alberta. J.L. Dodd (unpublished data) reported heavy use of burned plains pricklypear by antelope on an area burned for cactus control in northeastern Wyoming. In southern Texas, Everitt and Gonzalez (1979) reported that pricklypear made up approximately 55 percent of the total diet of whitetailed deer in Hidalgo county but only about 4% in Kennedy and Willacy counties. The latter two counties had a large preponderance of forbs on the range while Hidalgo county had few forbs.

Pricklypear as Animal Feed

One of the values of pricklypear is its utility as feed for livestock. A common idea is that this practice started during the drought of the 1930's and that the cactus was fed only as emergency feed when no other forage was produced. However, Griffiths (1905, 1906) reported cattle feeding trials with nopal pricklypear, *O. lindheimeri* Engelm., in south Texas. Pricklypear and cottonseed meal produced steer gains equivalent to other feeds. Milk cows fed pricklypear and sorghum gave adequate milk, and quality or flavor of milk were not adversely affected. Woodward et al. (1915) concluded that pricklypear is a good and palatable feed for dairy cows. Forsling (1924) reported that pricklypear had been used in Texas and the southwest for many years as emergency feed and also for fattening livestock.

Hare (1908) reported dry matter digestibility of pricklypear by cattle was 66% and protein digestibility was 57%. Mixing pricklypear with cured fodder or grain was recommended because digestibility was increased and animals scoured quite badly when fed pricklypear alone.

Growing pricklypear as a crop for livestock feed was explored in the U.S. early in the century. Griffiths (1908) reported that the French in North Africa had been propagating cactus through cuttings for some time. Some of his early studies involved propagating both the native nopal pricklypear and spineless pricklypear, *O. ficus-indica* (L.) Mill., in Texas. Spineless pricklypear is now widely grown in Mexico, Brazil, other Central and South American countries, South Africa, Australia, India and other countries for livestock feed.

Compared to other forages, the amount of calcium in pricklypear generally is quite high (3-6%), and nitrogen and phosphorus are quite low (N=0.7-1.3%; P=0.10-.25%) (Daniel 1935, Shoop et al. 1977, Everitt and Gonzalez 1979). In northeastern Colorado, Shoop et al. (1977) found crude protein levels in plains pricklypear (5.3%) were less than one-third that of alfalfa hay (16.8%) but about equal to grass-alfalfa hay pellets (5.7%). Pricklypear was about equal to alfalfa in gross energy, digestible soluble carbohydrates, and dry matter digestibility. They concluded that plains pricklypear is a palatable and nutritious cattle feed that could be used as winter supplement on shortgrass ranges. A ration of pure pricklypear would require a protein supplement.

Gonzalez (1986) reported that fertilization of nopal pricklypear with nitrogen on the Rio Grande Plain of Texas increased crude protein content from 5.4% (0.86% N) to as high as 9.8% (1.44% N). Phosphorus fertilization increased P content from 0.08% to as high as 0.19%. Dry matter production increased from 29.1 metric tons/ha on non-fertilized areas to 53.5 metric tons/ha on areas fertilized with both N and P.

METHODS FOR CONTROLLING PRICKLYPEAR

Even though the benefits of controlling pricklypear are not consistent and not well documented, a great deal of research has been done on control methodology.

Biological Control

The control of pricklypear in Australia is a classic example of effective biological control by insects (Vallentine 1971). Pricklypear was introduced into Australia for use as hedges. By 1920, the cactus had spread over an area 1300 by 500 km with many areas solid masses of impenetrable growth. Biological control was investigated because of the prohibitive cost of any other control measure. A moth from Argentina, *Cactoblastis cactorum* Berg, was chosen in 1925. By 1940 the moth had reduced cactus populations so much that the entire area was re-opened to settlement. Dodd (1940) indicated that the first investigations in the United States using *Cactoblastis* were at Uvalde, Texas by J.C. Hamlin, but success was considerably less than expected.

Biological control studies on pricklypear in Canada used *Cactoblastis doddi* Heinrich and *C. bucyrus* Dyar, both from Argentina. The insects entered injured cactus pads but tended not to move to other pads. Maw and Molloy (1980) concluded that their utility for biological control was limited.

In South Africa, *C. cactorum* Berg was reported to keep spineless pricklypear regrowth under control after chemical control (Zimmerman and Malan 1980). Imported cochineal insects, *Dactylopius opuntiae* Cockrell, were reported to be effective on spineless pricklypear after hand felling of the woody stems (Zimmerman and Moran 1982).

Mechanical Control

Various mechanical methods were among the first tried to control plains pricklypear in the U.S. In northeastern Colorado (Costello 1941), hand grubbing, railing, and scraping with a road grader resulted in almost complete control of cactus on level sites. Costello (1941) cautioned that unless plants are picked up and removed from the area, the uprooted plants will take root and the stand will re-establish itself. An example was given of greater cactus density three years after railing where the detached plants were not picked up.

Hyde et al. (1965) tested various methods of controlling plains pricklypear in northeastern Wyoming. Blading unfrozen ground with a road patrol during the winter of 1961-1962 windrowed a lot of soil with the pricklypear. The pricklypear rooted and the windrows became raised areas of almost solid cactus. Blading on frozen soil left little soil in the windrows, which resulted in little rooting of the pricklypear and 100% control.

A mechanical harvester that removed plains pricklypear from experimental plots was tested in northeastern Colorado by Laycock et al. (1986). The machine removed 77-92% of the cactus. Cactus biomass (dry weight) was reduced from 1170 kg/ha on the untreated plots to 190-270 kg/ha on the machine harvested plots and 15-55 kg/ha on plots where the cactus was removed by grubbing. Ground cover of cactus was 7.7% in the untreated area, 1.6-1.8% in the harvested areas and 0.2-0.6% in the grubbed areas.

In Texas, Englemann, nopal, and plains pricklypear have been controlled by mechanical or hand grubbing or by scraping with a grader blade, a bulldozer blade attachment, or railroad irons used as a drag (Hoffman et al. 1964). Railing two ways during the dry season rubbed off many spines, and cattle concentrated right after railing ate much of the cactus which reduced the number of detached pads left to root.

After chaining for mesquite control in Arizona, density of Englemann pricklypear remained unchanged (Martin et al. 1974). Dragging or railing reduced pricklypear density.

On the Rio Grande plains in southern Texas, Dodd (1968) found that rootplowing, either alone or combined with rootraking, reduced the density of mesquite and other woody species, but increased the density of pricklypear (*Opuntia*, species not given).

Herbicide Control

Many different herbicides have been used to control pricklypear.   Early herbicides used in Texas included ammonium sulfocyanate, arsenic pentoxide, arsenic trioxide, sodium arsenate, and calcium chlorate (Hoffman et al. 1964).   These were replaced by 2,4,5-T, Silvex, TCA, and dinitro compounds which achieved better kills.   Later, 2,4,5-T and Silvex were withdrawn and are no longer available for use.   Various formulations of picolinic acid (picloram) have been used in more recent years with good success. Tordon 22K (100% picloram), Tordon 2K (pelleted 100% picloram). Tordon 212 (picloram and 2,4-D), and Tordon 225E (picloram and 2,4,5-T) have been the main compounds used.

Application of 1% 2,4,5-T in an oil carrier was most effective in controlling Engelmann and nopal pricklypear in Texas (Hoffman et al. 1964).   In northeastern Wyoming, Hyde et al. (1965) treated cactus in 1/10 bloom with Trinox, Silvex, Tordon (formulation not given), 94A, TD446, and TD440 at 1.1, 2.2 4.5 kg/ha.   Silvex and Tordon were the only herbicides found to be effective at these rates; however, Tordon at 4.5 kg/ha injured the native grasses. Silvex was cheaper than Tordon.

In east-central Wyoming, Alley and Lee (1969) reported that application at early bloom stage of 2,4,5-T at 13.5 kg/ha and all rates of 2,3,6-TBA (2.8, 5.6, and 11.2 kg/ha) resulted in good control of plains pricklypear, but severely damaged the native herbaceous vegetation.   At 2.2-4.5 kg/ha, 2,4,5-T resulted in 45-85% cactus control with no apparent damage to the native vegetation.   Silvex effectively controlled cactus (90-99% mortality) at low rates (2.2-4.5 kg/ha) with no injury to associated vegetation.

In northeastern Colorado, Sims (1973) found that 80-90% kill of plains pricklypear could be obtained within four years after treatment with fourteen different herbicides.   Tordon 212 at 1.7 kg/ha or Tordon 225 at 0.6 kg/ha gave nearly 100% control the first year after treatment.   Tordon 22K with Kuron, or Esteron 245, 99, or 99C, resulted in first year cactus kill of 50%.   Kuron at 3.4 kg/ha reduced cactus 90%, while Tandex at 1.1-2.2 kg/ha killed 70-100% of the cactus.   Nopalmate, an arsenic compound, killed 86-96% of the cactus at 1.1, 2.2, and 4.5 kg/ha.   Harrowing prior to treatment to bruise the pads increased the first year mortality with some herbicides, but was found to be unnecessary for the Tordon compounds.

In southeastern Wyoming, Tordon 22K at 0.6 kg/ha and Silvex at 2.2 kg/ha applied to plains pricklypear in full bloom resulted in 80 and 70% first year control, respectively, with no apparent damage to forage species (Univ. Wyoming 1979). In another area, Tordon 22K at 0.3 and 0.6 kg/ha gave 90% or better control of pricklypear two years after initial application (Univ. Wyoming 1980). In Texas, Hoffman (1967) found that Tordon (formulation not given) achieved nearly 100% control and was superior to 2,4,5-T or Silvex at the same or lower rates in controlling pricklypear. Greater control was achieved with all herbicides when some sort of mechanical bruising of the pads, such as railing, was conducted prior to herbicide application.

In New Mexico, best control of Engelmann, nopal, and plains pricklypear occurred when spraying was conducted before or at the full bloom stage and when the pads were mechanically damaged by dragging light rails or cultipacking (McDaniel 1980). Best control was attained with 2,4,5-T in a water carrier but Tordon 225E also was found to be effective. Cost varied with cactus density and the application technique.

In Nebraska, treatment of plains pricklypear with Silvex and Dicamba resulted in 95% control (Wicks et al. 1969). Pricklypear control two years after spraying with Dicamba was 100% but some injury to the grasses was noticed. Application of picloram (formulation not given) at 4.5 kg/ha in May killed 100% of the pricklypear. Both May and June application of picloram at 2.2 and 4.5 kg/ha eliminated 50% of the native grass. Rotary hoeing increased the control of pricklypear with Dicamba, 2,4,5-T, Silvex, and picloram. Silvex at 2.2 and 4.5 kg/ha and picloram at 0.6 and 1.1 kg/ha applied after rotary hoeing controlled 98% of the pricklypear. In South Dakota, Johnson et al. (in press) reported that 0.3, 0.4, and 0.6 kg/ha of Tordon 22K reduced plains pricklypear 70-86%, and 0.6 and 0.8 kg/ha of Tordon 2K reduced pricklypear 67-89% compared to untreated areas.

Schuster (1971) reported that night spraying of phenoxy herbicides (2,4,5-T and Silvex) in Texas increased the kill of plains pricklypear. Nighttime opening of the cactus stomata may allow easier entry of the herbicide into the cactus pads. However, about 100% control within two years occurred for both the day and night treatments.

If the main advantages to pricklypear control with herbicides is increase in availability of forage, then the effects may not be realized immediately. After spraying, the dead cactus pads remain intact and even though the pads are dead, the spines continue to be a deterrent to grazing. In South Dakota, many of the pads had collapsed two years after treatment with Tordon 22K and 2K. Pad deterioration was nearly complete three years after treatment (J. Johnson, pers. comm.).

Fire

Much of the research on the effect of fire on pricklypear has been done in the southwest. On a desert grassland-shrub range in southern Arizona, June burning with 670 kg/ha fuel killed 28% of the Engelmann pricklypear (Reynolds and Bohning 1956). Cable (1967) found that June burning with 670 kg/ha of fuel reduced Engelmann pricklypear 32% the first year after treatment. No change in the density of Engelmann pricklypear was found following a single June burn treatment on a site with 340 kg/ha of fuel.

In a Texas tobosagrass-mesquite community, burning resulted in an average mortality of 55% for brownspine pricklypear, *O. phaeacantha* Engelm. (Neuenschwander et al. 1978). Mortality of the cactus was attributed to interactions between drought and biological activities following burning. Mortality was as high as 98% during a wet year and as low as 13% during a dry year following burning.

In the southern mixed prairie of Texas, Bunting et al. (1980) found that spring burning caused 70% mortality of brownspine pricklypear by the third year. Following fire, insect, rodent, and rabbit predation increased on the cactus pads. Mortality of tasajillo, *O. leptocaulis* DC., was 65% the first year after burning, with little resprouting. Engelmann pricklypear was not seriously damaged by fire because small mammals had reduced litter accumulations around the plants.

A fall burn in east-central Wyoming with 630 kg/ha of fine-fuels reduced plains pricklypear cover by 50% within four years after treatment (Smith et al. 1985). In another study in the same area, nearly 100% kill of pricklypear was achieved with a fall burn on a rangeland with 4,500 kg/ha dry-weight of pricklypear and 560 kg/ha of fine-fuel (Smith et al. 1985). Mortality of the pricklypear was attributed to dehydration of the cactus plants following removal of the waxy cuticle by fire, rather than heat damage. Burns on wet soils, during cool, damp weather, and preceding precipitation events by less than one month, resulted in low pricklypear mortality.

Fall burning in east-central Wyoming reduced pricklypear biomass 90 and 95% the first year after treatment in a low-fuel (460 kg/ha) and high-fuel (720 kg/ha) grassland site, respectively (J.L. Dodd et al. unpublished data). Little resprouting has occurred on these burns two years after treatment. Spring burning on similar sites reduced live cactus biomass 60-77% the first year, but resprouting and incomplete mortality of cactus plants resulted in near pretreatment levels of cactus or greater the following year. Shoop and Laycock (1983) found that grazing of singed-in-place plains pricklypear could reduce its abundance, but it was difficult to get cattle to eat very much of the singed cactus plants.

CONCLUSIONS

There is little evidence to support the theory that the abundance of pricklypear in the Great Plains today is the result of past overgrazing. Pricklypear historically has been an important component of North American rangelands and its abundance is regulated more by environmental factors such as drought, fire, soil type, and insect predation than by grazing. Pricklypear removal from rangelands on the central Great Plains only rarely increases total forage production unless competitive species such as sagebrush are also removed. In the northern Great Plains, production may be increased by reduction or removal of pricklypear (Johnson et al. in press).

Dense pricklypear stands can create serious problems for rangeland livestock producers including reduced animal performance and production, distribution problems, and reduced forage availability. In these cases, pricklypear may have to be controlled. The method chosen must be quite inexpensive for the control to be economically feasible. Mechanical methods, herbicides, and prescribed burning have all been found to be effective in controlling pricklypear on rangelands under specific conditions. Choosing the best method for a particular application should be based on: 1) economics of the control method and the expected returns, 2) the abundance of the pricklypear and the degree of control desired, 3) the labor and equipment available, and 4) how the control method will fit in with the overall land management plan.

After most weed or brush species are removed, some or all of the water and/or nutrients used by those species becomes available for the remaining species, and herbage production subsequently increases. On the central Great Plains, herbage production does not increase when pricklypear is removed. Possible explanations for the lack of response include: 1) pricklypear does not compete strongly with the grasses for water or nutrients, 2) pricklypear modifies the environment of the site so that growing conditions are more favorable for the grasses, or 3) the predominance of short warm-season grasses (blue grama and buffalo grass), which do not respond to pricklypear removal.

Removal of pricklypear may result in an increase in grass production on the northern Great Plains (Johnson et al. in press). Part of the reason for the different response may be that these ranges are dominated by taller cool-season grasses which may respond to pricklypear removal.

The extremely shallow root system of pricklypear does utilize water from very small rainfall events that the grasses cannot effectively utilize. However, pricklypear also uses moisture from larger events and therefore must compete with the grasses for this moisture.

Some evidence does exist that pricklypear modifies the micro-environment of the site in shortgrass areas. During snow drifting events in the winter, each plant or clump of pricklypear and associated ungrazed vegetation acts as a miniature snowfence, causing a small snowdrift on the leeward side (Laycock et al. 1986). This adds some water to the site during each snow drifting event in the winter. Air-borne soil particles and nutrients could be trapped in the same manner and evapotranspiration may also be reduced in the vicinity of the pricklypear plants. This modification of the microclimate of the site requires further study.

REFERENCES

Alley, H.P., and G.A. Lee. 1969. Chemical control of plains pricklypear in Wyoming. Wyoming Agr. Exp. Sta. Bull. 497.

Bement, R.E. 1968. Plains pricklypear: relation to grazing intensity and blue grama yield on central Great Plains. J. Range Manage. 21:83-86.

Bunting, S.C., H.A. Wright, and L.F. Neuenschwander. 1980. Long-term effects of fire on cactus in the southern mixed-prairie of Texas. J. Range Manage. 33:85-88.

Cable, D.R. 1967. Fire effects on semidesert grasses and shrubs. J. Range Manage. 20:170-176.

Cook, C.W. 1942. Insects and weather as they influence growth of cactus on the Central Great Plains. Ecology 23:209-214.

Costello, D.F. 1941. Pricklypear control on short-grass range in the Central Great Plains. USDA Leaflet 210.

Daniel, H.A. 1935. The total calcium, phosphorus, and nitrogen content of native and cultivated plants in the high plains of Oklahoma and a study of the mineral deficiencies that may develop in livestock when emergency feeds are fed. Panhandle Agr. Exp. Sta. Bull. 56.

Dodd, A.P. 1940. The biological campaign against prickly pear. Commonwealth Prickly Pear Board, Brisbane, Australia.

Dodd, J.D. 1968. Mechanical control of pricklypear and other woody species on the Rio Grande Plains. J. Range Manage. 21:366-370.

Dodd, J.L., and W.K. Lauenroth. 1975. Responses of *Opuntia polyacantha* to water and nitrogen perturbations in the shortgrass prairie, p. 229-240. *In:* M.K. Wali (ed.) Prairie: a multiple view. Univ. North Dakota Press, Grand Forks, ND.

96

Dorn, R.D. 1986. The Wyoming landscape, 1805-1878. Mountain West Publishing, Cheyenne, WY.

Dougherty, R.L. 1986. The soil water resource of *Opuntia polyacantha* in semiarid grassland. Unpublished PhD Dissertation, Colorado State Univ.

Everitt, J.H., and C.L. Gonzalez. 1979. Botanical composition and nutrient content of fall and early winter diets of white-tailed deer in south Texas. Southwest. Nat. 24:297-310.

Forsling, C.L. 1924. Saving livestock from starvation on southwestern ranges. USDA Farmers Bull. 1428.

Glendening, G.E. 1952. Some quantitative data on the increase of mesquite and cactus on a desert grassland range in southern Arizona. Ecology 33:319-328.

Gonzalez, C.L. 1986. Pricklypear (*Opuntia lindheimeri* Engelm.) can be used as cattle feed on regular basis. Abstracts of papers, 39th Ann. Mtg., Soc. Range Manage., Kissimmee, FL, No. 130.

Griffiths, D. 1905. The prickly pear and other cacti as food for stock. USDA Bur. Plant Ind. Bull. 74.

Griffiths, D. 1906. Feeding prickly pear to stock in Texas. USDA Bur. Animal Ind. Bull. 97.

Griffiths, D. 1908. The pricklypear as a farm crop. USDA Bur. Plant Indus. Bull. 124.

Griffiths, D. 1912. The thornless pricklypears. USDA Farmers' Bull. 483.

Hare, R.F. 1908. Experiments on the digestibility of pricklypear by cattle. USDA Bur. Animal Ind. Bull. 106.

Harvey, A.D. 1936. Rootsprouts as a means of vegetative reproduction in *Opuntia polyacantha*. J. Am. Soc. Agron. 28:767-768.

Hoffman, G.O. 1967. Controlling pricklypear in Texas. Down to Earth 23(1):9-12.

Hoffman, G.O., A.H. Walker, and R.A. Darrow. 1964. Pricklypear - good or bad. Texas Agr. Ext. Serv. Bull. 806.

Holscher, Clark E. 1944. Controlling the pricklypear. West. Livestock J. June 15.

Houston, W.R. 1963. Plains pricklypear, weather, and grazing in the northern Great Plains. Ecology 44:569574.

Hyde, R.M., A.D. Hulett, and H.P. Alley. 1965. Chemical and mechanical control of plains pricklypear in northeastern Wyoming. Univ. Wyoming Agr. Ext. Serv. Cir. 185.

Hyder, D.N., R.E. Bement, E.E. Remmenga, and C. Terwilliger, Jr. 1966. Vegetation-soils and vegetation-grazing relations from frequency data. J. Range Manage. 19:11-17.

Johnson, J.R., W.L. Tucker, C.E. Stymiest, and E.J. Bowker. Pricklypear cactus control in South Dakota. In: Beef cattle report, Dec. 1986, South Dakota State Univ., Dept. Animal and Range Sci. (in press).

Keasey, M.S. 1981. Prickly pear. Pac. Discov. 34:1-8.

Klipple, G.E., and D.F. Costello. 1960. Vegetation and cattle responses to different intensities of grazing on short-grass ranges on the Central Great Plains. USDA Tech. Bull. 1216.

Laycock, W.A. 1983. Hail as an ecological factor in the increase of pricklypear cactus, p. 359-361. In: J.A. Smith and V.W. Hays (eds.) Proc. XIV inter. grassland congress. Westview Press, Boulder, CO.

Laycock, W.A., D.M. Mueller, and M.C. Shoop. 1986. Removal of plains pricklypear has little effect on herbage production, p. 299-300. In: P.J. Joss, P.W. Lynch, and O.B. Williams (eds.) Rangelands: a resource under siege. Proc. second inter. rangeland congress. Aust. Acad. Sci., Canberra.

Lewis, M. 1903. History of the expedition under the command of Captains Lewis & Clark to the sources of the Missouri, Across the Rocky Mountains, down to the Columbia River to the Pacific in 1804-6. (A reprint of the edition of 1814). A.S. Barnes and Co., New York.

Lundgren, G.K., R.E. Whitson, D.E. Ueckert, F.E. Gilstrap, and C.W. Livingson, Jr. 1981. Assessment of the pricklypear problem on Texas rangelands. Texas Agr. Exp. Sta., Texas A&M Univ., MP-1483.

Martin, A.C., H.S. Zim, and A.L. Nelson. 1961. American wildlife plants. A guide to wildlife food habits. Dover, New York.

Martin, S.C., J.T. Thames, and E.B. Fish. 1974. Changes in cactus numbers and herbage production after chaining and mesquite control. Prog. Agr. Ariz. 26(6):3-6.

Martin, S.C., and F.H. Tschirley. 1969. Changes in cactus numbers after cabling. Prog. Agr. Ariz. 21(1):16-17.

Maw, M.G., and M.M. Molloy. 1980. Pricklypear cactus on the Canadian prairies. Blue Jay 38:209-211.

McDaniel, K. 1980. Methods of controlling pricklypear cactus. New Mexico State Univ. Coop. Ext. Serv. Guide B-801.

Merrill, L.B., C.A. Taylor, R. Dusek, and C.W. Livingston. 1980. Sheep losses from range with heavy pricklypear infestation, p. 91. In: D.N. Ueckert and J.E. Huston (eds.) Rangeland resources research: consolidated progress report. Texas Agr. Exp. Sta. No. 3665.

Neuenschwander, L.F., H.A. Wright, and S.C. Bunting. 1978. The effect of fire on a tobosagrass-mesquite community in the rolling plains of Texas. Southwest. Nat. 23:315-338.

Peden, D.G. 1976. Botanical composition of bison diets on shortgrass plains. Am. Midl. Nat. 96:225-229.

Platt, K.B. 1959. Plant control - some possibilities and limitations in the challenge to management. J. Range. Manage. 12:64-68.

Reed, M.J., and R.A. Peterson. 1961. Vegetation, soils and cattle responses to grazing on northern Great Plains range. USDA Tech. Bull. 1252.

Reynolds, H.G., and J.W. Bohning. 1956. Effects of burning on a desert grass-shrub range in southern Arizona. Ecology 37:769-777.

Schuster, J.L. 1971. Night applications of phenoxy herbicides on plains pricklypear. Weed Sci. 19:585-587.

Shoop, M.C., E.J. Alford, and H.F. Mayland. 1977. Plains pricklypear is a good forage for cattle. J. Range Manage. 30:12-17.

Shoop, M.C., and W.A. Laycock. 1983. Plant succession after burning range to singe plains pricklypear. Abstracts of paper, 36th Ann. Mtg., Soc. Range Manage., Albuquerque, NM. No. 210.

Sims, P.L. 1973. Effect of herbicides on the control of pricklypear cactus in northeastern Colorado. Colorado State Univ. Exp. Sta. Progress Report 73-2.

Smith, M.A., J.L. Dodd, and J.D. Rodgers. 1985. Prescribed burning on Wyoming rangeland. Univ. Wyoming Agr. Ext. Serv. Bull. 810.

Sparks, D.R. 1968. Diet of black-tailed jack rabbits on sandhill range in Colorado. J. Range Manage. 21:203-208.

Speirs, D.C. 1978. The cacti of western Canada. Nat. Cact. Succ. J. 33:83-84.

Stelfox, J.G., and H.G. Vriend. 1977. Prairie fires and pronghorn use of cactus. Can. Field-Nat. 9:282-285.

Stoddart, L. A., and A.D. Smith. 1943. Range management. McGraw-Hill, New York.

Thomas, G. W., and R.A. Darrow. 1956. Response of prickly-pear to grazing and control measures. Texas Agr. Exp. Sta. Prog. Rep. 1837.

Timmons, F.L., and L.E. Wenger. 1940. Jackrabbits and cactus team up; present serious problem in 30 counties. Kans. Farmer 77(8):5,22.

Turner, G.T., and D.F. Costello. 1942. Ecological aspects of the pricklypear problem in eastern Colorado and Wyoming. Ecology 23:419-426.

University of Wyoming. 1979. Research in weed science. Univ. Wyoming Agr. Exp. Sta. Res. J. 137.

University of Wyoming. 1980. Research in weed science. Univ. Wyoming Agr. Exp. Sta. Res. J. 147.

Vallentine, J.F. 1971. Range development and improvements. Brigham Young Univ. Press, Provo, UT.

Vaughan, T.A. 1967. Food habits of the northern pocket gopher on shortgrass prairie. Am. Midl. Nat. 77:176-189.

Weaver, J.E. 1920. Root development in the grassland formation. Carnegie Inst. Wash. Publ. 292.

Weaver, J.E., and F.W. Albertson. 1956. Grasslands of the Great Plains. Johnsen Publishing Co., Lincoln, NE.

Weaver, J.E., and F.E. Clements. 1938 (2nd ed.). Plant Ecology. McGraw-Hill, New York.

Wicks, G.A., C.R. Fenster, and O.C. Burnside. 1969. Selective control of plains pricklypear in rangeland with herbicides. Weed Sci. 17:408-411.

Woodward, T.E., W.F. Turner, and D. Griffiths. 1915. Pricklypear as feed for dairy cows. J. Agric. Res. 4:405-450.

Zimmermann, H.G., and D.E. Malan. 1980. Present status of prickly pear (*Opuntia ficus-indica* (L.) Mill.) control in South Africa, p. 79-85. *In:* S. Neser and A.L.P. Cairns (eds.) Proc. third nat. weeds conf. S. Afr. A.A. Balkema, Rotterdam.

Zimmermann, H.G., and V.C Moran. 1982. Ecology and management of cactus weeds in South Africa. S. African J. Sci. 78:314-320.

8. ECOLOGY AND MANAGEMENT OF BROOM SNAKEWEED

Kirk C. McDaniel
Department of Animal and Range Sciences
New Mexico State University, Las Cruces, New Mexico 88003

L. Allen Torell
Department of Agricultural Economics and Agricultural Business
New Mexico State University, Las Cruces, New Mexico 88003

Broom snakeweed, *Gutierrezia sarothrae* (Pursh) Britt. & Rusby, is an evergreen suffrutescent shrub common on western rangeland from Canada to Mexico (Figure 8.1, Solbrig 1960, Lane 1985). This plant is known by a variety of common names including turpentine weed, rubberweed, rockweed, stinkweed, yellowtop, matchweed, perennial broomweed, or snakeweed. Broom snakeweed is similar in appearance to threadleaf snakeweed, *G. microcephala* (DC) Gray, and they are often collectively referred to as perennial snakeweed.

Within the western United States and Great Plains grazing regions, species of the genus *Gutierrezia* occur on an estimated 57 million ha of rangeland (Platt 1959). The presence of broom snakeweed can vary widely from one location to another, and from one year to the next. Broom snakeweed populations are cyclic with high numbers of plants growing when there is enough soil moisture, and decreasing dramatically during drought or stress. There is no well-defined pattern in the population cycle, but above-average fall, winter, and spring precipitation following drought years appears to be related to the establishment of the plants on blue grama, *Bouteloua gracilis*, range (McDaniel et al. 1982).

Dense broom snakeweed stands cause significant economic losses to ranchers in the plains, prairie, and desert areas of central and southwestern United States (Torell et al. in press). The greatest plant numbers and most severe management problems exist on the southern High Plains and the Canadian-Pecos valleys of west Texas and eastern New Mexico. In this region, about 3.5 million ha of rangeland are classified as densely infested with snakeweed, and an additional 9.7 million ha are moderately infested by the weed (McDaniel et al. 1984).

FIG. 8.1 *Gutierrezia sarothrae* growing in eastern New Mexico.

Native grass production is often 100 to 800% higher on snakeweed-free rangeland compared to yield under a dense snakeweed stand (Ueckert 1979, McDaniel et al. 1982). This loss in herbage growth may be a more serious problem than poisonous properties associated with the species. The plant has a low forage value and it is seldom grazed unless other forage supplies are low. Livestock poisoning is reportedly worse in the spring when the plant is sprouting new leaves (Dollahite and Anthony 1957). In laboratory tests, smaller amounts of new growth than later growth are needed to cause toxic effects in cattle.

The main poisonous principle of broom snakeweed is saponin (Kingsbury 1964), which is a non-cardioactive steroid glycoside (Dollahite and Anthony 1957). Poisoned sheep and cattle exhibit listlessness, anorexia, an uneven and dull coat, diarrhea or constipation, fecal mucus, vaginal discharge, and occasionally hematuria. Rusted, peeling muzzles, along with excessive nasal discharges, have also been noted in cattle poisoned with broom snakeweed (Sperry et al. 1977). Sheep may exhibit a minor jaundice. Gastroneteritis lesions occur and there is a degeneration of the liver and kidneys. Death often occurs in cattle, goats, and sheep when they consume as little as 10 to 20% of their body weight in fresh broom snakeweed (Dollahite and Anthony 1957).

Consumption of broom snakeweed is believed to be a main cause of cattle abortions in certain areas of the Southwest (Kingsbury 1964). Calf losses on snakeweed-free sandy range sites in eastern New Mexico are reported to average 3 to 4%, whereas on snakeweed infested rangeland, losses can approach 60% in some herds (Sperry et al. 1977). When live calves are born to poisoned mothers, the calves are usually underweight and often die. The mother often retains the placenta and needs veterinary care.

Selenium poisoning may also occur from broom snakeweed plants growing in soils with a high selenium content (Hamilton and Beath 1963, Kingsbury 1964, Schickedanz 1977). Broom snakeweed accumulates this poisonous mineral in its tissues. Five ppm of affected plant material is sufficient to cause poisoning. Poisoned animals stop eating, breathe with difficulty, and urinate frequently. Ingestion of selenium-containing plants will cause blind staggers, causing animals to wander aimlessly or stumble. According to Schickedanz (1977), if an animal ingests plant material with 5 to 40 ppm of selenium for more than one month, alkali disease may result; this is manifested as deformed hooves, severe lameness, erosion of joints, or deformed offspring.

BOTANICAL DESCRIPTION AND DISTRIBUTION

Recent examinations of *Gutierrezia* have generated much morphological, anatomical, cytogenetic, and biochemical data that have created an interesting and sometimes controversial taxonomic history (Solbrig 1960, 1964, 1970, 1971; Ruffin 1971, 1974, 1977; Lane 1980, 1982, 1985).

According to Lane (1985), *Gutierrezia* is a new world genus comprised of seven woody shrubs, three herbaceous perennials, and six suffrutescent annuals in North America. Solbrig (1960) described 11 additional species in South America.

Gutierrezia sarothrae is the most widely distributed North American species and is the only perennial species common throughout the central plains and prairie regions (Figure 8.2). This taxon grows in the cold-temperate climates of Alberta and Manitoba Canada southward to subtropical areas in Nuevo Leon, Durango, Chihuahua, San Luis Potosi, and Baja California Norte, Mexico (Lane 1985). The species generally occurs west of the

100th meridian to the Sierra-Cascade axis near the Pacific coast (Solbrig 1960, 1964; Lane 1980, 1985). It is found at elevational ranges from 50 to more than 2900 m. Broom snakeweed tolerates macroclimates with mean annual air temperatures ranging from about 4 to 21°C, a growing season length from 100 to 300 days, and an annual rainfall from less than 20 to 50 cm (Solbrig 1960, Lane 1985). The plant normally blooms between August to November, but Lane (1985) noted it can bloom any month of the year in which there is sufficient moisture.

Broom snakeweed has three to five ray florets and three to five disk florets in each capitulum, and the involucre is 3.2 to 4.2 mm high and 1.2 to 2.0 mm wide. The plant is densely branched with a leafless woody main stem bearing numerous leafy branches above. Leaves are alternate, mostly linear, and 5 to 50 mm long and 0.5 to 1.0 mm wide. Flowers are yellow and are produced simply or in compact clusters of two to five flowers in the upper periphery of the canopy.

REPRODUCTION AND GROWTH

Broom snakeweed regenerates from an abundant seed source produced each fall. According to Ragsdale (1969), an average broom snakeweed plant produces 9,717 seeds, with a range of 1,263 to 21,853 seeds per plant. The environmental mechanisms that allow seeds to germinate, and the length of time seeds remain viable in the soil, are not well understood (Mayeux and Leotta 1981). Snakeweed populations have been described as being cyclic (Jameson 1966). That is, new seedlings may appear with favorable growing conditions, survive to maturity, then completely die out in later years because of insect damage, old age, drought, or by reasons unknown. When seedlings emerge, their mortality is usually high with 25 to 100% dying the first year, depending on available soil moisture. However, once established, a snakeweed's life expectancy is 3 to 10 years.

The phenological development of mature broom snakeweed can be divided into four annual events: initiation of perennating bud growth, stem elongation and leaf growth, development of flower buds, and flowering and seed set. When these phenological events occur varies according to the age and vigor of the plant.

For a mature, vigorous broom snakeweed plant growing in central Texas, growth of perennating buds usually begins in November and December, immediately after flowering and seed set (Ragsdale 1969). At this time, leaves change color from green to pale yellow or brown before shedding, and small perennating bud clusters of new leaves form along the main branches. Elongation of stems and new leaves begins from the perennating bud clusters in late January to March. Seedling plants begin vegetative growth over a longer period than mature vigorous plants, and mature nonvigorous plants usually begin vegetative growth later in the spring. Vegetative growth continues throughout summer until formation of flower buds, which usually occurs between August and November. Flowering and seed development sequentially follow formation of flowering buds at approximately 3-week intervals. Seed scatter usually precedes or coincides with the beginning of perennating bud growth.

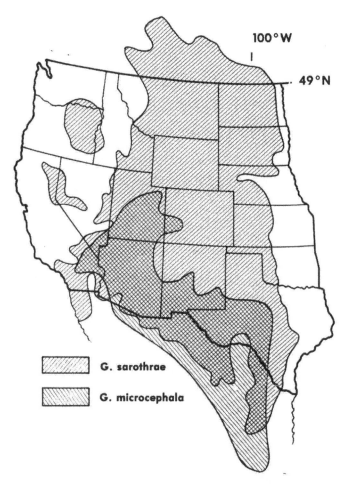

FIG. 8.2 Distribution of *Gutierrezia sarothrae* and *G. microcephala* in North America (adapted from Lane 1985).

ECOLOGICAL INTERACTIONS

Broom snakeweed competes with more valuable forage for moisture, space, nutrients, and light. Of these factors, moisture competition is probably most important, especially in the arid Southwest where rainfall is limiting.

A competitive advantage is derived for broom snakeweed from numerous adventitious (fibrous) roots produced in the upper 15 cm of soil depth, especially on shallow and loamy soils. Because the plant remains evergreen, it uses most of the available soil water during the fall and winter when warm season grasses are dormant; thus, little remains for the spring green up of grasses.

Broom snakeweed is a component of the natural ecosystem, but under pristine conditions it probably contributed about 5 to 10% of the species composition (Sosebee 1985). Under undisturbed conditions, the amount present would fluctuate over time according to climatic conditions. However, with the introduction of livestock the densities of these plants have increased on many southwestern rangelands.

Barnes reported as far back as 1913 that: "There is a great area of this weed (snakeweed) in eastern Colorado. From La Junta west along the Santa Fe road, one can see from the car windows great sweeps of lovely green prairies which delight the eye. On closer examination, the green is the green of snakeweed, and on acres and acres the grama and buffalo grass that once covered the whole country are gone, with the exception of an occasional struggling bunch grass" (p. 20).

As broom snakeweed populations increased on western rangeland it was reported that the plant was found mainly on rocky ridges, gravelly slopes, and infertile soils (Parker 1939). Later it became more noticeable around watering holes, trails and other areas where livestock tended to congregate. The species aggressively invades disturbed ranges that are heavily grazed, and some suggest the plant is a valuable range condition indicator (Barnes 1913, Wooten 1915, Jardine and Forsling 1922, Talbot 1926, Dayton 1931, Costello and Turner 1941, Green 1951, Nichol 1952, Humphrey 1955, Schmutz et al. 1968). However, because broom snakeweed populations increase and decrease in a cyclic pattern according to climatic conditions, others suggest the plant is not necessarily an indicator of overgrazing (Jameson 1970, Vallentine 1971, Ueckert 1979, McDaniel et al. 1984).

It was an early belief that conservative grazing practices would restore misused rangeland and prevent broom snakeweed infestations. We now know that this is a slow and unrewarding process in the Southwest because of the harsh environmental conditions. On the Ft. Stanton Experimental Ranch in central New Mexico, broom snakeweed populations increased following droughts in 1970-71, 1974, and 1976 on moderate and heavily grazed pastures. The degree of infestation was slightly less on the moderately grazed pasture (18 plants/m^2) compared to the heavily grazed pasture (25 plants/m^2), but associated grass production was highly suppressed under both situations (McDaniel et al. 1982). While grazing management can lead to increased productivity on broom snakeweed rangeland, other control methods such as burning, application of herbicides, and biological control yield more immediate results.

CONTROL AND MANAGEMENT

Foliar Sprays

The amount of broom snakeweed controlled with commercial herbicide sprays is largely related to growing conditions, especially soil moisture, before and at the time of treatment (McDaniel 1984, 1985). Above normal rainfall in March, April, or May stimulates new vegetative growth. When this occurs, the plant can be controlled with foliar sprays. However, broom snakeweed is more often controlled commercially with fall application of foliar herbicides after peak bloom. In eastern New Mexico, this occurs between late September to December. When the plant is stressed because of low soil moisture, it should not be treated with a foliar spray.

During acceptable growing conditions, 0.25 to 0.50 pound of liquid picloram per acre effectively controls snakeweed. The lower rate is recommended for fall application and the higher rate should be used in spring. Another herbicide, metsulfuron, is as effective as picloram when applied at 0.03 pounds of active ingredient per acre. Adding 2,4-D to picloram does not improve snakeweed control, but it may provide a broader spectrum of control for other broadleaf weeds.

Broom snakeweed can also be controlled with a foliar spray of dicamba alone or mixed equally with picloram. A tank mix of 0.25 pound of dicamba and 0.25 pound of picloram per acre is effective in the fall. If dicamba is sprayed alone, a rate of 0.75 pound per acre should be applied. A commercial formulation, or a tank mix of 3:1 ratio of 2,4-D and dicamba, applied at 2.0 pound per acre, is also effective.

Foliar sprays will control broom snakeweed present on an area being treated and will reduce the number of plants that may subsequently reinvade from seed. However, treatment life of herbicides varies, and the possibility for complete snakeweed eradication with a single application is remote, especially where a population is extremely dense before treatment. It may take two or more applications over a 5 to 10-year span to achieve the long-term goal of suppressing snakeweed.

Pelleted Herbicides

Soil-active herbicides in pelleted formulations are slow acting compared to foliar sprays. The pellets lie on the soil surface until dissolved by rain or snow, and the chemical is leached into the soil profile. The herbicide is taken up by plant roots, synthesized, and becomes toxic to the plant. A primary advantage of soil-applied herbicides over foliar sprays is that the chemical can persist in the soil for more than 1 year, extending the length of time snakeweed is controlled. The stage of plant growth has little or no effect on the herbicide's activity. A disadvantage is the indeterminate length of time required before the herbicide becomes activated in the soil and subsequently kills the weed. Additionally, a pelleted herbicide costs more than a comparable liquid, and a higher rate of active ingredient is usually required to control a dense stand of snakeweed.

Applied during or before periods of rainfall, tebuthiuron or picloram pellets at 0.5 to 1.0 pound of active ingredient per acre effectively control snakeweed. The pellets must be broadcast uniformly by aircraft or ground equipment.

Application just before seasonal rainfall will give the most rapid response. One advantage is that picloram and tebuthiuron pellets will control a number of different woody and broadleaf plants in addition to snakeweed. The pellets are particularly desirable in a mixed brush stand where snakeweed is an understory component.

Mechanical Methods

Broom snakeweed and other weed infestations often occur after mechanical treatments such as grubbing, disking, or dry land farming practices. Mechanical brush control treatments designed to kill woody plants other than snakeweed, or reseeding rangeland, often cause a high level of soil disturbance and provides an ideal seedbed for snakeweed invasion. These treatments are rarely beneficial for snakeweed control.

In native or disturbed pastures with nearby susceptible crops, herbicide applications may not be safe or practical. In these areas, an effective mechanical control method is the best alternative. Shredding or mowing can be done where brush, rocks and other obstacles do not make the treatment impractical. New stems often develop from the remaining branches when top growth is removed, which makes repeated mowings necessary. Mowing should be done when vegetative growth has reached a maximum, and soils are hot and dry (June to August). Shredding before flowering or seed set eliminates an annual seed crop and reduces the number of seedlings that may reinvade on a site.

Biological Control

More than 300 species of insects are reportedly associated with broom snakeweed, but all do not actually feed on the plant (Foster et al. 1981, Wangberg 1982). The most common destructive insects can be grouped according to their activity as (1) root borers, (2) plant juice feeders on roots, (3) leaf tiers and (4) plant juice feeder on above-ground vegetation. The root borer *Crossidius pulchellus* (LeConte) causes significant mortality to older broom and threadleaf snakeweed plants, and the grasshopper *Hesperotettix viridis* (Scudder) defoliates plants of all sizes in localized areas (Richman and Huddleston 1979, Parker 1985). In the case of *Crossidius*, the beetles lay their eggs on the broom snakeweed root crowns. When the larvae hatch, they bore their way into the fleshy tap root of mature plants, and eventually kill the root system. Entomologists suggest that drought conditions favor control by these beetles, whereas locally wet conditions help the broom snakeweed survive the rootborers (Richman and Huddleston 1979).

Mealybugs (*Eriococcus*, *Phenacoccus*, and *Spelococcus* species) also severely damage broom snakeweed root systems and crowns (Box et al. 1962). The nymphs and adult females have piercing-sucking mouth parts that let them feed on plant juices. It is not known whether the mealybugs kill broom snakeweed directly or by serving as a disease vector. The greatest potential for biological control of broom snakeweed is thought to be from the introduction of species from abroad (De Loach 1978). Argentina is reported to have several insects that attack and destroy indigenous *Gutierrezia* stands. Few of these insect species identified in Argentina are also found in the United States.

Prescribed Fire

Prescribed burning has been used to control many rangeland weeds, but has not been employed widely for broom snakeweed control. Research in New Mexico reported burning broom snakeweed in October, April, and June resulted in kills of 35, 45, and 96%, respectively. Burning in January increased broom snakeweed 25%. A successful burn can effectively reduce the standing crop of broom snakeweed, but generally does not reduce after-burn seedling populations. The effectiveness of fire is highly dependent upon the amount of fine fuel (usually grass) at the time of the burn. Often, sufficient fine fuel to carry a fire is not present when a snakeweed stand is dense; therefore, fire can be used only in selected areas.

Recommended Control Strategies and Economics

High priority should be given to rangeland where broom snakeweed numbers and yield are high, and where desirable grasses can be reestablished after control.   Removing a snakeweed stand makes soil moisture and nutrients more available to preferred forage.   The results can be dramatic.   Younger stages of a snakeweed infestation should be given early attention.   The greater the amount of grass species in the understory, the more satisfactory the results will be after control.   However, effectiveness of control treatments, subsequent rainfall, and follow-up grazing management contribute to the amount of forage response that may be expected.   Because grasses growing within dense snakeweed stands are often in low vigor, deferred grazing for one or two growing seasons after herbicide treatment is recommended.   Growing season deferment allows grasses to renew vegetative growth and root development that was suppressed because of snakeweed competition.   Dormant season grazing can be coupled with a growing season deferment regime for maximum range improvement.   Periodic maintenance control efforts may be needed to prevent snakeweed reinfestation.

The economic benefits from controlling broom snakeweed will vary depending upon beef prices, production costs, treatment costs, treatment life, and treatment success.   An economic evaluation of any range improvement must consider production costs and production levels both with and without the improvement.   Timing of benefits and costs is also important (Workman 1986).

Many of the potential benefits from controlling snakeweed have been identified, i.e., increased forage production, reduced animal poisoning, improved animal health, and improved livestock production.   More research is needed to quantify these parameters for an economic evaluation.   At present, only estimates from knowledgeable researchers and ranchers are available.

We have developed a general estimate of the economic potential for controlling a dense stand of broom snakeweed.   Table 8.1 summarizes the level of livestock production and economic variables considered to be representative for an eastern New Mexico or west Texas cow/calf producer, both with and without a heavy snakeweed infestation.   An average level is specified for each parameter based on livestock budgets prepared in both the panhandle of Texas and the northeastern corner of New Mexico (Torell et al. 1986, Texas A&M 1984).   Because of the uncertainty of many of the production parameters, a conservative estimate of

TABLE 8.1
Livestock and forage production parameters with and without heavy perennial snakeweed infestations, and prices received by ranchers for beef in northeastern New Mexico in 1985, 1980 and 1979.

Description	Heavy snakeweed infestation	Little or no snakeweed	% change
I. Livestock Production Parameters[a]			
Rangeland Carry Capacity:			
Animal units yearlong per section (AUYL/section)	10.0	20.0	100%
Cows yearlong per section (cows/section)	7.7	15.4	100%
Calf crop	75%	85%	10%
Replacement rate	15%	15%	0%
Cow to bull ratio	20:1	20:1	0%
Cow death loss	2%	1%	1%
Calf death loss	4%	3%	1%
Steer calf selling weight	428	475	10%
Heifer calf selling weight	405	450	10%
Cull cow selling weight	855	950	10%
Cull bull selling weight	1080	1200	10%

	1985	1980	1979
II. Economic Parameters			
Beef Prices ($/cwt)[b]			
Steer calves (400-500 lb)	$73.13	$82.08	$94.80
Heifer calves (400-500 lb)	60.41	69.97	81.48
Cull cows	38.50	45.91	51.02
Cull bulls	48.54	53.72	60.02

Discount rate	7%
Treatment life	5 years
Deferred grazing	1 year
Additional production expenses	$105/cow added

[a] Specification of livestock parameters are based on livestock budgets for northeastern New Mexico and the panhandle of Texas (Torell et al. 1986, Texas A&M 1984).

[b] Annual average prices quoted for the Clovis, New Mexico auction by the U.S. Department of Agriculture, Agricultural Marketing Service.

Table 8.2
Economic analysis of controlling rangeland heavily infested with snakeweed.

	($/Section) — Beef Price Situation —		
	1985 prices	1980 prices	1979 prices
I. GROSS BENEFITS:			
Increased Livestock Sales	$2,882	$3,298	$3,793
Increased Annual Livestock Production Costs[1]	808	808	808
Net Annual Benefit	$2,074	$2,490	$2,986
Present value of Net Annual Benefits (5 year uniform benefit, discounted at 7%)[2]	$7,947	$9,541	$11,441
II. PROJECT COSTS AND RETURNS:			
Spraying with .25 lb/acre of Picloram @ $9.40/acre (includes herbicide and application costs)	$6,016	$6,016	$6,016
Net cost of additional breeding stock[3]	$1,027	$1,027	$1,027
Total investment costs	$7,043	$7,043	$7,043
Net Present Value (NPV)	904	2,498	4,398
Benefit/Cost Ratio	1.07:1	1.20:1	1.36:1
Internal Rate of Return (IRR)	9.40%	13.41%	17.87%

[1] Includes miscellaneous expenses for feed, veterinary and medicine, fuel and repairs, and others (see Torell et al. 1986).

[2] All discounting starts at year 2 because of the assumed 1 year of grazing deferment.

[3] Cow numbers are increased by about 8 head with the additional carrying capacity. This represents a cost when the control treatment is initiated but about $3,077 is added to discounted benefits at the end of the planning period. The net cost of the breeding livestock is $1,027/section.

forage and livestock benefits was used. Substantially higher increases in livestock production after controlling snakeweed have been observed in some cases.

In our economic evaluation, three levels of beef prices were considered; 1985 prices, which are representative of the current depressed cattle market; 1979 prices, which reflect record high levels; and 1980 prices which were intermediate (Table 8.1). Three standard economic criteria commonly used in analyzing range improvement investments, including net present value (NPV), the benefit/cost ratio, and the internal rate of return (IRR), are presented in the economic evaluation of snakeweed control (Table 8.2). A more complete discussion of these three measures of investment profitability is given by Workman (1986) and other financial management textbooks. In general, an economically feasible investment is indicated by an IRR greater than the prevailing market interest rate, the NPV being positive, or a benefit/cost ratio greater than 1.

Even under relatively low beef prices, as was experienced in 1985, the economics of controlling a heavy infestation of broom snakeweed is estimated to be positive (Table 8.2). As (expected) beef prices increase the economics of snakeweed control improves. If record high 1979 beef prices occurred over a 5-year period, livestock benefits from controlling snakeweed would exceed costs by 36% and the internal rate of return would be a very competitive rate of nearly 18% (Table 8.2).

Snakeweed control should be approached from both a biological and economic perspective. The ultimate success of this range improvement practice is measured by the land manager in term of economic net return to the ranching enterprise and improved range condition. Increasing the value of these factors determines the ultimate success of the practice.

REFERENCES

Barnes, W.C. 1913. Western grazing grounds and forest ranges. Sanders Publishing Co., Chicago.

Box, T.W., E.W. Huddleston, D. Ashdown, and T. Copeland. 1962. Biological die-off of perennial broom snakeweed. Reports on Agr. Ind., Texas Tech. Col. 3:2.

Costello, D.F., and G.T. Turner. 1941. Vegetational changes following exclusion of livestock from grazed ranges. J. For. 39:310.

Dayton, W. A. 1931. Important western browse plants. USDA Misc. Pub. 101.

De Loach, C. J. 1978. Consideration in introducing foreign biotic agents to control native weeds on rangelands, p. 39-50. *In*: T.E. Freeman (ed.) Proc. IV internat. symp. on biol. control of weeds. Aug. 30-Sept. 2, 1976. Gainesville, FL.

Foster, D.E., D.N. Ueckert, and C.J. De Loach. 1981. Insects associated with broom snakeweed and threadleaf snakeweed in west Texas and eastern New Mexico. J. Range Manage. 34:446-454.

Greene, R.L. 1951. Utilization of winter forage by sheep. J. Range Manage. 4:233.

Hamilton, J.W., and O.A. Beath. 1963. Uptake of available selenium by certain range plants. J. Range Manage. 16:261.

Humphrey, R.R. 1955. Arizona range resources. II. Yavapui County, Ariz. Arizona Agr. Exp. Sta. Bull. 229.

Jameson, D.A. 1970. Value of broom snakeweed as a range condition indicator. J. Range Manage. 23:302-304.

Jameson, D.A. 1966. Competition in a blue grama-broom snakeweed-actinia community and responses to selective herbicides. J. Range Manage. 19:121-124.

Jardine, J.T., and C.L. Forsling. 1922. Range and cattle management during drought. USDA Bull. 1031.

Kingsbury, J.M. 1964. Poisonous plants of the United States and Canada. Prentice-Hall Inc., Englewood Hills, NJ.

Lane, M.A. 1980. Systematics of *Amphiachyris, Greenella, Gutierrezia, Gymnosperma, Thurovia,* and *Xanthocephalum* (Compositae: Astereae). Unpublished Ph.D. Dissertation, Univ. Texas.

Lane, M.A. 1982. Generic limits of *Xanthocephalum, Gutierrezia, Amphiachyris, Gymnosperma, Greenella,* and *Thurovia* (Compositae: Astereae). System. Bot. 7:405-416.

Lane, M.A. 1985. Taxonomy of *Gutierrezia* (Compositae: Astereae) in North America. Syst. Bot. 10:7-28.

Mayeux, H.S., Jr., and H. Leotta. 1981. Germination of broom snakeweed (*Gutierrezia sarothrae*) and threadleaf snakeweed (*G. microcephalum*) seed. Weed Sci. 29:530-534.

McDaniel, K.C. 1984. Snakeweed control with herbicides. New Mexico State Univ. Agr. Exp. Sta. Bull. 706.

McDaniel, K.C., R.D. Pieper, L.E. Loomis, and A.A. Osman. 1984. Taxonomy and ecology of perennial snakeweeds in New Mexico. New Mexico Agr. Exp. Sta. Bull. 711.

McDaniel, K.C. 1985. Control perennial snakeweed. Coop. Ext. Serv. Guide B-815. New Mexico State Univ., Las Cruces, NM.

Nichol, A.A. 1952. The natural vegetation of Arizona. Arizona Agr. Exp. Sta. Tech. Bull. 127:187-230.

Parker, K.W. 1939. The control of snakeweed in the southwest. Southwest For. and Range Exp. Sta. Res. Note 76.

114

Parker, M.A. 1985. Size-dependent herbivore attack and the demography on an arid grassland shrub. Ecology 66:850-860.

Platt, K.B. 1959. Plant control -- some possibilities and limitations. I. The challenge to management. J. Range Manage. 12:64-68.

Ragsdale, B.J. 1969. Ecological and phenological characteristics of perennial broomweed. Unpublished Ph.D. Dissertation, Texas A & M Univ.

Richman, D.B., and E.W. Huddleston. 1981. Root feeding by the beetle *Crossidius pulchellus* Le Conte and other insects on broomsnakeweed in eastern and central New Mexico. Environ. Entomol. 10:53-57.

Ruffin, J. 1971. Morphology and anatomy of the genera *Amphiachyris, Greenella, Gutierrezia, Gymnosperma, Thurovia,* and *Xanthocephalum* (Compositae). Unpublished Ph.D. Dissertation, Kansas State Univ.

Ruffin, J. 1974. A taxonomic re-evaluation of the genera *Amphiachyris, Amphipappus, Greenella, Gutierrezia, Gymnosperma, Thurovia,* and *Xanthocephalum* (Compositae). SIDA 5:301-333.

Ruffin, J. 1977. Polyphyletic survey of the genera *Amphiachyris, Amphipappus, Greenella, Gutierrezia, Gymnosperma* and *Xanthocephalum.* Contrib. Gray Herb. 207:117-131.

Schickedanz, J. 1977. Poisonous range plants of New Mexico. New Mexico State Univ. Coop. Ext. Serv. Pub. 400 B-1.

Schmutz, E.M., B.N. Freeman, and R.E. Reed. 1968. Livestock poisoning plants of Arizona. Univ. Arizona Press, Tucson.

Solbrig, O.T. 1960. Cytotaxonomic and evolutionary studies in the North American species of *Gutierrezia* (Compositae). Contrib. Gray Herb. 188:1-63.

Solbrig, O.T. 1964. Infraspecific variations in the *Gutierrezia sarothrae* complex (Compositae-Astereae). Contrib. Gray Herb. 193:67-115.

Solbrig, O.T. 1970. The phylogeny of *Gutierrezia*: an eclectic approach. Brittonia 22:217-229.

Solbrig, O.T. 1971. Polyphyletic origin of tetraploid populations of *Gutierrezia sarothrae* (Compositae). Madrono 21:21-25.

Sosebee, R.E. 1985. Timing -- the key to herbicidal control of broom snakeweed. Texas Tech. Management Note 6. Texas Tech Univ.

Sperry, O.E., J.W. Dollahite, G.O. Hoffman, and B.J. Camp. 1977. Texas plants poisonous to livestock. Texas A & M Ext. Ser. Rep. B-1028. College Station, TX.

Talbot, M.V. 1926. Indicators of southwestern range conditions. USDA Farmers' Bull. 1782.

Texas Agricultural Extension Service. 1984. Texas livestock enterprise budgets - Texas high plains. Region III. Texas A & M Univ. B-1241(L03).

Torell, L.A., B.A. Brockman, K.M. Garrett, and H.W. Gordon. 1986. 1985 costs and returns for a medium cow/calf enterprise, northeastern New Mexico. New Mexico State Univ. Coop. Ext. Serv. CRE85-LA-NE2, Las Cruces, NM.

Torell, L.A., H.W. Gordon, K.C. McDaniel, and A. McGinty. Economic impacts of broom snakeweed infestations. Proc. ecol. econ. impact poisonous plants on livestock prod. Westview Press, Boulder, CO. (in press).

Vallentine, J.F. 1971. Range development and improvements. Brigham Young Univ. Press, Provo, UT.

Wangberg, J.K. 1982. Destructive and potentially destructive insects of snakeweed in western Texas and eastern New Mexico and a dioristic model of their biotic interactions. J. Range Manage. 35:235-238.

Wooten, E.O. 1915. Factors affecting range management in New Mexico. USDA Bull. 211.

Workman, J.P. 1986. Range economics. MacMillan Pub. Co., Riverside, NJ.

9. ECOLOGY AND MANAGEMENT OF MORMON CRICKET, Anabrus simplex HALDEMAN

Charles M. MacVean
Department of Entomology
Colorado State University, Fort Collins, Colorado 80523

The Mormon cricket, *Anabrus simplex* Haldeman, is a flightless, shield-backed grasshopper which occurs primarily in the Great Plains and sagebrush-dominated regions of western United States and Canada. It is a gregarious insect and is probably best known for its huge migratory aggregations, or bands. These typically develop in permanent breeding areas in broken, mountain habitat and then spread, by walking, to surrounding areas, including agricultural lowlands and valleys (Wakeland and Shull 1936). Dating to the early encounter in 1848 between hordes of this insect and Mormon settlers in the Salt Lake Valley - from which the name "Mormon crickets" stems - sporadic outbreaks of crickets have caused severe damage to crops, especially wheat and alfalfa (Cowan 1929, Wakeland 1959, Evans 1985). Though crickets normally feed on a wide diversity of rangeland plants, crops are highly preferred (Swain 1944). Homesteaders were forced to abandon farming in northwest Colorado due to the yearly invasions of crickets during the 1920's. Damaging numbers of crickets persisted into the late thirties, with the peak of the epidemic occurring in 1938.

Damage by crickets to rangeland plants has been much more difficult to assess than crop damage (Swain 1940, 1944). While crickets do feed on range grasses, particularly the inflorescences, they clearly prefer broad-leaf, succulent species of lesser forage value when these are present (Cowan 1929, Swain 1944, Wakeland 1959). However, Wakeland (1959) and Wakeland and Shull (1936) claimed that the economic importance of Mormon crickets arose primarily from their destruction of range grasses and the subsequent impact on cattle grazing. While true for certain areas during the 1930's, the available evidence suggests that serious range damage occurred only during a relatively short time when drought and overgrazing were severe. Despite a scarcity of quantitative studies, the reputed destruction of rangeland has lead to extensive control campaigns in the past, and is responsible for recent control efforts in northwestern Colorado and northeastern Utah in the vicinity of Dinosaur National Monument, where crickets have again become abundant in the last 6 to 8 years. A great deal of controversy and confusion surrounds the suppression campaigns in these areas due to ranchers' fears of a new cricket plague, and opposition from conservation agencies involved in protection of local endangered species.

Because knowledge of Mormon cricket biology is so limited, a comprehensive ecological review is impossible. Most of the voluminous body of literature on Mormon crickets is old (pre-1960, reviewed by Wakeland (1959)) and is primarily concerned with describing the life cycle and control measures. However, recent studies have advanced our understanding of diet composition, mating behavior, sexual selection, and biological and chemical control. These areas are highlighted in the first part of this chapter. The second part addresses the economic importance of Mormon crickets, stressing the inadequacy of existing data and methods for a sound determination of pest status and the need for control.

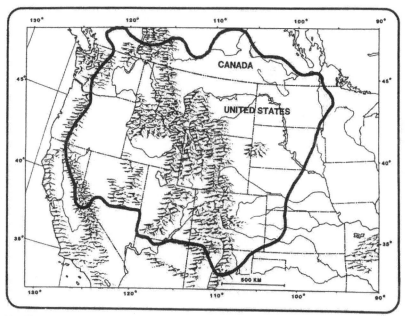

FIG. 9.1   Approximate geographic distribution (area enclosed by dark border) of the Mormon cricket, *Anabrus simplex* Haldeman (modified from Wakeland 1959).

TAXONOMY AND DISTRIBUTION

Anabrus is one of 22 North American genera of shield-backed grasshoppers (Tettigoniidae: Decticinae). The name "Mormon cricket" applies to *Anabrus simplex* Haldeman, though it has also been used for congeners, due to their great similarity in appearance and biology (Wakeland 1959). Presently, 4 species of *Anabrus* are recognized: *simplex* Haldeman, *cerciata* Caudell, *longipes* Caudell, and *spokan* Rehn and Hebard (Caudell 1907, Gurney 1939, Rentz and Birchim 1968).

While all are potential pests, *A. simplex* is the most widespread, occurring throughout much of western North America, and also has caused the most damage. In addition to the area indicated in Figure 9.1, specimens have been reported from Tennessee, though this record may be in error (Goodwin and Powders 1970). Mormon crickets commonly occur between 1300 m and 2400 m elevation (Caudell 1907, Schweis et al. 1939), but have been found as high as 4000 m in Colorado (Scudder 1898). Their habitat includes mixed, shortgrass, and sagebrush-dominated grasslands, and a variety of broadleaf and coniferous forests.

The other three species of *Anabrus* are confined to the northwestern United States. The related genus *Peranabrus* includes only *P. scabricollis* Thomas, the "coulee cricket", which occurs primarily in Washington (Caudell 1907). This species is extremely similar to *A. simplex* in appearance and habits, and has periodically achieved pest status (Snodgrass 1905, Melander and Yothers 1917).

BEHAVIOR AND ECOLOGY

Detailed descriptions of the life history of gregarious Mormon crickets are given by Gillette (1905), Corkins (1922, 1923), Criddle (1926), and Cowan (1929), so only major features will be included here. Aside from mating behavior, Mormon cricket biology has been studied only in high-density, gregarious populations. Given that mating in gregarious crickets is quite different than in solitary ones, other differences in life history traits may exist and warrant further investigation.

Banding and Migration

Mormon crickets are univoltine, early-season insects, hatching in early spring (March to May, depending on elevation and weather) from eggs laid singly in the ground during the preceding summer. The nymphs develop through 7 stadia before the final molt to adults, with a total lifespan of about 100 days. In high-density populations, crickets display gregarious behavior throughout most of their lifetime. After the first 2 or 3 molts they aggregate into dense migratory bands which may cover several square kilometers. During daylight hours, these large groups of insects walk in fixed directions, with individuals within the band pausing occasionally to feed (Cowan and Shipman 1943). Feeding and migrating behaviors are closely linked and generally occur under the same weather conditions: clear, sunny skies with air temperatures ranging from ca. 10° to 35°C and ground temperatures of 24° to 45°C (LaRivers 1941, 1944, Wakeland 1959). However, directional movement and feeding can occur under cloudy conditions if ambient temperature is within the specified range. At dusk, or in adverse weather (cold, rain, snow), the army-like migrations cease and crickets form dense clusters around rocks or on branches of sage and other shrubs. Under very hot conditions, crickets will climb into the plant canopy or up the culms of grasses to roost, apparently escaping high ground temperatures which often reach 50°C during the summer. A typical daily activity cycle includes morning and evening periods of movement and feeding, a mid-day period of inactive roosting, and cluster formation at night. The occurrence and timing of these events are variable and seem to be highly influenced by weather conditions. Further temporal partitioning of behavior is seen soon after the final molt to adults (end of May to mid-June) when crickets will engage in mating only during morning hours and oviposition mostly in the afternoon.

At any given moment, bands can be found criss-crossing an area in all compass directions. They may coalesce into larger units, but have also been observed maintaining their respective headings while "flowing" through each other at the intersection of their paths (Sorenson and Jeppson 1940, MacVean per. obs.), suggesting band-specific orientation cues rather than generalized movement by an aggregation of insects. No research has been conducted to determine the causes or functions of migratory behavior, nor is the mechanism(s) which produces and maintains a band understood. Orientation to sun, wind, and conspecifics have been suggested (Cowan 1929, Swain 1944, Wakeland 1959), but none of these has been tested. The size and distribution of bands in a geographic area, as well as density within bands, are also problems which remain open for investigation. Bands of 2 to 16 km in length and 1 to 2 km in width have been reported and rates of travel of 1 to 2 km per day have been estimated.

It is not understood how or why a transition occurs from widely scattered, sedentary individuals to densely-aggregated, migratory bands of insects (Wakeland 1959, Cowan and Shipman 1943). The process bears a striking resemblance to phase transformations in the African plague locusts, *Schistocerca gregaria* (Forskal) and *Locusta migratoria migratorioides* (Reiche and Fairmaire). Historically, it accounts for the transition from an entirely benign state to outbreak and pest status. It has been noted that solitary individuals tend to be smaller and more lightly colored than their gregarious counterparts (Gwynne 1984, MacVean per. obs.). The latter are usually black as adults, whereas solitary crickets vary from solid emerald green (except for white venter) to tan and light purple with white mottling on the abdomen.

Mating

One of the few aspects of cricket ecology which has been well studied using current ethological and evolutionary approaches is mating behavior. Gwynne (1981, 1984) has found an unusual role reversal between the sexes where, contrary to the pattern seen in most animals, males actively select among potential mates, rejecting the smaller females in the population. Females, in turn, compete among themselves for access to males. Accepted females are significantly heavier and possess a higher number of mature eggs than their rejected counterparts. Thus, it appears that males increase their fitness by mating preferentially with the more fecund females in the population (Gwynne 1981, 1984).

The evolution of this role reversal is apparently related to the unusually large investment that males make in each copulation in the form of a large spermatophore which they extrude and transfer to the female's genital opening. This large white "sac", often noted by early investigators and considered a rather awkward insemination device (Gillette 1904, Snodgrass 1905, Caudell 1908), represents up to 27% of the male's body weight, a substantial energy expenditure (Gwynne 1981, 1984). It is composed of two small sperm ampules and a much larger proteinaceous bulb. While the sperm are draining from the ampules into the female's spermathecae, she eats the proteinaceous mass. Subsequently, the nutrients are used in somatic tissues and for egg production. Gwynne (1984) found that the number of eggs produced is directly related to the number of spermatophores received in successive copulations. It is also possible that ovipositional stimulants, such as prostaglandins, are transferred with the spermatophore (Stanley-Samuelson et al. 1986). Female reproduction appears to be limited by access to spermatophore-bearing males, which results in female-female competition for this resource.

Role reversal is typical of high-density, gregarious, populations (5-10 adults/m^2 or higher) but is not shown by crickets at low densities (less than 1/m^2). Gwynne (1984) found that "solitary-phase" males, while producing large spermatophores, did not reject females and in fact competed for them; that is, they displayed the typical male role of most species. Also, a greater proportion of the males in a low-density population possessed well-developed spermatophores than in a high-density population. Gwynne (1984) suggested that high-density populations may be more food-limited and that the males with mature spermatophores are a limiting component of female fitness. In this scenario, sexual selection has favored large, aggressive females that compete successfully for the available males, which in turn select the larger females. Whether food is indeed limiting in high-density populations and not in the low-density ones, and how role reversal develops in the transition from solitary to gregarious states, are not known.

Food Habits

Mormon crickets are omnivorous insects. Food items include plants, conspecifics, other insects, livestock manure, and carrion (Wakeland 1959). Over 400 species of food plants were reported by Swain (1944), ranging from grasses and small forbs to the foliage of large shrubs and trees. Although practically every plant species in the crickets' habitat is fed upon at one time or another, definite preferences are discernible. In their native habitat, crickets prefer succulent, herbaceous species over grasses or woody plants. Most crops, especially wheat and alfalfa, are readily eaten in preference to range vegetation (Gillette 1905, Corkins 1922, 1923, Cowan 1929, Mills 1939, Ueckert and Hansen 1970). In the most comprehensive study of cricket feeding habits conducted to date, Swain (1944) assigned all known food plants to 3 categories, ranging from most to least preferred. Forbs such as wild onion, *Allium* spp., arrowleaf balsamroot, *Balsamorhiza sagittata* (Pursh) Nutt., crucifers, *Sisymbrium altissimum* L. and *Brassica* spp., and lupine, *Lupinus* spp., were among the most highly utilized plants, while slender wheatgrass, *Agropyron pauciflorum* (Schwein), cheatgrass, *Bromus tectorum* L., and Sandberg bluegrass, *Poa secunda* Presl., were preferred grasses. Other authors have reported similar preferences in Colorado (Corkins 1923), Montana (Cowan 1929), and Nevada (Schweis et al. 1939).

Crickets also exhibit marked preferences for flowers and seeds over vegetative tissue. In this regard, cricket injury to crops can often be distinguished from that of *Melanoplus* grasshoppers (Acrididae). In areas where crickets and grasshoppers occurred together, Swain (1944), Wakeland (1959), and Cowan (1929) found that grasshopper damage to wheat resulted in plants denuded of leaves but with intact culms, while the opposite was true for damage due to crickets.

Ueckert and Hansen (1970) and Hansen and Ueckert (1970) examined the crop contents of Mormon crickets from a population near Red Feather Lakes, Colorado, and found that plants comprising large portions of the diet made up a small proportion of the total available herbage and were thus actively selected by crickets. Grasses, clubmoss, and grasslike plants made up only 8% of the diet. Forbs represented about 50%, and fungi composed 16% of the diet. The authors did not determine whether fungus consumption was the incidental result of feeding on plants bearing fungal growth, but suggested that fungus-infected plants might be preferred due to higher carbohydrate and protein content. The remaining 21% of the diet was made up of arthropod parts, most of them apparently the remains of small insects and other Mormon crickets. Diet composition varied during the season, with arthropods increasing from 10 to 20% between July and September, and forbs decreasing from about 60 to 30% in the same period. This may reflect an increased protein requirement for mating and egg production.

These studies confirm the general observations of many authors, particularly the propensity for cannibalism among Mormon crickets. Injured, weakened individuals, or those rendered vulnerable while molting, are common targets of cannibalistic attacks. The cricket literature abounds with descriptions of "road slicks" caused by successive automotive slaughters of crickets congregating in the roadway to feed on the previous rash of victims. (An especially vivid description for the curious reader is given by Snodgrass (1905)).

Crickets are also known to prey upon other insects. In the earliest account of predation, Thomas (1872) reported crickets feeding on cicadas roosting in shrubs. Ueckert and Hansen (1970) found remains of aphids, ants, and lepidopterans in the crops of Mormon crickets. However, aside from these reports, predatory behavior has received no formal attention. With respect to the pest status of Mormon crickets, their cannibalistic and predatory nature, as well as scavenging on feces or carrion, deserve further study, since these components of the diet may serve to offset damage to range vegetation.

The available evidence suggests that Mormon crickets are opportunistic feeders that consume succulent, high-protein tissues or animal products without any strict regard to the species involved. Food preference patterns are variable and dependent upon the taxonomic composition of the community in which crickets occur, as well as the phenological states of potential food items. However, because Mormon crickets prefer succulent, weedy species to range grasses, they are generally predisposed not to compete with livestock for forage.

POPULATION REGULATION

The factors controlling population fluctuations are poorly understood. Though numerous predators and a few parasites and pathogens are known to occur, their function as regulating agents is largely unknown.

Predators

Probably the best studied natural enemy is *Palmodes laeviventris* (Cresson) (Sphecidae), a solitary digger wasp. After stinging and paralyzing a cricket, the female wasp drags it into its burrow where it serves as a food item for a developing larva (LaRivers 1944, 1945). Gwynne and Dodson (1983) found that the size of crickets taken by *P. laeviventris* was positively correlated with the size of the wasps, and that the sex ratio among prey was female-biased. They suggested that females may be more vulnerable to predation because they are more active and conspicuous during the morning hours when they are responding to calling males and the wasps are foraging. The sphecid wasp, *Tachysphex semirufus* (Cresson), is also known to provision its nest with Mormon crickets (Kurczewski and Evans 1986).

Predation by vertebrates, especially birds, has long been considered an important mortality factor. A great deal of literature and folklore surrounds the plague of crickets that invaded the Mormon settlers' crops and was allegedly arrested by huge flocks of California gulls in 1848 (Bancroft 1889, McAtee 1926, Henderson 1931, Tanner 1940, Evans 1985). Since then, the list of species known to feed on crickets has grown considerably and includes about 50 species of birds, rodents, and reptiles (Kalmbach 1918, Cowan 1929, Knowlton 1937, 1941, 1943, 1948, Knowlton et al. 1946, Wakeland 1959). Most records of predation are of nymphs and adults, though excavation of cricket egg beds by rodents and birds, presumably to feed on the eggs, has also been noted. An interesting consequence of the banded nature of cricket populations is that birds such as kestrels hover over the band and "track" it as it migrates (Wakeland 1959).

Recent aerial spraying against Mormon crickets in northwestern Colorado has stimulated new interest in avian predators because of potential indirect effects on the endangered Peregrine Falcon. This species is the subject of an intense recovery program in Dinosaur National Monument (NPS 1983). Declines in populations of passerine prey species following aerial application of insecticides have been noted, probably as a result of emigration from the site (Moulding 1976). In addition, toxic effects could arise through consumption of insecticide-treated crickets.

Aside from their role as a food item for wild bird species, crickets have been utilized as food for domestic fowl. Homesteading farmers found that crickets trapped in ditches (built to keep the insects out of crop fields) made a very good diet for turkeys and chickens (Cowan 1929). More recently, the nutritional value of crickets as chicken food has been investigated and found to be high (DeFoliart et al. 1982). However, commercial harvesting of crickets on open rangeland has not yet attracted the attention of industrial feed companies.

Lastly, the role of Mormon crickets in human nutrition should be mentioned. American Indians were known to herd large numbers of crickets into corrals made of sagebrush and greasewood, or to catch them in baskets as they floated down rivers, then dry and grind them. A type of flour was made from the crushed cadavers, which yielded a pasty winter food (Bancroft 1889, LaRivers 1944, Wakeland 1959, Evans 1985).

Parasites

Mormon crickets are notably free of insect parasites. Although two hymenopteran species are known to attack the eggs, their impact on population dynamics appears to be small. *Sparaison pilosum* Ashmead (Scelionidae) occurs throughout the cricket's range, but overall percent parasitism averaged less than 3% in a 1939 survey (Wakeland 1959). However, Cowan (1929) found that in western Montana parasitism was as high 50%, and was associated with an apparent reduction in hatch. Two additional parasites have been reported. Gahan (1942) described the wasp *Oencyrtus anabrivorus* (Encyrtidae), reared from cricket eggs collected in the Big Horn Mountains of Wyoming. Mature larvae of the flesh fly *Sarcophaga harpax* Pandelle (as *S. tuberosa*) (Sarcophagidae) were reported from adult Mormon crickets (LaRivers 1944, 1945). Nothing is known of the distribution or importance of these two species as regulating agents.

A horsehair worm, *Gordius robustus* Leidy (Nematomorpha), is known to infect crickets and can be abundant in localized areas near standing water. However, its efficacy as a control agent is severely limited by the scarcity of water in much of the Mormon cricket's geographic range (Thorne 1940). A nematode parasite, *Agamospirura anabri,* was described by Christie (1929) from crickets collected in Montana, but its biology is unknown.

Pathogens

The role of diseases in natural cricket population cycles is virtually unknown. The few observations of disease-related mortality found in the literature are vague and contradictory.

Riley (1894) mentioned that "crickets died off by millions from disease" in the vicinity of the Snake River in Idaho. However, Wakeland and Shull (1936) claimed that "no disease is known to occur" in Mormon crickets. Attempts made to introduce *Entomophaga grylli* Fresenius, a fungus found in many orthopterans, for cricket control were unsuccessful (Doten 1904, Ball 1915).

More recent applied studies have focused on a group of pathogens which appears to hold some promise for cricket management: the microsporidians (Protozoa: Microsporida: Nosematidae). A number of authors have suggested that *Nosema locustae* Canning, the best-studied microsporidian from grasshoppers, holds great potential for long-term suppression of acridids (Henry and Oma 1974, 1981, Ewen and Mukerji 1980, Henry and Onsager 1982, Erlandson et al. 1985, 1986) and Mormon crickets (Henry and Onsager 1982). This pathogen is typically acquired through transovarial infections or by ingestion of spores. *N. locustae* attacks the fat body, neural, and pericardial tissues of grasshoppers, but it appears to be confined to gut tissue in Mormon crickets (Canning 1962a, b, Henry and Onsager 1982). *N. locustae* can be easily propagated in the laboratory and applied to bran bait, which can then be spread in the field to inoculate a wild population with high levels of spores. Infected insects become weakened, resulting in slower growth, less feeding, reduced fecundity, or death. Infected insects may also be more susceptible to cannibalism, which should serve as an excellent means of transmission (Henry and Oma 1981).

After initial isolation of *N. locustae* from adult Mormon crickets in 1974 (Henry and Oma 1981), susceptibility of third instars to infection via application of treated bran bait was confirmed in a large-scale field test in northwestern Colorado (Henry and Onsager 1982). However, due to movement of bands into and out of the treated area and the lack of good controls, it was impossible to discern the true impact of *N. locustae* on the cricket population. Additional research is needed to evaluate further the effects of *N. locustae* on Mormon cricket survival and reproduction, and its potential as a management tool.

A new species of microsporidian found in Mormon crickets in 1985 (tentatively *Vairimorpha* sp., Henry, per. comm., 1986) appears to hold greater potential for cricket suppression. Unlike *N. locustae,* it builds up extremely high spore levels in many tissues, is common in the cricket population of northwestern Colorado, and is sometimes associated with sluggish behavior and high mortality (MacVean and Capinera, unpublished data). Thus, it may be more important as a natural regulator of cricket populations.

Weather

The role of weather in cricket population dynamics is a matter of conjecture. Circumstantial evidence suggests that cold, wet conditions can adversely affect survival. Despite the crickets' ability to withstand inclement weather by clustering in protected sites (Corkins 1923), reductions in numbers have been observed during the early nymphal period (instars 1 to 4) following prolonged periods (several weeks) of rain or snow and daily lows near freezing. Other mortality factors, such as predators and parasites, were not observed during these periods, leading to the conclusion that mortality was weather-induced (Schweis et al. 1939, Wakeland 1959). The later instars appear to be less susceptible to inclement weather. Severe winter conditions appeared to be of little consequence to egg viability, as indicated by consistently high populations (Wakeland and Shull 1936). However, little can be gleaned from the sketchy observations in the literature, and the role of weather remains unclear.

PEST STATUS AND MANAGEMENT

Since Mormon crickets first acquired pest status by damaging wheat fields in Utah in 1848, crop losses have been recorded throughout the range of the insect at one time or another up to about 1960. From this time to the beginning of the current outbreak (ca. 1980) crickets have occurred as relatively small, localized populations, causing little damage.

Crop acreages, dollar amounts lost to crickets, and the impact on local economies for the pre-1960 era are available in the literature (Corkins 1922, Cowan and McCampbell 1929, Cowan 1932, Wakeland et al. 1939, Swain 1944, Wakeland and Parker 1952, Wakeland 1959). Northwestern Colorado, an area with the most prolonged cricket infestation on record (1918-1938; Wakeland 1959) witnessed a reduction in the number of farms from 420 in 1920 to 258 in 1927, due to ravages by crickets (Cowan 1929, 1932). Significant crop losses have also been documented in Montana (Strand 1937, Mills 1939, Cowan and Shipman 1943, Morrill 1983) and Utah (Sorenson and Thornley 1938).

In contrast, forage loss on rangeland and the associated monetary costs are much more difficult to document. Only two studies exist in the literature which attempt to quantify range damage due to crickets. They suggest that while total consumption by crickets is potentially damaging, only a few relatively small areas have experienced serious losses. Since evidence of the Mormon cricket's damage potential in croplands is well documented, and because the current controversy over cricket control centers on damage to rangelands, the remainder of this discussion will focus on the latter problem.

Based on consumption studies with caged insects, Cowan and Shipman (1947) concluded that during a 4-month period (the approximate lifespan of a cricket) crickets at a constant density of $12/m^2$ would consume 4.4 times the amount of forage taken by cattle under proper stocking rates, i.e., 4 head of cattle per 100 ha for 9 months (10 head of cattle/section for 9 months). Furthermore, they pointed out that the plant species most frequently consumed by crickets, *Bromus tectorum*, *Lupinus caudatus* Kellogg, *Poa* sp., and *Balsamorhiza sagittata*, were also the most important forage species in Nevada (although lupine is generally considered a poor forage plant). Thus, the likelihood for competition between crickets and livestock was high, and had in fact forced ranchers to move livestock out of infested areas.

However, Swain (1944) found that severe damage due to crickets was rare and localized. Using visual estimations of consumption (calibrated with hand balances) in replicated, $9.3m^2$ open plots in the areas of worst cricket infestations in Nevada and Idaho, total amounts of herbage removed by crickets were calculated. The relative losses of forage for livestock, based on "proper-use factors" and species composition, were also computed. Total dry weight removed by the end of the season ranged from 1 to 56% of the total current year's growth, and averaged 15% (averages of data given by Swain (1944)). Loss in forage available for livestock (i.e., the fraction of the total year's growth which can be removed without overgrazing) ranged from 1 to 100%, but averaged 35% for cows and horses, and 40% for sheep and goats. In 7 of 36 transects (each transect consisting of ten $9.3m^2$ plots) relative losses approached 100%. Three of these transects were dominated by broadleaf trees which possess little forage value, but where even moderate feeding by crickets can remove 100% of the utilizable plant growth. Excluding these locations, only 4 of 36 study sites, or 11%, experienced loss of all the forage available for livestock.

Clearly, even the most severely damaged areas in Swain's study suffered much less damage than Cowan and Shipman's (1947) data would have predicted, i.e., removal of all utilizable plant growth in any heavily infested area. The discrepancy can probably be explained by two major factors: cricket density over time (cricket-days) and diet composition. While a few transect sites experienced cricket densities of $6-24/m^2$ during the entire season, most areas were occupied and injured by crickets for only 3-4 successive days (Swain 1944). Cowan and Shipman (1947) claimed that bands of crickets covering a given area at a density of $12/m^2$ for an entire season (required for their damage projections) were not uncommon, but gave no supporting data.

More likely, as suggested by Swain's study (1944), such infestations are rare. Secondly, Cowan and Shipman (1947) estimated forage loss by extrapolating consumption of lettuce or alfalfa leaves by caged crickets. As pointed out earlier, the predatory, cannibalistic, and scavenging components of cricket feeding behavior may well reduce forage consumption, and are probably reflected in the lower consumption values found by Swain (1944).

Swain (1944) also discussed cricket injury to range plants with respect to major vegetation types. The transect studies were conducted in "northern desert shrub" areas (Oregon, Nevada, Idaho, Utah), where the worst cricket outbreaks were occurring in 1938-39 and where estimates of total available plant biomass per unit area were much lower than in the "mixed prairie grassland" in the eastern portion of the crickets' range (portions of South Dakota, Nebraska, Montana, Wyoming). In the latter areas, Swain (1944) estimated green clipped weight at ca. 400 g/m^2, compared to 55 g/m^2 in typical sagebrush vegetation of Nevada. Based on transect data, losses in the mixed prairie areas were barely detectable (less than 5% of total biomass) with weedy species suffering most of the damage (*Taraxacum, Tragopogon, Penstemon*). These plants made up a small proportion of the total vegetation, but were preferred over grasses such as *Stipa comata* Trin. and Rupr., *Festuca* spp. and *Agropyron* spp. Swain (1944) thus pointed out the importance of total biomass availability and species composition in determining the pest status of Mormon crickets. When succulent forbs are available, most feeding will be confined to them, and to a lesser extent, the inflorescences of grasses. However, in areas of high shrub density, crickets will utilize grass foliage and inflorescences more heavily.

Cricket densities have generally remained lowest in the eastern part of the species' range. While economic damage has occurred in eastern Colorado, the Dakotas, Nebraska, and Montana, (Cowan and McCampbell 1929, Swain 1944, Wakeland 1959, Morrill 1983), the major outbreaks have always been reported west of the shortgrass and mixedgrass regions.

While quantitative assessments of cricket damage to rangeland were scarce, control campaigns were widespread throughout the infested area for many years (Fig. 9.1). Wakeland (1951, 1959) provides an excellent review of the evolution of control measures, from mechanical barriers such as trenches, sheet metal fences, or oil-on-water traps designed to divert or capture migrating bands of crickets, to arsenite dusts, baits, and finally to aerial sprays. In part, control measures were conducted in the immediate vicinity of crop lands, but most of the effort was directed at cricket control on open rangeland, where infested acreages were much greater (Wakeland and Parker 1952). In 1939, almost 19 million acres were estimated to be infested (Wakeland 1951).

The claim was often made that cricket injury to rangeland and loss of stock carrying capacity was significant and warranted control (Schweis et al. 1939, Cowan and Shipman 1943, Wakeland 1959, Cowan and Wakeland 1962). While undoubtedly true for certain areas, it seems likely that damage was overestimated in many instances. Evidence for this view is seen in the limited amounts of damage in many of Swain's (1944) study sites, despite their location in areas of worst cricket infestations. Even in agricultural areas with higher monetary value than open rangeland, control campaigns were sometimes conducted at an expense greater than the value of the crop (Wakeland and Shull 1936, Wakeland 1959). However, the psychological impact of hordes of large, black crickets traversing not only the range but also invading houses and barns (Johnson 1905, Corkins 1922), destroying vegetable gardens, and contaminating well water with thousands of dead bodies (Wakeland 1959) cannot be underestimated.

Drought conditions during the 1930's and overgrazing on most western rangelands (Wakeland and Shull 1936, Voigt 1976) certainly worsened the impact of cricket herbivory, making control measures necessary where they might not have been required due to crickets alone. The worst cricket outbreak peaked in 1938-39, soon after the Dust Bowl drought years (LaRivers 1944, Wakeland 1951, Navarra 1979). It also coincided with a time when overgrazing by sheep and cattle was at its height and just beginning to draw attention by land management agencies (Wakeland and Shull 1936, Foss 1960, Voigt 1976). Prior to the Dust Bowl era, it appears that Mormon crickets had no significant detrimental effects on rangeland, even during outbreak years. Doten (1904) could find no trace of feeding damage despite the fact that "crickets fairly covered the higher mountain slopes" in Nevada. Corkins (1922, 1923), working in Colorado, reported that "so long as this insect confines itself to its native habitat, the sagebrush covered hills, little harm results." In the early thirties Cowan (1932) also found that "crickets do not make very appreciable inroads on range grass" in northwestern Colorado. Thus, it appears that during the outbreak of the late thirties, drought and overgrazing aggravated an otherwise tolerable, perhaps inconsequential, level of herbivory by crickets with respect to livestock carrying capacities.

The current outbreak in northwestern Colorado and northeastern Utah has again triggered strong action by ranchers and government to control crickets. In 1985, the Animal and Plant Health Inspection Service (APHIS) conducted aerial ULV applications of Sevin-4-Oil (carbaryl) on 23,700 ha of rangeland. The campaign immediately came under legal scrutiny for potential damage to endangered species, primarily the American peregrine falcon, and no spraying was conducted in 1986. Control efforts in 1985 were partly in response to claims that range damage was occurring, partly for prevention of crop damage by bands of crickets migrating from nearby uplands, and perhaps most significantly from apprehension that an uncontrolled population would reach epidemic proportions, as in the 1930's.

Damage to rangeland in infested areas is often slight or imperceptible. Wakeland (1959) described cricket damage as varying from scalloped leaf margins, or holes in leaves, to destruction of whole plants. Currently, only the first two types are occurring, resulting in biomass losses that are practically immeasurable through range survey methods which rely on vegetation weight estimates (e.g., Pechanec and Pickford 1937, Ahmed et al. 1983, Cabral and West 1986, MacVean and Capinera, unpublished). Since quantitative records of cricket population density do not exist, it is difficult to compare the current outbreak with that of the 1930's. However, numerous photographs in the literature depict bands of migrating crickets which appear very similar in density to those found today in northwest Colorado. The total area covered by such bands is obviously much smaller at present than during past outbreaks, but the levels of herbivory in infested areas would be expected to be similar. Why then, is range damage so slight at present, given that severe defoliation was occurring in the same area, and others with similar plant composition, during the 1930's? It seems likely that at least two factors are involved: weather and grazing pressure. In contrast to conditions prevailing at that time, recent years have seen above-average precipitation in Colorado (Karl et al. 1983, Doesken et al. 1987) and stocking rates are more carefully monitored on public lands. Thus, forage availability is probably higher now than during previous outbreaks, thereby reducing the relative impact of consumption by crickets. As discussed earlier, crickets are selective feeders where an abundance of forbs and grasses exists, consuming only succulent tissues and inflorescences. If current damage levels are truly representative of an area receiving favorable precipitation and moderate grazing pressure, then the Mormon cricket's economic importance is negligible.

MANAGEMENT PROBLEMS AND OPTIONS

Currently, an understanding of pest status and the basis for management decisions suffer from at least two fundamental deficiencies: the lack of appropriate sampling methods and lack of economic injury levels. Three levels of sampling are required to determine the total cricket population in a given area:

within-band density, band size, and band abundance (e.g. number of bands per 100 km^2). Furthermore, the size and distribution of bands are likely to change during the season due to movement and possible coalescing of bands, such that the cricket density found at a given site over time (cricket-days) could be highly variable.

The relationship between cricket density and plant injury is little known, aside from Cowan and Shipman's (1947) laboratory consumption study. Although cricket damage to rangeland appears to be slight at present, further research is necessary to understand damage potential in relation to local plant species composition and annual productivity.

It is realistic to expect that the concerns of ranchers and government for cricket control will persist. Throughout the history of Mormon cricket control, prevention of damage to crops by timely destruction of migratory bands on rangeland has been considered essential (Wakeland 1959). However, while perhaps a valid approach for protection of farms in the 1920's, this view is inappropriate today. By virtue of the high, rapid cricket mortality produced by low concentrations of carbaryl in bran or rolled-wheat baits (Foster et al. 1979), these can be used in a reactive sense to protect agricultural areas and need not be applied further than the immediate surroundings of the crop field. More than one application may be necessary, but treatment of localized areas would result in obvious reductions in cost and pesticide loads in the environment. Biocontrol agents, such as spores of microsporidians, which can be incorporated into baits, may also provide a highly selective suppression tactic well suited for protected areas such as national park lands. While slow-acting in nature, such pathogens may persist from one generation to another and be transmitted horizontally through cannibalism. Further research is required to determine the effectiveness of these agents for use on rangeland.

ACKNOWLEDGMENTS

I wish to thank the National Park Service and the Colorado Agricultural Experiment Station for financial assistance; Howard Evans, Boris Kondratieff, Dave Horton, Rick Redak, and Sarah Klahn for reviewing the manuscript; and, in particular, John Capinera for providing the opportunity to write this chapter, and for his enduring support, criticism, and encouragement.

REFERENCES

Ahmed, J., C.D. Bonham, and W.A. Laycock. 1983. Comparison of techniques used for adjusting biomass estimates by double sampling. J. Range Manage. 36:217-221.

Ball, E.D. 1915. How to control the grasshoppers. Utah Agr. Coll. Exp. Sta. Bull. 138:116.

Bancroft, H.H. 1889. History of Utah, p. 262, 279-281. *In:* The works of Hubert Howe Bancroft, Vol. 26. The History Co., San Francisco.

Cabral, D.R., and N.E. West. 1986. Reference unit-based estimates of winterfat browse weights. J. Range Manage. 39:187-189.

Canning, E.U. 1962a. The life cycle of *Nosema locustae* Canning in *Locusta migratoria migratorioides* (Reiche and Fairmaire), and its infectivity to other hosts. J. Insect Pathol. 4:237-247.

Canning, E.U. 1962b. The pathogenicity of *Nosema locustae* Canning. J. Insect Pathol. 4:248-256.

Caudell, A.N. 1907. The Decticinae (a group of Orthoptera) of North America. Proc. U.S. National Museum 32:285-290, 351-369.

Caudell, A.N. 1908. An old record of observations on the habits of *Anabrus*. Entomol. News 19:44-45.

Christie, J.R. 1929. Notes on larval nemas from insects. J. Parasitol. 16:250-252.

Corkins, C.L. 1922. Notes on the habits and control of the Western or Mormon cricket, *Anabrus simplex* Hald. Colorado Office of State Entomol. Cir. 36.

Corkins, C.L. 1923. Mormon cricket control. Colorado Office of State Entomol. Cir. 40.

Cowan, F.T. 1929. Life history, habits, and control of the Mormon cricket. USDA Tech. Bull. 161.

Cowan, F.T. 1932. Mormon cricket control in Colorado. Colorado Office of State Entomol. Cir. 57.

Cowan, F.T., and S.C. McCampbell. 1929. The Mormon cricket and its control. Colorado Office of State Entomol. Cir. 53.

Cowan, F.T., and H.J. Shipman. 1943. Mormon crickets and their control. USDA Farmers' Bull. 1928.

Cowan, F.T., and H.J. Shipman. 1947. Quantity of food consumed by Mormon crickets. J. Econ. Entomol. 40:825-828.

Cowan, F.T., and C. Wakeland. 1962. Mormon crickets - how to control them. USDA Farmers' Bull. 2081.

Criddle, N. 1926. The life history and habits of *Anabrus longipes* Caudell (Orthop.). Can. Entomol. 58:261-265.

DeFoliart, G.R., M.D. Finke, and M.L. Sunde. 1982. Potential value of the Mormon cricket (Orthoptera: Tettigoniidae) harvested as a high-protein feed for poultry. J. Econ. Entomol. 75:848-852.

Doesken, N., and T. McKee. 1987. Colo. climate summary, water-year series (Oct. 1985-Sept. 1986). Climatol. Report 87-2. Colorado Climate Center, Dept. Atmosph. Sci., Colorado State Univ.

Doten, S.B. 1904. The Western cricket. Nevada Agr. Exp. Sta. Bull. 56.

Erlandson, M.A., M.K. Mukerji, A.B. Ewen, and C. Guillot. 1985. Comparative pathogenicity of *Nosema acridophagus* Henry and *Nosema cuneatum* Henry (Microsporida: Nosematidae) for *Melanoplus sanguinipes* (Fab.)(Orthoptera: Acrididae). Can. Entomol. 117:1167-1175.

Erlandson, M.A., A.B. Ewen, M.K. Mukerji, and C. Guillot. 1986. Susceptibility of immature stages of *Melanoplus sanguinipes* (Fab.)(Orthoptera: Acrididae) to *Nosema cuneatum* Henry (Microsporida: Nosematidae) and its effects on host fecundity. Can. Entomol. 118:29-35.

Evans, H.E. 1985. The pleasures of entomology, portraits of insects and the people who study them. Smiths. Inst. Press, Washington, D.C.

Ewen, A.B., and M.K. Mukerji. 1980. Evaluation of *Nosema locustae* (Microsporida) as a control agent of grasshopper populations in Saskatchewan. J. Invertebr. Pathol. 35:295-303.

Foss, P.O. 1960. Politics and grass, the administration of grazing on the public domain. Univ. Washington Press, Seattle.

Foster, R.N., C.H. Billingsley, R.T. Staten, and D.J. Hamilton. 1979. Field cage tests for concentrations of carbaryl in a bait and its application rates for control of Mormon cricket. J. Econ. Entomol. 72:295-297.

Gahan, A.B. 1942. Descriptions of five new species of Chalcidoidea, with notes on a few described species (Hymenoptera). Proc. U.S. National Museum 92 (3137):41, 49-51.

Gillette, C.P. 1904. Copulation and ovulation in *Anabrus simplex* Hald. Entomol. News 15:321-325.

Gillette, C.P. 1905. The western cricket, life history and remedies. Colorado Agr. Exp. Sta. Bull. 101:1-10.

Goodwin, J.T., and V.N. Powders. 1970. A range extension for the Mormon cricket. Ann. Entomol. Soc. Am. 63:623-624.

Gurney, A.B. 1939. Aids to the identification of the Mormon and coulee crickets and their allies (Orthoptera; Tettigoniidae, Gryllacrididae). USDA Bur. Entomol. and Plant Quar. E-479.

Gwynne, D.T. 1981. Sexual difference theory: Mormon crickets show role reversal in mate choice. Science 213:779-780.

Gwynne, D.T. 1984. Sexual selection and sexual differences in Mormon crickets (Orthoptera: Tettigoniidae, *Anabrus simplex*). Evolution 38:1011-1022.

Gwynne, D.T., and G.N. Dodson. 1983. Nonrandom provisioning by the digger wasp, *Palmodes laeviventris* (Hymenoptera: Sphecidae). Ann. Entomol. Soc. Am. 76:434-436.

Hansen, R.M., and D.N. Ueckert. 1970. Dietary similarity of some primary consumers. Ecology 51:640-648.

Henderson, W.W. 1931. Crickets and grasshoppers in Utah. Utah Agr. Exp. Sta. Cir. 96:7-15, 33-38.

Henry, J.E., and E.A. Oma. 1974. Effects of infections by *Nosema locustae* Canning, *Nosema acridophagus* Henry, and *Nosema cuneatum* Henry, (Microsporida: Nosematidae) in *Melanoplus bivittatus* (Say)(Orthoptera: Acrididae). Acrida 3:223-231.

Henry, J.E., and E.A. Oma. 1981. Pest control by *Nosema locustae* a pathogen of grasshoppers and crickets, p. 573-586. *In:* H.D. Burges (ed.) Microbial control of pests and plant diseases 1970-1980. Academic Press, N. Y.

Henry, J.E., and J.A. Onsager. 1982. Experimental control of the Mormon cricket *Anabrus simplex* by *Nosema locustae* (Microspora: Microsporida), a protozoan parasite of grasshoppers (Ort.: Acrididae). Entomophaga 27:197-201.

Johnson, S.A. 1905. Distribution and migrations of the Mormon cricket (*Anabrus simplex* Hald.) in Colorado. USDA Bur. Entomol. Bull. 52:62-66.

Kalmbach, E.R. 1918. The crow and its relation to man. USDA Bull. 621:19-21.

Karl, T.R., L.K. Metcalf, M.L. Nicodemus, and R.G. Quayle. 1983. Statewide average climatic history, Colorado 1888-1982. National Oceanic and Atmospheric Administration, National Climatic Data Center, Asheville, NC.

Knowlton, G.F. 1937. Utah birds in the control of certain insect pests. Proc. Utah Acad. Sci. Arts and Letters 14:159-161.

Knowlton, G.F. 1941. California gull and insect control in Utah. J. Econ. Entomol. 34:584-585.

Knowlton, G.F. 1943. Raven eats Mormon cricket eggs. Auk 60:273.

Knowlton, G.F. 1948. Vertebrate animals feeding on the Mormon cricket. Am. Midl. Nat. 39:137-138.

Knowlton, G.F., D.R. Maddock, and S.L. Wood. 1946. Insect food of the sagebrush swift. J. Econ. Entomol. 39: 382-383.

Kurczewski, F.E., and H.E. Evans. 1986. Correct names for species of *Tachysphex* observed by Evans (1970) at Jackson Hole, Wyoming, with new information on *T. alpestris* and *T. semirufus* (Hymenoptera: Sphecidae). Proc. Entomol. Soc. Wash. 88:720-721.

LaRivers, I. 1941. Response of *Anabrus simplex* to temperature. J. Econ. Entomol. 34:121-122.

LaRivers, I. 1944. A summary of the Mormon cricket (*Anabrus simplex*)(Tettigoniidae: Orthoptera). Entomol. News 55:71-77, 97-102.

LaRivers, I. 1945. The wasp *Chlorion laeviventris* as a natural control of the Mormon cricket. Am. Midl. Nat. 33:743-763.

McAtee, W.L. 1926. The role of vertebrates in the control of insect pests, p. 421-422. *In:* Smiths. Inst. Ann. Rpt. for 1925.

Melander, A.L., and M.A. Yothers. 1917. The coulee cricket. Washington State Coll. Agr. Exp. Sta. Bull. 137.

Mills, H.B. 1939. Montana insect pests for 1937 and 1938. Montana Agr. Exp. Sta. Bull. 366:17-21.

Morrill, W.L. 1983. Early history of cereal grain insect pests in Montana. Bull. Entomol. Soc. Am. 29:24-28.

Moulding, J.D. 1976. Effects of a low-persistence insecticide on forest bird populations. Auk 93:692-708.

Navarra, J.G. 1979. Atmosphere, weather and climate: an introduction to meteorology. W.B. Saunders Co., Philadelphia.

NPS (National Park Service). 1983. Natural resources management plan and environmental assessment. Dinosaur National Monu., Colo. and Utah.

Pechanec, J.F., and G.D. Pickford. 1937. A weight estimate method for the determination of range or pasture production. J. Amer. Soc. Agron. 29:894-904.

Rentz, D.C., and J.D. Birchim. 1968. Revisionary studies in the Nearctic Decticinae. Mem. Pacific Coast Entomol. Soc. 3:1-173.

Riley, C.V. 1894. The western or great plains cricket. USDA Div. Entomol. Ann. Rpt.:202-203.

Schweis, G.G., L.M. Burge, and G.M. Shogren. 1939. Mormon cricket control in Nevada, 1935-38. Nevada Sta. Dept. Agr. Bull. 1 and 2.

Scudder, S.H. 1898. The alpine Orthoptera of North America. Appalachia 8:19-20.

Snodgrass, R.E. 1905. The coulee cricket of central Washington (*Peranabrus scabricollis* Thomas). J. New York Entomol. Soc. 13:74-83.

Sorenson, C.J., and L.R. Jeppson. 1940. Some insect pests of farm crops in the juniper-pinon belt of Utah - the Mormon cricket. Proc. Utah Acad. Sci. Arts and Letters 17:49-52.

Sorenson, C.J., and H.F. Thornley. 1938. Mormon crickets and their control in Utah since 1923. Proc. Utah Acad. Sci. Arts and Letters. 15:63-70.

Stanley-Samuelson, D.W., J.J. Peloquin, and W. Loher. 1986. Egg-laying in response to prostaglandins injections in the Australian field cricket, *Teleogryllus commodus*. Physiol. Entomol. 11:213-219.

Strand, A. L. 1937. Montana insect pests for 1935 and 1936. Montana Agr. Exp. Sta. Bull. 333:18-24.

Swain, R.B. 1940. A field method for estimating Mormon cricket injury to forage. J. Kansas Entomol. Soc. 13:124-127.

Swain, R.B. 1944. Nature and extent of Mormon cricket damage to crop and range plants. USDA Tech. Bull. 866.

Tanner, V.M. 1940. The gulls and crickets. Great Basin Nat. 1:49-54.

Thomas, C. 1872. Notes on the saltatorial Orthoptera of the Rocky Mountain regions. U. S. Geol. Survey of Montana and Portions of Adjacent Territories, Ann. Rpt. 5:428-431, 438-441.

Thorne, G. 1940. The hairworm, *Gordius robustus* Leidy, as a parasite of the Mormon cricket, *Anabrus simplex* Haldeman. J. Wash. Acad. Sci. 30:219-231.

Ueckert, D.N., and R.M. Hansen. 1970. Seasonal dry-weight composition in diets of Mormon crickets. J. Econ. Entomol. 63:96-98.

Voigt, W., Jr. 1976. Public grazing lands, use and misuse by industry and government. Rutgers Univ. Press.

Wakeland, C. 1951. Changing problems and procedures in grasshopper and Mormon cricket control. J. Econ. Entomol. 44:76-82.

Wakeland, C. 1959. Mormon crickets in North America. USDA Tech. Bull. 1202.

Wakeland, C., and J.R. Parker. 1952. The Mormon cricket, p. 605-608. *In:* USDA Yearbook of Agriculture, Washington DC.

Wakeland, C., and W.E. Shull. 1936. The Mormon cricket with suggestions for its control. Idaho Ext. Serv. Bull. 100.

Wakeland, C., W.B. Mabee, and F.T. Cowan. 1939. Practical methods of Mormon cricket control. USDA Bur. Entomol. and Plant Quar. E-470.

10. BEHAVIORAL RESPONSES UNDERLYING ECOLOGICAL PATTERNS OF RESOURCE USE IN RANGELAND GRASSHOPPERS

Anthony Joern
School of Biological Sciences
University of Nebraska, Lincoln, Nebraska 68588

Coexisting grasshopper species in North American rangeland typically feed on different plants (Mulkern et al. 1969, Joern 1979a, 1983, Pfadt and Lavigne 1982), are found in different microhabitats (Anderson 1964, Joern 1982, 1986), and are phenologically shifted throughout the year (Joern 1979b). Recognition of behavioral mechanisms underlying observed patterns is required before population and community level processes in grasshopper assemblies can be clarified. Understanding such processes may allow range managers to assess better the impact of grasshopper consumption of forage and recognize previously unappreciated points of vulnerability for control programs. Development of noninvasive, cultural controls which act in concert with biological processes may be the ultimate economic gain of such studies. In this paper, complex interactions among behavioral responses which underlie patterns of resource use are outlined.

When describing patterns of forage loss from grasshoppers, one is also describing temporal and spatial patterns of plant and grasshopper aggregation on a species-specific basis. Localized spatial distributions and patterns of diet selection by grasshoppers are the result of varied individual responses to many environmental pressures. Compromises concerning selection of diet, microhabitat, or other resource are typically mediated by behavioral choices constrained within evolved, species-specific limits. Several very basic factors strongly influence individual behavior (Figure 10.1) and have been examined in grasshoppers. However, the sometimes subtle and intricate interactions among the environmental pressures and grasshopper responses, which are not specifically indicated in Figure 10.1, may be more important in determining behavioral responses than direct influences. For

138

example, food plants which are high in energy (e.g., carbohydrate reserves in leaf tissue, see Rittenhouse and Roath, Chapter 3) may be low in nitrogen or heavily defended by some toxic chemical (see Redak, Chapter 4). Even though some plant in a patch may contain the optimal mix of nutrients with few chemical defenses, the grasshopper must be able to locate this plant against a heterogeneous background. If it cannot, it must either use suboptimal resources or else move around and sequentially select plants to meet nutritional needs. Similarily, high quality host plants may exist in microhabitats which expose a feeding grasshopper to significant risk of predation/parasitism or thermal stress. In such cases, clear trade-offs exist and responses represent compromises imposed by these interactions. Spatial distributions of other, co-occurring grasshoppers may also affect individual behavior since predator behavior or species-specific mate location are often density- or frequency-dependent. Many other indirect, but clearly important, interactions among such environmental pressures can be envisioned.

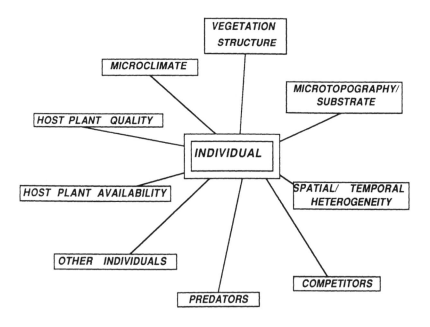

FIG. 10.1 Environmental pressures which direct behavioral responses in patterns of resource use.

AN ECOLOGICAL/BEHAVIORAL HIERARCHY FOR UNDERSTANDING RANGE GRASSHOPPERS

The underlying ecological, behavioral and physiological mechanisms which determine which plants will be eaten, and to what extent, represent a hierarchy of interactions. This hierarchy (Table 10.1) includes questions concerning where in the habitat an individual will feed in addition to which plant will be chosen and how much will be eaten. For the most part, such a classification is for the benefit of biologists attempting to untangle complex interactions, as individual grasshoppers may not face such a range of "decisions" in their ecological lifetime.

Habitat/grasshopper relationships are often biogeographical in origin, as evidenced by the fact that different taxonomic mixes of species are often associated with different habitats (Uvarov 1977, Otte 1981). Sometimes, habitats are interspersed in a patchy mosaic. This increases the degree of interspersion of particular taxa among available habitats. At present, little work has been done to establish the broad patterns of association between grasshopper taxa and habitat specificity, let alone underlying biogeographic and ecological mechanisms.

TABLE 10.1
Hierarchical levels are presented beginning with the most general and ending with the most specific.

Hierarchical Level	Level of Resolution
Habitat	Biogeographical
Patch	Ecological
Microhabitat within patch	Ecological
Choice of leaf from among those available	Behavioral
Actual assessment (accept/ reject)	Behavioral/ Physiological

At ecotonal boundaries, the problem of habitat selection is probably the same as patch selection. A habitat is usually very heterogeneous to a grasshopper, but may contain patches of relative homogeneity. The normal grasshopper has the physical capability of encountering a variety of patches if generally moving about in an unconstrained (physical or behavioral) fashion. Behavioral responses to a wide variety of environmental pressures

140

are involved in selection of patches (Anderson 1964). Aspect of the habitat, such as structure or overall microclimate (sun vs. shade), may be largely responsible (Joern 1986) for patch selection. Although patches are difficult to delineate in the field, grasshopper species respond in predictable ways (Joern 1982). Given the opportunity, species presumably move until they find an appropriate patch which fit predetermined, species-specific criteria. The utility of understanding relationships among habitat, patch and microhabitat are illustrated in managed range systems where grasshopper populations may shift in response to grazing by large vertebrate herbivores (Capinera, Chapter 11).

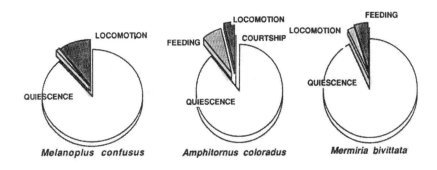

FIG. 10.2 Time/activity budgets of three grasshopper species for general classes of behavior.

ACTIVITY

Time / Activity Budgets

Few activity budgets have been compiled for grasshoppers. Time/activity budgets (daylight hours) for three coexisting species from western Nebraska sandhills grassland are shown in Figure 10.2 (Joern et al. 1986). In all cases, quiescent activity predominated (ca. 90% of daylight period). During night hours (ca. 1930-0600h), all species crawled up on vegetation and were primarily quiescent. No movement was observed in several marked individuals. Quiescent activity should not be perceived as non-activity since individuals were thermoregulating, minimizing exposure to predators, etc. The small percentage of time spent feeding is consistent with some estimations (e.g., *Chorthippus parallelus* Zett., Bernays and Chapman 1970a) but lower than others; *Locusta migratoria* (L) nymphs in the lab spent about 15%

of the time feeding (Ellis 1951) and *Nomadacris septemfasciatus* Serville in the field spent 30-35% of the time feeding (Chapman 1957).

Thermoregulation

Control of body temperature is central to many physiological activities (May 1979). For example, developmental rates (Hardman and Mukerji 1982), food processing abilities (Baines et al. 1973), reproductive activity (Loher and Wiedenmann 1981), life cycle characteristics (Orshan and Pener 1979, Pener and Orshan 1980), and metabolic activity (Chappell 1983) are temperature-dependent processes. Developmental rate increased 5.6-fold and more eggs were produced in caged *Chorthippus brunneus* (Thunberg) when provided with light bulbs which allowed basking and elevated T_b (Begon 1983).

Insect body temperatures (T_b) are greatly influenced by microclimate, especially incoming solar radiation, wind, air temperature, and humidity (Porter and Gates 1969, May 1979). Body color, size, orientation to incoming radiation, and microhabitat selection act in concert to mediate the amount of radiation received and the subsequent T_b. Body temperature is only negligibly affected by metabolism and evaporation in resting insects (Anderson et al. 1979).

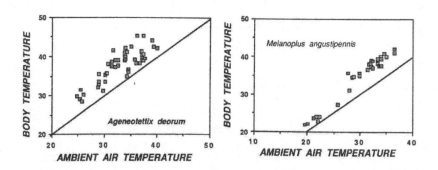

FIG. 10.3 Relation between body and ambient air temperatures.

In general, T_b is correlated with air temperature although typically 2-10°C higher than ambient through the day (Figure 10.3). Such general relationships have been observed for both ground and vegetation dwelling species (Anderson et al. 1979,

Kemp 1986). Coexisting species sometimes exist in very different thermal niches (Anderson et al. 1979, Gillis and Possai, 1983) although not always (Kemp 1986). Given the extensive impact of air temperature on T_b and the extreme temperatures observed during the day in most ground level microhabitats (>60°C), it is not surprising that mechanisms for moderating such conditions exist. In addition to merely avoiding extremes, many species actively maintain T_b within a narrow range over a wide range of environmental conditions (Anderson et al. 1979).

Behavior (posturing and microhabitat selection) has been repeatedly shown to be important in regulating T_b in grasshoppers (Chapman 1959, Anderson et al. 1979, Joern 1981, Gillis and Possai 1983). Individuals orient the longitudinal axis of the body either perpendicular or parallel to the incoming solar rays in response to body temperature and microclimatic conditions (Anderson et al. 1979). A varied repertoire of postures in addition to orientation to the sun has been described whereby individuals position themselves to either increase T_b rapidly or else minimize further increases. Extreme postures include "crouched" positions where the body hugs the substrate (typical of cool conditions with a warm substrate) to "stilting" where the individual maximally extends legs so that the body is as far above the substrate as possible (typical of hot conditions). Body temperatures may vary by several degrees based on posturing alone. When conditions become too extreme, grasshoppers move to different microhabitats (e.g., into shade, or up on vegetation). Regulation of T_b through posturing seems more prevalent in ground dwelling species; vegetation dwelling species appear to move around and select sites with suitable microclimates (Anderson et al. 1979, Joern 1981).

Local Movement

Local spatial distributions and the attendant mosaic of densities within a habitat fluctuate in response to individual movement. Dispersal, where an individual either enters or leaves a habitat or patch, is perhaps the best studied example of movement (Uvarov 1977) and clearly has important implications for generally understanding population fluctuations. However, environmental conditions within a habitat or patch are neither homogeneous nor constant on either a daily or longer time scale. In response to such variation, individual grasshoppers may track suitable microenvironments within an area without actually leaving. In return, mechanisms which encourage localized movement may greatly contribute to the heterogeneity of both resource availability and local patterns of grasshopper dispersion.

Localized movements also contribute to the actual population structure of grasshoppers within a site. Is there much or little gene exchange among individuals found within an area? What are the consequences of the genetic structure of populations? As an example, substrate matching by individuals of some grasshopper species is sometimes very precise (Isley 1938, Rowell 1971, Gillis 1982). In some populations, however, a mosaic of backgrounds exists and grasshoppers with different coloration and patterning coexist (Gillis 1982). To the degree that such coloration patterns are genetically determined, patterns of gene flow may restrict or enhance the development of such a pattern, possibly mediated through assortative mate choice as well as likelihood of moving.

Under natural conditions (i.e., availability of vegetation rather than plowed field), most grasshoppers do not move great distances (Johnson et al. 1986). In addition, grasshoppers do not move unidirectionally; i.e., estimates based on only a few days cannot be easily extrapolated. For example, *Cordillacris crenulata* (Bruner) moves an average of 4.9 m/day but probably spends a lifetime within a area with a radius of 60-70 m (Joern 1983). Some species such as *Hypochlora alba* (Dodge) appear restricted to a patch of Louisiana sagewort, *Artemisia ludoviciana* Nutt. (Smith, per. comm.). A cautionary note concerns the nature of studies used to obtain such estimates; values are based only on recaptured individuals and may sometimes underestimate actual average movement. Individuals which move farthest are least likely to be recaptured.

What factors influence small-scale movement? Few studies exist which specifically examine this question. Dempster (1955) concluded that small-scale movements of *C. brunneus* and *Chorthippus parallelus* (Zett.) were primarily influenced by vegetation height. Vegetation structure clearly influences microhabitat choice (Anderson 1964, Joern 1982; discussed below). Host plant quality may also play a role, especially in affecting activity level (Bernays and Chapman 1970b, Mulkern 1969). As discussed below, more grass-feeding grasshoppers were captured in patches of the grass *Calamovilfa longifolia* fertilized with nitrogen than in unfertilized patches (Heidorn and Joern unpublished).

MICROHABITAT SELECTION

A microhabitat is a specific spatial location which may be not much larger than an individual. Selection of particular microhabitats by grasshoppers represents a compromise among multiple factors including both biotic and abiotic components. The list of possible pressures which might determine the suitability of a local site is varied. For grasshoppers, factors known to influence microhabitat selection include: vegetation structure, number of plant species, microclimate, soil characteristics, availability of suitable food plants, oviposition site availability, substrate characteristics which render an individual cryptic, aggressive interactions deriving from territoriality (Otte and Joern 1975, Greenfield and Shelly 1985), and relative heterogeneity within a patch which may influence the ability to locate an appropriate microhabitat (Gould and Stinner 1984). Which of these microhabitat characteristics are most important to grasshoppers? Does the relative importance of each factor vary among grasshopper species? And, how do these factors interact to determine observed patterns of microhabitat selection? Most of these questions cannot yet be answered.

Vegetation Structure

Vegetation structure is often a dominant cue in grasshopper microhabitat selection (Dempster 1955). Several studies have shown that grasshoppers have the behavioral capacity and visual acuity to respond to vegetation structure (Williams 1954, Wallace 1958, Bernays and Chapman 1970b, Mulkern 1969).

Grasshoppers respond to spatial cues. Grass-inhabiting *Gomphocerippus rufus* (L.) and *C. parallelus* moved toward vertical rather than horizontal or near horizonal stripes (Williams 1954). *C. parallelus* showed similar tendencies to select vertical objects when given choices among vertical and horizontal wires in a 3-dimensional setting (Bernays and Chapman 1970b). *Melanoplus keeleri* (Thomas) and *M. femurrubrum* (DeGeer) nymphs responded to vertical lines projected on a ground glass screen by crawling upward along the edges of the lines; horizontal lines inhibited upward movement (Mulkern 1969). Presumably, ground-dwelling species are less responsive to such cues although the critical studies have not been done.

Substrate Matching

Several studies have indicated that grasshoppers often select backgrounds against which they best blend (Isely 1938, Rowell 1971, Gillis 1982). For example, two syntopic color morphs of the cryptic *Circotettix rabula* (Rehn & Hebard) actively selected backgrounds according to the best match to body color (Gillis 1982). Red individuals in which grey rings had been painted around the eye chose grey backgrounds while grey individuals with red rings around the eye chose red backgrounds; red/red and grey/grey controls responded appropriately. Further experiments eliminated responses due to reflectance as an explanation.

FEEDING BEHAVIOR

A wide-ranging variety of studies on food use by grasshoppers exists including: biochemical studies of nutritional needs (Dadd 1960, 1963), physiological investigations concerning sensory capabilities (Chapman and Thomas 1978, Stadler 1982), digestion and the role of nutrition and defensive chemicals in mediating host plant acceptance or rejection (Williams 1954, Chapman, 1974, Bernays and Chapman 1978), and ecological studies which assess preference among plant species and patterns of host plant use by various groups of grasshoppers in a variety of ecological settings (Gangwere 1961, Mulkern 1967, Otte and Joern 1977, Joern 1979, Heidorn and Joern 1984). Combined insights lead to a series of hypotheses for understanding behavioral mechanisms which direct host plant selection.

At the final levels in the foraging hierarchy, selection of a host plant for assessment may be a response to visual, olfactory, or tactile stimuli, or may merely represent a host plant in easy reach when hungry (Bernays and Chapman 1970b). *C. parallelus* found in dense grassland appears initially to select grasses based on arbitrary encounter (Bernays and Chapman 1970b). Overall plant abundance may be the primary criterion for selection at this level as the most abundant plant species would be encountered most frequently. Finally, assessment is largely a physiological response to chemical cues (phagostimulants or antifeedants) in the plant tissue (Bernays and Chapman 1978). If suitable cues exist, the grasshopper eventually feeds. Otherwise, the presence of antifeedants or insufficient positive information results in rejection of that leaf and often plant, and the grasshopper continues its search for acceptable food (Blaney et al. 1985).

Detailed examination of responses of the first three levels of the foraging hierarchy (Table 10.1) have been discussed elsewhere since they involve movement, microhabitat selection, and so forth. These factors clearly play an integral role in the final patterns of forage loss in natural systems. This section concentrates on the behavioral mechanisms which underly actual choice within a microhabitat.

General Feeding Patterns

General patterns of feeding behavior are known for a large number of grasshopper taxa from a variety of North American grassland types. Most species exhibit noteworthy species-specific selectivity, even those with seemingly wide-ranging diets. Comparisons among species, however, indicate that significant differences in actual selection of diets exist in terms of plant taxonomic identity, number of plants eaten, and category (grass/forb) eaten (Gangwere 1961, Mulkern 1967, Otte and Joern 1977, Joern 1979). North American range species tend to be either grass or forb feeders; only a small number of species can be classified as mixed-feeders (Mulkern et al. 1969, Joern 1983a).

Phylogenetic affinities in feeding behaviors of North American species also exist. Gomphocerines and oedipodines are primarily grass feeders while melanoplines are primarily forb feeders. Grass-feeding species, in general, have significantly lower diet breadths than do forb feeders. For example, diet breadth (weighted by relative frequency of taxa included in the diet) of gomphocerines from a sandhills grassland in Nebraska (Joern 1983a) was 4.7 compared to 7.5 for oedipodines and 11.4 in melanoplines; melanopline species from this site averaged a total of 17.1 plant taxa in the collective diet compared to 8.0 for gomphocerines. True specialist feeders are forb-feeders, leading to the interesting result that monophagous and polyphagous species are forb-feeders while grass-feeding species are typically oligophagous.

Diet breadth also varies in association with habitat characteristics. Vegetation-dwelling species tend to have a lower diet breadth than do ground-dwelling species (Mulkern 1967, Otte and Joern 1977, Joern 1979, 1983a). In Nebraska sandhills grassland (Joern 1983a), vegetation-dwelling species have a diet breadth of 5.5 compared to 10.4 for ground-dwelling species. Food plants taken by *N. septemfasciatus* varies as it moves throughout the vegetation during the day in response to microclimatic changes (Chapman 1957). Grasshoppers from disturbed areas may have larger diet breadths than observed in grasshoppers from undisturbed sites although the result is tenuous (Joern 1983a). Mean diet breadth of coexisting species also varies according to grassland type (Joern 1983a) ranging from relatively low values in shrub steppe and shortgrass prairie (Pawnee Site, B=4.2; Pfadt and Lavigne 1982) to significantly higher diet breadths in sandhills grassland (northeastern Colorado, B= 8.2; Ueckert and Hansen 1971). A significant, positive correlation exists between mean annual rainfall and mean diet breadth of grasshoppers from a range of North American grassland sites (Joern in press).

Do behavioral preferences correlate with suitable measures of success? At a general level, the answer is clearly yes. Most grasshoppers perform well on preferred plant species (Mulkern 1967). Deterrent chemicals often prevent or reduce feeding even when no alternative food sources are present and subsequent growth, survival and reproduction are reduced. Grasshoppers typically avoid these species (Chapman 1974, Bernays and Chapman 1978). Approximate digestibility, a measure of feeding efficiency, tends to correlate with preference rankings in two *Melanoplus* species (Bailey and Mukerji 1976). Some exceptions exist as well. The very hairy leaves of *Tribulus terrestris* (Zygophyllaceae) deters feeding in *Schistocerca americana* (Drury) although nymphs forced to feed solely on this plant survive and grow exceedingly well (Otte 1975). *Melanoplus bivittatus* (Say) preferred alfalfa in feeding trials although individuals raised solely on alfalfa exhibited reduced survivorship and fecundity (Pfadt 1949).

General Feeding Behavior

Predictable behavioral sequences during host plant evaluation by grasshoppers have been observed (Blaney and Chapman 1970, Mordue 1979). Four stages exist: palpation, biting, nibbling, and feeding. Rejection may occur at any stage. A grasshopper probes the leaf surface with maxillary palps during palpation. Contact chemoreceptors on the palp tips and within the buccal cavity record the sensory stimuli upon touching the leaf. Rejection of a leaf can occur at this stage (Blaney and Chapman 1970, Mordue 1979, Blaney et al. 1985). If either insufficient or slightly positive information concerning suitability is received during palpation, grasshoppers bite into the leaf and release constitutive material from the cells. Again, chemoreceptors respond to the mix of phagostimulants and antifeedants. If not rejected at this level, the grasshopper may nibble and ultimately continue to feed. Such behaviors, including "random biting" followed by immediate rejection and continued search have been repeatedly observed under field conditions (Williams 1954, Gangwere 1961), although it is now known that biting is not indiscriminant but follows palpal examination.

Intrinsic Plant Qualities

Physical and chemical qualities intrinsic to individual plants greatly mold and modify host plant selection by grasshoppers. Although some phagostimulatory chemicals have been identified (Dadd 1960, Cook 1977, Bernays and Chapman 1978), most cues which influence host plant choice and meal size are deterrents (Chapman 1974, Bernays and Chapman 1978, see also Chapter 4) including: toxic secondary chemicals, hairiness, hardness, water content, etc. (Table 10.2). While the majority of the detailed studies of these factors have concerned locusts (especially *L. migratoria* and *Scistocerca gregaria* (Forsk.)), a sufficient number of studies on other grasshopper species, including some from North American rangeland, indicate that the results may be general.

Nutritional Chemicals. Nutritional needs of grasshoppers are similar to those of most insects and include minimal levels of protein, carbohydrates, lipids (especially sterols and small amounts

TABLE 10.2
Brief summary of attributes which influence types and amount of host plants taken by grasshoppers.

Attribute	Response	Reference
INTRINSIC FACTORS IN PLANT		
Nutritional		
Sugar	Hexose and disaccharide sugars are stimulatory	Cook 1977
Nitrogen	Amino acids (L-proline, L-serine)	Bernays & Chapman 1978
	Grasshoppers concentrate on fertilized grass	Cook 1977
	May affect movement	Heidorn & Joern unpublished
Quantitative	Affected digestive efficiency	Bernays 1982
	Moved away from low quality	Mordue & Hill 1970
Water Content	Affects latency to feed, meal length, decreased switching, growth, water balance	Parker 1984
		Lewis & Bernays 1985
Wilting	Alters preference ranking	Roessingh et al. 1985
Antifeedant Chemicals	Deter feeding, reduce survival, alters reproduction	Lewis 1979, 1982
		Bernays & Chapman 1978
Physical Attributes		
Hardness	Mandibles often worn, early nymphs often cannot eat	Williams 1954
		Bernays & Chapman 1970b
Texture (trichomes)	Reduced feeding, sometimes only affects early instars	Knutson 1982
Leaf Thickness	Deters feeding by early instars	Bernays & Chapman 1970b
		Bernays & Chapman 1970b

TABLE 10.2 (continued)

Attribute	Response	Reference
Leaf Shape	No effect	Williams 1954
		Bernays & Chapman 1970b
Color	No effect	Williams 1954
Leaf Damage	Amount eaten, growth rate, oocyte production, mortality	McCaffery 1982
GRASSHOPPER ATTRIBUTES		
Starvation	Less selective and eats broader range of plants	Gangwere 1961
		Kaufman 1965
		Hill et al. 1968
Age	Quality eaten decreases with age	Hill & Goldsworthy 1968
Haemolymph Osmotic Pressure	Increase reduces meal size	Bernays & Chapman 1973, 1974
Phylogeny	General patters of grass- vs. forb-feeding	Williams 1954
		Otte & Joern 1977
Habituation	Decreased inhibitory affect of antifeedants	Gill 1972 (cited in Bernays 1983)
Associative Learning	Increased rate and mode of rejection of unpalatable host	Blaney et al. 1985
		Blaney & Simmonds 1985

of fatty acids), some water soluble vitamins, and presence of inorganic salts at low concentrations (Dadd 1963, Bernays and Chapman 1978). Many of these needs are probably easily obtained from food plants (e.g., inorganic salts), but others such as protein may be often limiting since protein levels in plants are generally quite low (Mattson 1980). Effects of sugars and nitrogen on host selection have been examined in most detail.

Initiation and maintenance of feeding typically requires phagostimulants, chemicals which induce feeding. Among the classes of nutrients known to be required, sugars and some amino acids have stimulatory roles while most other classes which have been tested (inorganic salts, water soluble vitamins, phospholipids, sterols) either had no effect or results are conflicting (Bernays and Chapman 1978). Leaf water content has demonstrated effects, but in complicated ways (Bernays and Chapman 1978, Lewis 1979, 1982, Lewis and Bernays 1985).

Many hexose sugars and disaccharides stimulate feeding in *L. migratoria* while pentose sugars either had no effect or were inhibitory at high concentrations (Cook 1977). However, only sucrose and fructose are probably common enough in most plants to be regarded as common phagostimulants. Sucrose has been shown to be a phagostimulant for all acridoid species tested irrespective of taxonomic status or feeding habits (Bernays and Chapman 1978). In addition, sucrose and fructose are additive in their influence (in association with a variety of other phagostimulants) on meal size in short term experiments (Bernays and Chapman 1978). If levels of these or other suitable sugars fluctuate with regularity or in response to various types of stress (Gershenzon 1984), they may play an even more important role in diet selection.

In general, proteins and amino acids seldom act as phagostimulants at naturally occurring concentrations. Purified wheat proteins had little effect as phagostimulants to *Camnula pellucida* (Scudder) or *M. bivittatus* (Thorsteinson and Nayar 1963). Of commonly occurring amino acids, only L-proline and L-serine elicited strong feeding responses in *L. migratoria*; other amino acids resulted in weak or no response, especially at naturally occurring levels observed in most plants (Cook 1977). Bernays and Chapman (1978) conclude that while amino acids are stimulating, relative proportions seem unimportant and they do not enhance the stimulating power of sucrose alone. As with sugars, concentration fluxes of soluble nitrogen or specific amino acids in response to environmental stress may result in short-term but important influences.

Nutrients may elicit other behavioral responses which affect forage loss, even if the nutrient does not act as a phagostimulant. Nutritionally poor diet may lead to increased activity and movement (Mulkern 1969, Bernays 1982) which may result in grasshoppers leaving or staying in a particular food patch. An increased number of grass-feeding (but not forb-feeding) grasshoppers from Nebraska sandhills prairie were found in N-fertilized portions of nearly monospecific *C. longifolia* clones (Heidorn and Joern unpublished). Foliar nitrogen levels increased with fertilization and all treatments were located with ca. 5 m or less of one another (portions of same clone). Yet, *Ageneotettix deorum* (Scudder) did not preferentially choose among leaves from three levels of fertilization in paired choice tests which suggests an inability to detect foliar nitrogen levels. It is possible that individuals which happened upon high quality foliage did not move on while those in low quality patches did.

Antifeedants. Many host plants are rejected on encounter because they contain deterrent chemical defenses (Redak, Chapter 4). In most situations, selection ends at palpation or sometimes biting as chemosensory palps are stimulated. Action of antifeedants is often additive (Adams and Bernays 1978, Bernays and Chapman 1978).

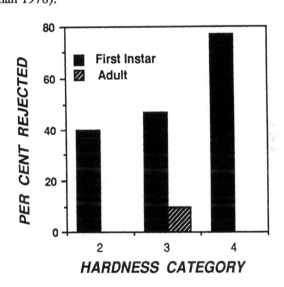

FIG. 10.4 Effect of leaf toughness on acceptability in nymphs and adults of *C. parallelus* (after Bernays and Chapman 1970b). Hardness increases with numerical score.

Rejection is not the only response to secondary chemicals. In some cases, plants are eaten but to a lesser degree (Bernays and Chapman 1978). Also, North American grasshoppers which include the most plant species in their diet are primarily forb feeders (Joern 1983a). These grasshoppers are exposed to a wide

152

range of different chemicals which must be detoxified. Interestingly, grass-feeding *L. gregaria* was sensitive to tannins while the polyphagous, forb-feeding *S. gregaria* was not (Bernays 1978). Rejection responses to plants including defensive chemicals reflect the underlying physiological capabilities to detoxify antibiotic constituents.

Other Factors. A wide range of non-chemical attributes of host plants affect feeding in some grasshopper species (Table 10.2). In general, the effects of these factors are less important than chemical stimulants or deterrents although the age-specific nature of many has great consequences for population processes and cannot be ignored. An example concerns plant hardness (Figure 10.4) in which early instar *C. parallelus* were unable to eat tough leaves although adults had no trouble (Bernays and Chapman 1970b).

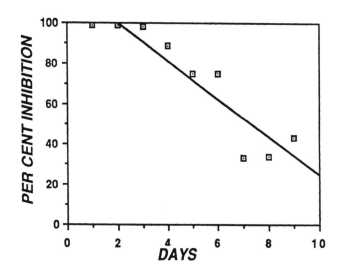

FIG. 10.5 Habituation to plant chemical defenses by *S. gregaria* feeding on artificial diet (after Gill 1972 cited in Bernays 1983).

Experience

Previous experience with host plants may influence willingness to feed and the time required for such decisions to be reached. Habituation has been demonstrated for *S. gregaria* feeding on artificial medium impregnated with the antifeedant chemical azadirachtin (Figure 10.5) (Gill 1972 cited in Bernays 1983). Nearly complete inhibition was initially observed but then dropped to 30-40% after 9 days. Grasshoppers were well fed with green, palatable food for 4 hours each day.

Learning may also be involved. Food selection behavior in *L. migratoria* nymphs was examined (Blaney and Simmonds 1985). Initially, non-preferred hosts were rejected at the biting stage, following palpation. On subsequent contacts, the mode of rejection switched to rejection at the palpation stage. Addition of alternative hosts during an ongoing experiment resulted in reversion to the original rejection mode (biting) for the new plant until the grasshopper encountered it repeatedly. Associative learning was implicated.

Presence of Alternate Host Plants

Grasshopper diet selection may be influenced by the range of available plants in addition to absolute chemical and physical attributes of specific plant taxa. Preferences among available taxa have been clearly demonstrated and selection is often chemically mediated as discussed above. However, to what extent is the actual selection of plants in the field modified by relative abundances or spatial distributions of other taxa? Association of particular plants with other individuals provides the context in which herbivores must find host plants; altered attack rates on particular plant species may result from density-dependent (Root 1973) or frequency-dependent (Atsatt and O'Dowd 1976) responses.

Arphia sulphurea (Fabr.) tended to prefer rare grasses to common ones in Missouri prairie (Landa and Rabinowitz 1983). Reexamination of feeding habits of grasshoppers from Michigan old-field habitat (Gangwere et al. 1976) suggested that the same pattern existed under field conditions (Landa and Rabinowitz 1983). Negative frequency-dependent selection among alternate hosts was observed in laboratory trials of *S. gregaria* feeding on five cultivars (Chandra and Williams 1983). Clear preferences between species pairs were observed when plants were presented at constant densities. Preference then shifted to the least preferred as the relative frequency of this species dropped. Both studies suggest that rare species, when encountered, are at greater risk than common species.

An opposite response was observed for the English grasshopper *Omocestus viridulus* (L.) when grasses were presented in varying frequencies (Cottam 1985). Both relative availability and palatability influenced choice; *O. viridulus* concentrated on the most abundant grass when it was palatable but less so when the most abundant was relatively unpalatable. Preference rankings between pairs of species were not consistent. This result suggests that grasshoppers were not responding to specific plant characters *per se* and that the palatability of a plant and the behavioral mechanism responsible for selection may be altered by surrounding plants.

Clearly, behavioral mechanisms in diet selection involve both the intrinsic qualities of individual plants as well as the overall aspect of the vegetation. Chemical defenses are probably of overriding significance in diet selection although explanation of choice among grasses has proved anomalous. Such mechanisms such as frequency-dependent selection may prove important in understanding selection among grasses. However, Redak (Chapter 4; see also Capinera et al. 1983, Roehrig and Capinera 1983) discusses the relative importance of previously underappreciated chemical defenses in grasses, so the general problem of selection among grasses remains unanswered.

CONCLUDING COMMENTS

Individual behavioral interactions underly most of the basic and applied problems which have been considered to date. Too little emphasis has been placed on these underlying mechanisms in understanding rangeland grasshoppers. As a result, potential opportunities for increased precision of estimating potential damage (or determining actual damage) or of uncovering weak links in the life cycles of problem grasshopper species have not been appreciated. Or, grasshoppers may behaviorally respond to chemical insecticides by hiding, which results in overestimates of efficacy of particular spray programs (Johnson et al. 1986). It is time to increase the implementation of research programs including such detailed studies.

Species-specific differences in how grasshoppers respond to almost any cue should be highlighted rather than ignored. Mulkern (1967) made a similar plea 20 years ago which, for all practical purposes, has gone unheeded. Understanding the diversity of responses to environmental cues and pressures as well as the similarities will provide the needed insights for true IPM.

Finally, it must be recognized that the underlying behavioral responses and resulting ecological patterns are truly complex. Under field conditions, it is not enough to examine merely the intrinsic attributes of plants to predict whether they are likely to be taken by grasshoppers. Where the plant is found, its neighbors, the microclimatic conditions in that microhabitat, the probability that a grasshopper will encounter it, and grasshopper responses to each of these conditions must be known as well. Also, understanding specific behavioral responses of grasshoppers and other insect herbivores may spawn the successful development of unique forms of crop and range protections such as is advocated by Bernays (1983). Clearly, much remains to be understood before this is possible.

ACKNOWLEDGMENTS

Linda S. Vescio kindly commented on the manuscript. Research has been supported by NSF Grant BSR-8408097 and USDA Competitive Grant 86-CRCR-1-1974. Logistical support from Cedar Point Biological Station is also greatly appreciated.

REFERENCES

Adams, C.M., and E.A. Bernays. 1978. The effects of combinations of deterrents on the feeding behaviour of *Locusta migratoria*. Entomol. Exp. Appl. 23:101-109.
Anderson, N.L. 1964. Some relationships between grasshoppers and vegetation. Ann. Entomol. Soc. Am. 57:736-742.
Anderson, R.V., C.R. Tracy, and Z. Abramsky. 1979. Habitat selection in two species of short horned grasshoppers: the role of thermal and hydric stress. Oecologia 38:359-374.
Atsatt, P.R., and D.J. O'Dowd. 1976. Plant defense guilds. Science 193:24-29.
Bailey, C.G., and M.K. Mukerji. 1976. Consumption and utilization of various food plants by *Melanoplus bivittatus* and *Melanoplus femurrubrum*. Can. J. Zool. 54:1044-1050.
Baines, D.M., E.A. Bernays, and E.M. Leather. 1973. Movement of food through the gut of fifth-instar males of *Locusta migratoria migratorioides* (R & F). Acrida 2:319-332.
Begon, M. 1983. Grasshopper populations and weather: the effects of insolation on *Chorthippus brunneus*. Ecol. Entomol. 8:361-370.
Bernays, E.A. 1978. Tannins: an alternative viewpoint. Entomol. Exp. Appl. 24:44-53.
Bernays, E.A. 1981. Plant tannins and insect herbivores: an appraisal. Ecol. Entomol. 6:353-360.
Bernays, E.A. 1982. The insect on the plant - a closer look, p. 3-17. *In:* J.H. Visser and A.K. Minks (eds.) Proc. 5th int. symp. on insect-plant relationships. Centre Agr. Publ. Documen., Wageningen.
Bernays, E.A. 1983. Antifeedants in crop pest management, p. 259-271. *In:* D.S. Woodhead and W.S. Bowers (eds.) Natural products for innovative pest management. Pergammon Press, Oxford.

156

Bernays, E.A., and R.F. Chapman. 1970a. Food selection by *Chorthippus parallelus* (Zetterstedt) (Orthoptera: Acrididae) in the field. J. Anim. Ecol. 39:383-394.

Bernays, E.A., and R.F. Chapman. 1970b. Experiments to determine the basis of food selection by *Chorthippus parallelus* (Zetterstedt) (Orthoptera: Acrididae) in the field. J. Anim. Ecol. 39:761-776.

Bernays, E.A., and R.F. Chapman. 1974a. Changes in haemolymph osmotic pressure in *Locusta migratoria* in relation to feeding. J. Entomol. A. 48:149-155.

Bernays, E.A., and R.F. Chapman. 1974b. The effects of haemolymph osmotic pressure on the meal size of nymphs of *Locusta migratoria* (L.). J. Exp. Biol. 61:473-480.

Bernays, E.A., and R.F. Chapman. 1977. Deterrent chemicals as a basis of oligophagy in *Locusta migratoria* (L.). Ecol. Entomol. 2:1-18.

Bernays, E.A., and R.F. Chapman. 1978. Plant chemistry and acridoid feeding behaviour. p. 99-141. *In:* J.B. Harborne (ed.) Biochemical aspects of plant and animal coevolution: annual proceedings of the phytochemical society of Europe. Academic Press, London.

Blaney, W.M., and R.F. Chapman. 1970. The functions of the maxillary palps of Acrididae (Orthoptera). Entomol. Exp. Appl. 13:363-376.

Blaney, W.M., and M.S.J. Simmonds. 1985. Food selection in locusts: the role of learning in rejection behaviour. Entomol. Exp. Appl. 39:273-278.

Blaney, W.M., C. Winstanley, and M.S.J. Simmonds. 1985. Food selection by locusts: an analysis of rejection behaviour. Entomol. Exp. Appl. 38:35-40.

Capinera, J.L., A.R. Renaud, and N.E. Roehrig. 1983. Chemical basis for host selection in *Hemileuca oliviae:* role of tannins in preference of C_4 grasses. J. Chem. Ecol. 9:1427-1437.

Chandra, S., and G. Williams. 1983. Frequency-dependent selection in the grazing behaviour of the desert locust *Schistocerca gregaria.* Ecol. Entomol. 8:13-21.

Chapman, R.F. 1957. Observations on the feeding of adults of the Red Locust (*Nomadacris septemfasciata* Serville). Br. J. Anim. Behav. 5:60-75.

Chapman, R.F. 1959. The behaviour of nymphs of *Scistocerca gregaria* (Forskal) (Orthoptera: Acrididae) in a temperature gradient, with special reference to temperature preference. Behaviour 24: 283-317.

Chapman, R.F. 1974. The chemical inhibition of feeding by phytophagous insects: a review. Bull. Entomol. Res. 64:339-363.

Chapman, R.F., and Thomas. 1978. The numbers and distribution of sensilla on the mouthparts of Acridoidea. Acrida 7:115-148.

Chappell, M.A. 1983. Metabolism and thermoregulation in desert and montane grasshoppers. Oecologia 56:126-131.

Cook, A.G. 1977. Nutrient chemicals as phagostimulants for *Locusta migratoria* (L.). Ecol. Entomol. 2:113-121.

Cottam, D.A. 1985. Frequency-dependent grazing by slugs and grasshoppers. J. Ecol. 73:925-933.

Dadd, R.H. 1960. Observations on the palatability and utilization of foods by locusts, with particular reference to the interpretation of performances in growth trials using synthetic diets. Entomol. Exp. Appl. 3:283-304.

Dadd, R.H. 1963. Feeding behaviour and nutrition in grasshoppers and locusts. Adv. Insect Physiol. 1:47-109.

Dempster, J.P. 1955. Factors affecting the small scale movements of some British grasshoppers. Proc. Royal Entomol. Soc., London (A) 30:145-150.

Ellis, P.E. 1951. The marching behaviour of hoppers of the African Migratory Locust (*Locusta migratoria migratorioides*) in the laboratory. Anti-Locust Bull.7.

Gangwere, S.K. 1961. A monograph of food selection in Orthoptera. Trans. Amer. Entomol. Soc. 87:67-230.

Gangwere, S.K., F.C. Evans, and M.L. Evans. 1976. The food habits and biology of Acrididae in an old-field community in southeastern Michigan. Great Lakes Entomol. 9:83-123.

Gershenzon, J. 1984. Changes in levels of plant secondary metabolites under water and nutrient stress. Rec. Adv. Phytochem. 18:273-320.

Gillis, J.E. 1982. Substrate colour-matching cues in the cryptic grasshopper *Circotettix rabula* (Rehn & Hebard). Anim. Behav. 30:113-116.

Gillis, J.E., and K.W. Possai. 1983. Thermal niche partitioning in the grasshoppers *Arphia conspersa* and *Trimerotropis suffusa* from a montane habitat in central Colorado. Ecol. Entomol. 8:155-161.

Gould, F., and R.E. Stinner. 1984. Insects in heterogeneous habitats. p. 427-450. *In:* C.B. Huffaker and R.L. Rabb (eds.) Ecological entomology. John Wiley & Sons, New York.

Greenfield, M.D., and T.E. Shelly. 1985. Alternative mating strategies in a desert grasshopper: evidence of density-dependence. Anim. Behav. 33:1192-1210.

Hardman, J.M., and M.K. Mukerji. 1982. A model simulating the population dynamics of the grasshoppers (Acrididae) *Melanoplus sanguinipes* (Fabr.), *M. packardii* Scudder, and *Camnula pellucida* (Scudder). Res. Pop. Ecol. 24:276-301.

158

Heidorn, T.J., and A. Joern. 1984. Differential herbivory on C_3 versus C_4 grasses by the grasshopper *Ageneotettix deorum* (Orthoptera: Acrididae). Oecologia 65:19-25.

Hill, L., and G.J. Goldsworthy. 1968. Growth, feeding activity and the utilisation of reserves in larvae of *Locusta*. J. Insect Physiol. 14:1085-1098.

Hill, L., A.J. Luntz, and P.A. Steele. 1968. The relationships between somatic growth, ovarian growth, and feeding activity in the adult desert locust. J. Insect Physiol. 14:1-20.

Isley, F.B. 1937. Seasonal succession, soil relations, numbers, and regional distribution of northeastern Texas acridians. Ecol. Monogr. 7:319-344.

Isley, F.B. 1938. Survival value of acridian protective coloration. Ecology 19:370-389.

Joern, A. 1979a. Feeding patterns in grasshoppers (Orthoptera: Acrididae): factors influencing diet specialization. Oecologia 38:325-347.

Joern, A. 1979b. Resource utilization and community structure in assemblages of arid grassland grasshoppers (Orthoptera: Acrididae). Trans. Amer. Entomol. Soc. 105:253-300.

Joern, A. 1981. Importance of behavior and coloration in the control of body temperature by *Brachystola magna* Girard (Orthoptera: Acrididae). Acrida 10:117-130.

Joern, A. 1982. Vegetation structure and microhabitat selection in grasshoppers (Orthoptera: Acrididae). Southwestern Nat. 27:197-209.

Joern, A. 1983a. Host plant utilization by grasshoppers (Orthoptera: Acrididae) from a sandhills prairie. J. Range Manage. 36:793-797.

Joern, A. 1983b. Small-scale displacements of grasshoppers (Orthoptera: Acrididae) within arid grasslands. J. Kansas Entomol. Soc. 56:131-139.

Joern, A. 1986a. Experimental study of avian predation on coexisting grasshopper populations (Orthoptera: Acrididae) in a sandhills grassland. Oikos 46:243-249.

Joern, A. 1986b. Resource partitioning by grasshopper species from grassland communities, p. 75-100. *In:* D.Nickle (ed.) Proceedings 4th triennial meeting, Pan American acridological society. Pan American Acridological Society.

Joern, A. Insect herbivory in the transition to California annual grasslands: did grasshoppers deliver the coup de grass? *In:* H.A. Mooney and L. Huenneke (eds.) California annual grasslands: a model system. Dr. W. Junk, The Hague, Netherlands (in press).

Joern, A., R. Mitschler, and H. O'Leary. 1986. Activity and time budgets of three grasshopper species (Orthoptera: Acrididae) from a sandhills grassland. J. Kansas Entomol. Soc. 59:1-6.

Johnson, D.L., B.D. Hill, C.F. Hinks, and G.B. Schaalje. 1986. Aerial application of the pyrethroid deltamethrin for grasshopper (Orthoptera: Acrididae) control. J. Econ. Entomol. 79:181-188.

Kaufman, T. 1965. Biological studies of some Bavarian Acridoidea (Orthoptera), with special reference to their feeding habits. Ann. Entomol. Soc. Am. 58:791-801.

Kemp, W.P. 1986. Thermoregulation in three rangeland grasshopper species. Can. Entomol. 118:335-343.

Knutson, H. 1982. Development and survival of the monophagous grasshopper *Hypochlora alba* (Dodge) and the polyphagous *Melanoplus bivittatus* (Say) and *Melanoplus sanguinipes* (F.) on Louisiana sagewort, *Artemisia ludoviciana* Nutt. Environ. Entomol. 11:777-782.

Landa, K., and D. Rabinowitz. 1983. Relative performance of *Arphia sulphurea* (Orthoptera: Acrididae) for sparse and common prairie grasses. Ecology 64:392-395.

Lewis, A.C. 1979. Feeding preference for diseased and wilted sunflower in the grasshopper *Melanoplus differentialis*. Entomol. Exp. Appl. 26:202-207.

Lewis, A.C. 1982. Leaf wilting alters a plant species ranking by the grasshopper *Melanoplus differentialis*. Ecol. Entomol. 7:391-395.

Lewis, A.C. 1984. Plant quality and grasshopper feeding: effects of sunflower condition on preference and performance in *Melanoplus differentialis*. Ecology 65:836-843.

Lewis, A.C., and E.A. Bernays. 1985. Feeding behavior: selection of both wet and dry food for increased growth in *Schistocerca gregaria* nymphs. Entomol. Exp. Appl. 37:105-112.

Loher, W., and G. Wiedenmann. 1981. Temperature-dependent changes in circadian patterns of cricket premating behaviour. Ecol. Entomol. 6:35-43.

Mattson, W.J., Jr. 1980. Herbivory in relation to nitrogen content. Annu. Rev. Ecol. Syst. 11:119-162.

May, M.L. 1979. Insect thermoregulation. Annu. Rev. Entomol. 24:313-343.

McCaffery, A.R. 1982. A difference in the acceptability of excised and growing cassava leaves to *Zonocerus variegatus* (L.). Entomol. Exp. Appl. 32:111-115.

Mordue, A.J. 1979. The role of the maxillary and labial palps in the feeding behavior of *Schistocerca gregaria*. Entomol. Exp. Appl. 25:279-288.

Mulkern, G.B. 1967. Food selection of grasshoppers. Annu. Rev. Entomol. 12:59-79.

Mulkern, G.B. 1969. Behavioral influences on food selection in grasshoppers (Orthoptera: Acrididae). Entomol. Exp. Appl. 12:509-523.

Mulkern, G.B., K.P. Pruess, H. Knutson, A.F. Hagen, J.B. Campbell, and J.D. Lambley. 1969. Food habits and preferences of grasshoppers. North Dakota State Univ. Agr. Exp. Sta. Bull. 481.

Orshan, L., and M.P. Pener. 1979. Repeated reversal of the reproductive diapause by photoperiod and temperature in males of the grasshopper, *Oedipoda miniata*. Entomol. Exp. Appl. 25:219-226.

Otte, D. 1975. Plant preference and plant succession: a consideration of the evolution of plant preference in *Schistocerca*. Oecologia 18:129-144.

Otte, D. 1977. Species richness patterns of old world desert grasshoppers in relation to plant diversity. J. Biogeo. 3:197-209.

Otte, D. 1981. The North American grasshoppers. Vol. 1. Acrididae. Gomphocerinae and Acridinae. Harvard Univ. Press, Cambridge, MA.

Otte, D. and A. Joern. 1975. Insect territoriality and its evolution: population studies of the desert grasshoppers on creosote bushes. J. Anim. Ecol. 44:29-54.

Otte, D., and A. Joern. 1977. On feeding patterns in desert grasshoppers and the evolution of specialized diets. Proc. Acad. Natural Sciences, Phila. 128:89-126.

Parker, M.A. 1984. Local food depletion and the foraging behavior of a specialist grasshopper, *Hesperotettix viridis*. Ecology 65:824-835.

Pener, M.P., and L. Orshan. 1980. Reversible reproductive diapause and intermediate states between diapause activity in male *Oedipoda miniata* grasshoppers. Physiol. Entomol. 5:417-426.

Pfadt, R.E. 1949. Food plants as factors in the ecology of the lesser migratory grasshopper (*Melanoplus mexicanus*). Wyoming Agr. Exp. Sta. Bull. 290.

Pfadt, R.E., and R.J. Lavigne. 1982. Food habits of grasshoppers inhabiting the Pawnee site. Univ. Wyoming Agr. Exp. Sta. Sci. Mono. 42.

Porter, W.P., and D.M. Gates. 1969. Thermodynamic equilibria of animals. Ecol. Mono. 39:227-244.

Roessingh, P., E.A. Bernays, and A.C. Lewis. 1985. Physiological factors influencing preference for wet or dry food in nymphs of *Scistocerca gregaria*. Entomol. Exp. Appl. 37:89-94.

Root, R.B. 1973. Organization of plant-arthropod association in simple and diverse habitats: the fauna of collards (*Brassica oleraceae*). Ecol. Mono. 43:95-124.

Rowell, C.H.F. 1971. The variable colouration of Acridoid grasshoppers. Adv. Insect Physiol. 8:145-198.

Stadler, E. 1982. Sensory physiology of insect-plant relationships - round table discussion, p. 81-92. *In:* J.H. Visser and A.K. Minks (eds.) Proc. 5th int. symp. insect-plant relationships. Centre for Agr. Publ. Docum. Wageninen, The Netherlands.

Thorsteinson, A.J., and J.K. Nayar. 1963. Plant phospholipids as feeding stimulants for grasshoppers. Can. J. Zool. 41:931-935.

Ueckert, D.N. and R.M. Hansen. 1971. Dietary overlap of grasshoppers on sandhill rangeland in northeastern Colorado. Oecologia 8:276-295.

Uvarov, B. 1977. Grasshoppers and locusts: a handbook of acridology. Vol. 2. Centre for Overseas Pest Research, London.

Wallace, G.K. 1958. Some experiments on form perception in the nymphs of the desert locust, *Schistocerca gregaria* Forskal. J. Exp. Biol. 35:765-775.

Williams, L.H. 1954. The feeding habits and food preferences of Acrididae and the factors which determine them. Trans. Royal Entomol. Soc. Lond. 105:423-454.

11. POPULATION ECOLOGY OF RANGELAND GRASSHOPPERS

John L. Capinera
Department of Entomology
Colorado State University, Fort Collins, Colorado 80523

Grasshopper populations on western grasslands historically have exhibited dramatic shifts in numerical abundance. Increases in abundance, alone or in combination with migration, have had devastating effects on the agricultural economy of the affected areas.

Population ecology of grasshoppers on the shortgrass plains is not particularly well known. Commonly it is assumed that all grasshopper species are quite similar, or that regional differences in community composition are not important. Here I review the important elements affecting the population ecology of western rangeland grasshoppers, and emphasize the different or unique aspects associated with the shortgrass region.

GRASSHOPPER POPULATION STRUCTURE

Many studies on grasshopper population ecology are restricted to *Melanoplus* or other crop-feeding species. Thus, there is surprisingly little information available on the numerical behavior of grasshoppers on rangeland. Examination of literature on grasshoppers in the Great Plains region of North America indicates that each of the three major grasshopper subfamilies is well represented throughout the region. Species distribution among subfamilies at locations which could be considered shortgrass (Arizona, Colorado, Montana, Wyoming, Texas) do not appear to differ from mixed to tallgrass locations (Alberta, Kansas, Nebraska, Oklahoma, South Dakota) (Figure 11.1).

Close examination of grasshopper population structure at these sites, however, suggests that species dominance varies from shortgrass to mixed and tallgrass locations (Table 11.1). Most shortgrass sites seem to be dominated by gomphocerines, while melanoplines predominate elsewhere. With so much of our

162

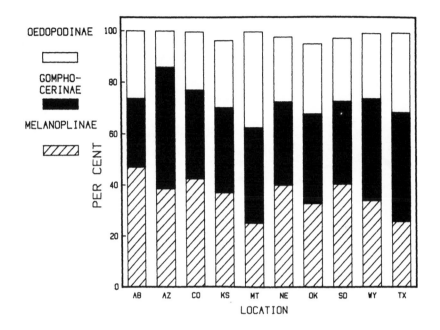

FIG. 11.1 Dominance of grasshopper subfamilies at various locations in the United States and Canada. Dominance is expressed as % of species in each of the 3 major subfamilies; sources are given in Table 11.1.

knowledge of grasshopper ecology and management based on studies of melanoplines, applicability to shortgrass rangeland management must be questioned.

Grasshopper phenology, in particular, varies with location due to differences in grasshopper population composition. Although key species in the northern Great Plains exhibit considerable synchrony in hatch and development, southern regions apparently have more late-developing species. Consequently, there is less synchrony in hatch, and management strategies that depend upon precise assessment of instar (see Chapter 13) are more difficult to institute.

There have been few studies of the temporal characteristics of grasshopper population structure. Pfadt (1977) monitored changes in species abundance and richness in Wyoming during a period of population increase following insecticide application. Capinera and Thompson (in press) conducted a similar study during a period of natural population change in Colorado. Finally, Joern

TABLE 11.1
Dominant species at various shortgrass, and mixed- or tallgrass, rangeland sites. Also designated is grasshopper subfamily: Gomphocerinae (G), Melanoplinae (M), Oedipodinae (O).

Shortgrass sites

Arizona[a]
G-*Eritettix simplex*
M-*Melanoplus gladstoni*
M-*Melanoplus desultorius*

Colorado[b]
G-*Opeia obscura*
G-*Aulocara elliotti*
G-*Ageneotettix deorum*
G-*Philibostroma quadrimaculatum*
M-*Melanoplus gladstoni*
O-*Trachyrhachys kiowa*

Montana[c]
G-*Ageneotettix deorum*
G-*Ageneotettix femoratum*
M-*Melanoplus sanguinipes*

Wyoming[d]
G-*Cordillacris occipitalis*
G-*Ageneotettix deorum*
G-*Aulocara elliotti*
G-*Philibostroma quadrimaculatum*
M-*Melanoplus occidentalis*

Texas[e]
G-*Cordillacris crenulata*
G-*Parapomala wyomingensis*
G-*Opeia obscura*
O-*Trachyrhachys kiowa*

Mixed or tallgrass sites

Alberta
G-*Chorthippus curtipennis*
M-*Melanoplus infantilis*
M-*Melanoplus dawsoni*
O-*Encoptolophus costalis*

Kansas[g]
G-*Ageneotettix deorum*
M-*Campylacantha olivacea*
M-*Melanoplus bivittatus*
M-*Melanoplus femurrubrum*

Nebraska[h]
G-*Ageneotettix deorum*
M-*Melanoplus angustipennis*
M-*Phoetaliotes nebrascensis*

Oklahoma[i]
M-*Melanoplus bivittatus*
M-*Melanoplus confusus*
M-*Melanoplus packardii*

South Dakota[j]
M-*Melanoplus confusus*
M-*Melanoplus occidentalis*
M-*Melanoplus sanguinipes*
M-*Phoetaliotes nebrascensis*

[a]Jepson (1985), [b]Capinera and Thompson (in press), [c]Anderson (1952), [d]Pfadt (1977), [e]Joern (1979), [f]Hardman and Smoliak (1982), [g]Campbell et.al. (1974), [h]Joern (1982), [i]Smith (1940), [j]Redak (unpublished).

and Pruess (1986) examined community structure in Nebraska. These latter studies included both increasing and decreasing periods of abundance.

A common pattern emerges from the three aforementioned studies. It appears that during periods of change in numerical abundance, many, but not all, species exhibit similar trends. A relatively small number of species tend to dominate the species complex at a location, and dominance tends to persist over long periods of time.

Knowledge of grasshopper population structure has rarely been utilized in past grasshopper control efforts, but this may change in the near future. With suppression programs based solely on liquid insecticide treatments, there was little concern about species identification as most species are quite susceptible. The only concerns raised were species-specific differences in egg hatch or oviposition, and occasionally the host plant preferences of target insects. However, interest in the use of the protozoan disease of grasshoppers, *Nosema locustae* Canning, raises new problems. Although not well studied, there appear to be great differences in susceptibility to infection. Henry (1971) reported that the Melanoplinae were most susceptible, with Oedipodinae intermediate and Gomphocerinae exhibiting a low susceptibility to infection. Even among three relatively susceptible *Melanoplus* spp., levels of infection differed markedly due to age at time of infection. Early instars were more susceptible. The acceptance of bran bait, which is used to apply *Nosema* and sometimes used to apply insecticide, also is species-dependent. While there is considerable variation among species within subfamilies, it appears that greatest to least acceptance of bran occurs in Gomphocerinae, Melanoplinae, and Oedopodinae, respectively (Onsager et al. 1980).

GRASSHOPPER POPULATION CYCLES

Any long-term collection of data on grasshopper occurrence demonstrates periods of great abundance (outbreak) interspersed by periods of rarity (subsidence) (Munro 1949, Mitchener 1953, Schlebecker 1953, Smith 1954). In many cases, outbreaks seem to occur at regular intervals, prompting association of cycles with regularly occurring phenomena such as weather cycles. Sunspots, in particular, have long been of interest; Archibald (1878), Swinton (1883), Criddle (1932), Dirsh (1974) and others have suggested a strong relationship between sunspot cycles and outbreaks (Figure 11.2). Sunspots do not directly affect grasshoppers, of course, but influence temperature and precipitation patterns (King 1973, 1975, White 1977). The nature of the 11- and 22-year solar influence is ameliorated by latitude and other factors which reduce

166

predictability (King 1975). Lunar cycles, which occur in 18.6 year intervals, also influence weather (Currie 1984), and may affect grasshopper populations.

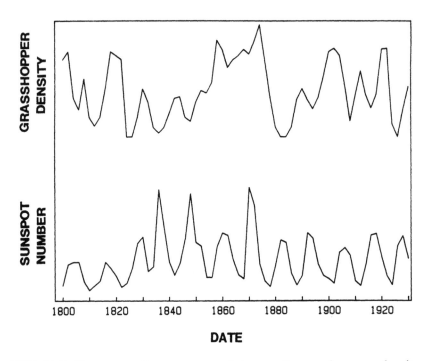

FIG. 11.2 Patterns of sunspot activity and grasshopper density (from Criddle 1932).

INFLUENCE OF WEATHER ON GRASSHOPPER POPULATIONS

Many attempts have been made to relate grasshopper abundance to weather. Grasshopper growth, survival, and reproduction generally are favored by warm, dry weather (Shotwell 1941, MacCarthy 1956, Wakeland 1958, Edwards 1960, Dempster 1963, Smith 1969, Gage and Mukerji 1977). Edwards (1960) reported that abnormally high July to September temperatures (and associated low rainfall) in Saskatchewan resulted in abnormally high grasshopper numbers. Since it takes more than a single season for grasshopper numbers to build to outbreak levels, grasshopper numbers were most highly correlated with the previous three years of weather. Gage and Mukerji (1977), also working in Saskatchewan, reported that while weather from the previous two or three years seemed to influence grasshopper abundance, the preceding year was most important. Additionally, individual species differed in their response to heat and precipitation.

The period of egg hatch for a given species varies considerably among locations, and among years. Warm temperatures are necessary for egg hatch (a minimum of 10° to 15^oC, with the optimum for nymphal growth usually in the range of 27° to 38^oC (Dempster 1963). Timing of hatch is important in that eclosion of young grasshoppers should be in synchrony with availability of food plants, and with favorable weather.

Embryonic development in most species occurs in the summer and autumn months, followed by a period of diapause. Additional development occurs in the spring following diapause. The maximum amount of development which occurs prior to diapause is species-specific. For example, in *Melanoplus sanguinipes* (F.) diapause follows attainment of 80% development, while in *Camnula pellucida* (Scudder) only 50% development occurs (Moore 1948).

When grasshopper eggs are laid late in the year, or when cold weather begins early, they are not exposed to adequate heat for the attainment of maximum prediapause development. Thus, additional development must occur during the spring months and egg hatch is delayed. Delayed hatch is detrimental to population development because maturation and reproduction are delayed, reducing potential fecundity (Shotwell 1941, Randell and Mukerji 1974, Mukerji and Gage 1978). Also, eggs laid later in the season may be less viable (Pickford 1976).

Moisture also influences embryonic development and hatching. Eggs are very resistant to drought, and can lose 1/3 to 2/3 of their moisture content without significant loss in viability (Salt 1952). However, grasshopper eggs usually absorb moisture, and development will be delayed or prevented at low soil moisture levels (Mukerji and Gage 1978). Since the critical level of soil moisture is quite low (for *Melanoplus sanguinipes* it is 13.5% or approximately the wilting point of plants) it is unlikely that moisture is limiting except during severe drought.

Grasshopper eggs may be exposed to extremely hot or cold temperatures, although soil generally acts to buffer temperature extremes considerably. The high temperatures (above 45° C) and low temperatures (less than -25° C) necessary to kill embryos apparently do not occur in undisturbed soil (Parker 1930), so the egg stage can be considered essentially immune to adverse temperatures.

Grasshopper body temperature is usually quite similar, but not identical, to the temperatures of the general environment. By basking in the sun, some grasshoppers acquire greatly elevated body temperatures (Uvarov 1966, Begon 1983, see also Chapter 10). Young nymphs seem to be particularly susceptible to inclement weather, especially cool, wet weather. Outbreak of fungal disease (*Entomophaga grylli* Fres.), stimulated by rainfall, commonly is implicated in mortality among grasshoppers (Parker 1930, Pickford and Riegert 1964) although nymphal mortality seems to be frequent even in the absence of disease.

The rate of food consumption by grasshoppers is affected by temperature. Parker (1930) found much higher rates of consumption associated with warmer temperatures. Other metabolic activities, particularly food utilization efficiencies, probably are affected in a similar manner, although this is poorly studied. Starvation may be a significant factor in the population ecology of grasshoppers when low temperatures are experienced.

Warm weather also promotes early maturation of grasshoppers, an important component of reproduction and total fecundity. There seems to be little variation in the number of eggs produced per egg pod, but considerable variation occurs in the number of pods produced per female. Grasshoppers hatching earlier, or developing more quickly, have the highest fecundity because they have a longer period of time in which to produce egg pods; they also experience more favorable temperatures during the oviposition period (Pfadt 1977, Hilbert and Logan 1981).

Weather also may affect grasshoppers indirectly by modifying the quality of their food. This concept is reviewed by White (1976), who suggested that plants stressed by excess or shortage of precipitation would become more suitable for grasshopper growth and survival due to an increase in nitrogen (protein) content. Visscher et al. (1979) also indicated that temperature might affect the quality of foliage and resultant grasshopper reproductive potential.

While warm dry conditions generally favor abundance, forage availability can be limiting in more arid regions of the world. Thus, in Africa and Arizona some grasshopper species increase in abundance following meteorological events which promote growth of forage plants early in the season (Nerney 1961, Sayer 1962); this also seems to be the case for Colorado (Capinera and Thompson in press). A key relationship in the southern shortgrass region, where the forage often is moisture-limited, may be this combination of adequate winter or spring precipitation which promotes forage growth, followed by warm, dry summer conditions which favors grasshopper survival, growth, and reproduction. Failure to observe this phenomenon in Canadian studies (e.g., Edwards 1960, Smith 1969, Gage and Mukerji 1977) probably reflects the different grasshopper complex and weather present in northern latitudes.

RESOURCE AVAILABILITY

Resources may be limiting more frequently than is commonly supposed. Grasshoppers have distinct food preferences (Mulkern et al. 1969, Capinera and Sechrist 1982a, Pfadt and Lavigne 1982), and survival rates and reproductive performance usually are highest on preferred hosts. At high densities grasshoppers commonly exhaust the supply of preferred host plants, and consume less suitable hosts (Pfadt 1949, Bailey and Mukerji 1976), which may lead to population decrease. Within-resource variation in quality also may occur, as discussed previously under "influence of weather."

Natural or artificial rangeland perturbations may affect rangeland grasshopper populations. Overgrazing (principally by livestock) stimulates change in plant community structure, and often allows invasion of weed flora which favors certain grasshoppers. Thus, high levels of livestock grazing are generally considered to be incompatible with effective grasshopper management. Recent studies (Capinera and Sechrist 1982b, O'Meilia et al. 1982), however, suggest otherwise. If forage plant communities are not excessively disrupted, grasshopper numbers are not favored by heavy livestock use.

Resource enrichment through application of fertilizer might be expected to influence grasshopper abundance. In laboratory studies, grasshoppers were favored by high levels of nitrogen (Smith and Northcott 1951) or low phosphorus (Smith 1960). However, under field conditions it has not been possible to observe such effects (Kirchner 1977, Capinera and Thompson in press). Other activities such as interseeding rangeland with legumes (Hewitt et al. 1982), rangeland renovation (Hewitt and Rees 1974), fire (Evans 1984), and sulfur dioxide emissions (McNary et al. 1981) also affect grasshopper abundance. The extent of these practices thus far is quite limited, so effects are localized, and often transient.

NATURAL ENEMIES OF GRASSHOPPERS

The natural enemies of grasshoppers, particularly predators and parasites, have long been of interest to entomologists. While fairly complete lists of these beneficial organisms exist (Greathead 1963, Lavigne and Pfadt 1966, Rees 1973), quantitative data on long term affects, or regulatory behavior, on grasshopper populations are infrequent.

Predators

Important insect predators of grasshopper eggs include meloids, bombyliids, and carabids. Principal insect predators of nymphs and adults are formicids, sphecids, and asilids; spiders and mantids sometimes also appear to be significant mortality factors. Vertebrates, particularly birds, coyotes, and foxes depend to a great extent on grasshoppers for food.

The most obvious predators in most shortgrass environments are asilids, which frequently are observed subduing grasshopper prey. The asilid *Protacanthus milbertii* Macquart was estimated to consume two (Joern and Rudd 1982) to seven (Dennis and Lavigne 1975) prey per day. Joern and Rudd (1982) estimated that these asilids could reduce grasshopper populations at a rate of up to 2% per day over a six to eight week period. Species generally were consumed in proportion to their abundance, with the possible exception of sedentary or flightless species. Not all asilids, however, are so beneficial (see parasites, below). On occasion, predation by sphecid wasps may be evident (Newton 1956).

Another fairly apparent mortality factor is avian predation. Exclusion of birds by Joern (1986), for example, indicated that grasshopper sparrows, meadowlarks, and other avifauna consumed over 27% of a Nebraska grasshopper population. Species diversity also was reduced. This is in contrast to some bioenergetics models which suggested little effect of avian predation on rangeland grasshopper populations (Wiens and Dyer 1975, Wiens 1977)(see also Chapter 22).

Parker and Wakeland (1957) examined destruction of egg pods by bombyliids, meloids, and carabids from throughout the western United States. On average, 18.8% of eggs were destroyed although average predation in some counties reached 80 - 100%. Meloid beetle larvae were the most effective predators (8.8%), followed by bombyliids (6.2%), and carabids (2.9%). In some cases, predation occurred in a density-dependent manner, suggesting regulatory activity. Predation seemed to vary with crop environment, but no consistent pattern was evident. Lavigne and Pfadt (1966) reported similar results from Wyoming.

Parasites

Parasites of rangeland grasshoppers include scelionids, nemestrinids, and tachinids, but most common are sarcophagids. Scelionids, which are egg parasites, are surprisingly uncommon in the shortgrass region (Pickford 1964, Mukerji 1987). The other parasites attack nymphal and adult grasshoppers. Nemestrinids generally are rare (Lavigne and Pfadt 1966), but occasionally they may have very significant effects on grasshopper abundance

(Prescott 1960). Mites also are associated with grasshoppers, both as egg predators and as nymphal and adult parasites.

The only serious attempt to assess the effects of grasshopper parasites was conducted by Smith (1958, 1964) in Canada. Over a 10-year period, Smith monitored parasitism of *Melanoplus sanguinipes* by several parasitoid species. Rates of parasitism were not high, never exceeding 17%. Secondary parasites, especially *Perilampus,* apparently affected parasitism. Nevertheless, there was a positive relationship between grasshopper density and numbers of grasshoppers parasitized. No evidence was found to support the idea that parasites affected trends in grasshopper abundance.

In a recent examination of the affects of parasitism on fecundity, Rees (1986) reported a reduction in the proportion of female *M. sanguinipes* which deposited egg pods, although egg number per pod and viability were not affected. Prescott (1960) also showed a reduction in fecundity in parasitized grasshoppers, due principally to reduced longevity.

Parasitism and predation are not always complementary processes. Asilid predators which feed on grasshoppers also may attack parasitic flies. When asilids were excluded from field cages, fly parasites lived longer, and caused higher levels of parasitism in grasshoppers (Rees and Onsager 1985).

Perhaps the most unusual parasite is the grasshopper mite, *Eutrombidium locustarum* (Walsh), which feeds on grasshopper eggs as a mature nymph and adult, but attacks grasshopper nymphs and adults as a young nymph. The young nymphs feed externally and seem to have little effect on the vigor of their host, while mature nymphs and adults each destroy several eggs (Severin 1944, Lavigne and Pfadt 1966).

Diseases

Important disease agents of rangeland grasshoppers include nematodes, fungi, and protozoa. Bacterial and viral diseases also occur, but are less well-known. Additional information on potential of microbial agents is given by Dempster (1963) and Streett (Chapter 14).

Grasshoppers are attacked by at least two species of nematodes, but only *Mermis nigrescens* Dujardin is found in western North America. While it may infect high proportions of grasshoppers under favorable conditions (i.e., high humidity) (Mongkolkiti and Hosford 1971, Bland 1976), in arid rangeland it does not seem to be an important mortality factor (Capinera 1987).

The fungus *Entomophaga grylli* Fres. is one of the most widespread grasshopper pathogens (MacLeod et al. 1980). Epizootics are dramatic, but usually are localized and sporadic

(Nelson et al. 1982). Occasionally large areas are affected (Pickford and Riegert 1964), especially in the northern Great Plains. Differential susceptibility among grasshopper species or infectivity among *E. grylli* strains, as well as environmental constraints (i.e., high moisture requirements for germination of spores and conidia), limit the effect of this disease in much of the shortgrass region.

The protozoan *Nosema locustae* Canning occurs in a wide variety of grasshoppers. While natural infections usually occur at low levels, sometimes high levels of infection are observed, particularly late in the season or in late-season species. The disease is transmitted through eggs, feces, and by cannibalism. The incidence of infection seems to vary among species, and possibly among habitats (Henry 1971, 1972, 1981). A delayed density-dependent infection response typical of regulatory agents has been observed (Henry and Oma 1981). Reductions in survival, oviposition, and consumption by populations inoculated with *N. locustae* have been reported (Ewen and Mukerji 1980, Henry and Onsager 1982, Abusief 1983, Johnson and Pavlikova 1986), and presumably this occurs in natural epizootics as well. Other, less-well-known microsporida, including other *Nosema* spp., also offer potential for grasshopper biocontrol (Streett and Henry 1984, Erlandson et al. 1985, 1986).

Influence of Natural Enemies on Grasshopper Population Dynamics

In an excellent review of population dynamics of grasshoppers and locusts, Dempster (1963) concluded that while numerical data on the effects of natural enemies were almost completely lacking, it seemed unlikely that natural enemies could regulate acridid abundance. At most, he suggested that "natural enemies do little more than damp the peaks in the population fluctuations."

Given the wealth of data supporting the importance of meteorological/abiotic factors in governing grasshopper abundance, and the lack of data on natural enemies/biotic factors, it is easy to underestimate the importance of the latter. However, there are numerous examples of individual natural enemies causing high levels of mortality (e.g., Parker and Wakeland 1957, Prescott 1960, Bland 1976, Joern 1986). The collective effects of mortality factors, which individually contribute small increments of additional mortality or reduced fecundity, may total to significant effects. Models of grasshopper population dynamics clearly illustrate how small changes in daily survival may result in large changes in generation survival (Onsager 1983, Mann et al. 1986). Indeed, models which do not include biotic mortality factors

(Hardman and Mukerji 1982) greatly overestimate actual abundance of grasshoppers observed in the field.

POPULATION QUALITY AND MIGRATION

Genetic and non-genetic variation among individuals contributes to the overall quality of populations (Capinera 1979). Population quality is reflected in different rates of such life history traits as survival, reproduction, and dispersal. Differential expression of life history traits results, in the extreme case, in development of "phases" in such insects as African locusts. Elements of phase polymorphism also occur in North American grasshopper populations, and affect population ecology principally through migration.

Shifts between light and dark forms of grasshoppers are well known, and this usually is related to temperature or crowding (Parker 1930, Shotwell 1941, Brett 1947, Kelly and Middlekauff 1961, Dingle and Haskell 1967). However, there is little to suggest that very marked changes in phase occur regularly. *Melanoplus sanguinipes* has undergone marked phase shift, and was known as Rocky Mountain Locust during its gregarious, migratory form during the 1800's (Faure 1933). Although Gurney and Brooks (1959) maintained that *M. sanguinipes* and Rocky Mountain Locust were different species, some of their species designations have been synonymized, and their arguments are unconvincing. Mormon cricket, *Anabrus simplex* Haldeman, similarly exhibits major elements of phase transformation, although this is poorly documented (see Chapter 6).

Even in the absence of phase transformation, migration can be a significant factor in the population ecology of some grasshoppers. Species of *Melanoplus*, *Camnula*, and other grasshoppers sometimes move long distances (Shotwell 1941, Parker et al. 1955, Wakeland 1958). Except for occasional situations such as the *Dissosteira longipennis* (Thomas) outbreak (Wakeland 1958), however, migration seems to be a greater problem in the northern Great Plains.

Small-scale displacement (dispersal) of grasshoppers is a widespread phenomenon, either within rangeland (Joern 1983), or between rangeland and cropland. However, gomphocerines are less likely to attack crops than are melanoplines, so movement of grasshoppers from rangeland to cropland is less of a problem in shortgrass regions.

CONCEPTUAL MODEL OF GRASSHOPPER POPULATION ECOLOGY

The factors governing grasshopper population ecology differ spatially (habitat) and temporally (phase of outbreak cycle). Reversal of the effects responsible for stimulating the outbreak phase do not necessarily lead to subsidence, or vice versa. Thus, any model which purports to treat population ecology comprehensively must consider both the spatial and temporal components.

Weather, undoubtedly, is the key factor in stimulating the increase phase of grasshopper outbreaks. Nymphal and adult survival rates usually are positively correlated with temperature, and negatively correlated with precipitation. In the southern, more arid regions of the Great Plains, however, grasshoppers appear to be food-limited, so excess precipitation prior to egg hatch may be a prerequisite to outbreak. In the more northern areas, moisture is less limiting but heat (day-degrees) frequently may be inadequate. Grasshoppers in southern areas are favored by warm weather after hatching, but high temperatures alone will not stimulate outbreak.

Phase polymorphism does not seem to be a significant factor in most North American grasshopper populations. Certain grasshopper assemblages, however, contain species which are more likely to exhibit migratory tendencies leading to local or regional redistribution.

Quantitative data on the effects of beneficial organisms and resource competition generally are lacking. From the inability of abiotic factors to explain fully population change, and from the limited quantitative data showing regulatory potential of biotic agents, I infer that predators, parasites, pathogens, and competition do contribute to population collapse. In more northern (and moist) locations fungal epizootics are especially important. At some locations, beneficial organisms and competition cause collapse of grasshopper populations even while environmental conditions favor increase in grasshopper abundance; this results in the mosaic of increasing and decreasing population densities commonly observable within a general regional outbreak. In other instances, beneficial organisms give the appearance of causing population collapse when adverse environmental conditions are largely responsible. A major challenge for grasshopper management is to reconcile the relative importance of biotic and abiotic factors, and to predict the direction of population trend in an accurate and economic manner.

ACKNOWLEDGMENT

Preparation of this manuscript was supported by Colorado Agricultural Experiment Station.

REFERENCES

Abuseif, S.M. 1983. Ecology, life history, and biological control of *Aulocara elliotti* (Thomas) and *Melanoplus sanguinipes* (F). Unpublished Ph.D Dissertation, Univ. Wyoming.

Anderson, N.L. 1952. Grasshopper investigations on Montana rangelands. Montana Agr. Exp. Sta. Bull. 486.

Archibald, E.D. 1878. Locusts and sun-spots. Nature 19:145-146.

Bailey, C.G., and M.K. Mukerji. 1976. Feeding habits and food preferences of *Melanoplus bivittatus* and *M. femurrubrum* (Orthoptera: Acrididae). Can. Entomol. 108:1207-1212.

Begon, M. 1983. Grasshopper populations and weather: the effects of insolation on *Chorthippus brunneus*. Ecol. Entomol. 8:361-370.

Bland, R.G. 1976. Effects of parasites and food plants on the stability of *Melanoplus femurrubrum* Environ. Entomol. 5:724-728.

Brett, C.H. 1947. Interrelated effects of food, temperature, and humidity on the development of the lesser migratory grasshopper, *Melanoplus mexicanus mexicanus* (Saussure) (Orthoptera). Oklahoma Agr. Exp. Sta. Tech. Bull. T-26.

Campbell, J.B., W.H. Arnett, J.D. Lambley, O.K. Jantz, and H. Knutson. 1974. Grasshoppers (Acrididae) of the Flint Hills native tallgrass prairie in Kansas. Kansas Agr. Exp. Sta. Res. Pap. 19.

Capinera, J.L. 1979. Qualitative variation in plants and insects: effect of propagule size on ecological plasticity. Am. Nat. 114:350-361.

Capinera, J.L. 1987. Observations on natural and experimental parasitism of insects by *Mermis nigrescens* Dujardin (Nematoda: Mermithidae). J. Kansas Entomol. Soc. 60:159-162.

Capinera, J.L., and T.S. Sechrist. 1982a. Grasshoppers (Acrididae) of Colorado: identification, biology and management. Colorado State Univ. Agr. Exp. Sta. Bull. 584S.

Capinera, J.L., and T.S. Sechrist. 1982b. Grasshopper (Acrididae)-host plant associations: response of grasshopper populations to cattle grazing intensity. Can. Entomol. 114:1055-1062.

Capinera, J.L., and D.C. Thompson. Dynamics and structure of grasshopper assemblages on shortgrass prairie. Can. Entomol. (in press).

176

Criddle, N. 1932. The correlation of sunspot periodicity with grasshopper fluctuation in Manitoba. Can. Field Nat. 46:195-199.

Currie, R.G. 1984. Evidence for a 18.6-year lunar nodal drought in western North America during the past millennium. J. Geophys. Res. 89:1295-1308.

Dempster, J.P. 1963. The population dynamics of grasshoppers and locusts. Biol. Rev. 38:490-529.

Dennis, D.W., and R.J. Lavigne. 1975. Comparative behavior of Wyoming robber flies (Diptera: Asilidae). II. Univ. Wyoming Agr. Exp. Sta. Sci. Mono. 30.

Dingle, H., and J.B. Haskell. 1967. Phase polymorphism in the grasshopper *Melanoplus differentialis*. Science 155:590-592.

Dirsh, V.M. 1974. Genus Schistocerca (Acridomorpha, Insecta). Dr. W. Junk B.V., The Hague.

Edwards, R.L. 1960. Relationship between grasshopper abundance and weather conditions in Saskatchewan, 1930-1958. Can. Entomol. 92:619-624

Erlandson, M.A., M.K. Mukerji, A.B. Ewen, and C. Gillott. 1985. Comparative pathogenicity of *Nosema acridophagus* Henry and *Nosema cuneatum* Henry Microsporida: Nosematidae) for *Melanoplus sanguinipes* (Fab.) (Orthoptera: Acrididae). Can. Entomol. 117:1167-1175.

Erlandson, M.A., A.B. Ewen, M.K. Mukerji, and C. Gillott. 1986. Susceptibility of immature stages of *Melanoplus sanguinipes* (Fab.) (Orthoptera: Acrididae) to *Nosema cuneatum* Henry (Microsporida: Nosematidae) and its effect on host fecundity. Can. Entomol. 118:29-35.

Evans, E.W. 1984. Fire as a natural disturbance to grasshopper assemblages of tallgrass prairie. Oikos 43:9-16.

Ewen, A.B., and M.K. Mukerji. 1980. Evaluation of *Nosema locustae* (Microsporida) as a control agent of grasshopper populations in Saskatchewan. J. Invertebr. Pathol. 35:295-303.

Faure, J.C. 1933. The phases of the Rocky Mountain locust *Melanoplus mexicanus* (Saussure). J. Econ. Entomol. 26:706-718.

Gage, S.H., and M.K. Mukerji. 1977. A perspective of grasshopper population distribution in Saskatchewan and interrelationships with weather. Environ. Entomol. 6:469-479.

Greathead, D.G. 1963. A review of the insect enemies of Acrididae (Orthoptera). Trans. Roy. Entomol. Soc., Lond. 114:437-517.

Gurney, A.B., and A.R. Brooks. 1959. Grasshoppers of the *Mexicanus* group, genus *Melanoplus* (Orthoptera: Acrididae). Proc. U.S. Natl. Mus. 110:1-93.

Hardman, J.M., and M.K. Mukerji. 1982. A model simulating the population dynamics of the grasshopper (Acrididae) *Melanoplus sanguinipes* (Fabr.)., *M. packardii* Scudder, and *Camnula pellucida* (Scudder). Res. Pop. Ecol. 24:276-301.

Hardman, J.M., and S. Smoliak. 1982. The relative impact of various grasshopper species on *Stipa-Agropyron* mixed prairie and fescue prairie in southern Alberta. J. Range Manage. 35:171-176.

Henry, J.E. 1971. Experimental application of *Nosema locustae* for control of grasshoppers. J. Invertebr. Pathol. 18:389-394.

Henry, J.E. 1972. Epizootiology of infections by *Nosema locustae* Canning (Microsporida: Nosematidae) in grasshoppers. Acrida 1:111-120.

Henry, J.E. 1981. Natural and applied control of insects by protozoa. Annu. Rev. Entomol. 26:49-73.

Henry, J.E., and E.A. Oma. 1981. Pest control by *Nosema locustae*, a pathogen of grasshoppers and crickets, p. 573-586. *In:* H.D. Burges (ed.) Microbial control of pests and plant diseases 1970-1980. Academic Press, London.

Henry, J.E., and J.A. Onsager. 1982. Large-scale test of control of grasshoppers on rangeland with *Nosema locustae*. J. Econ. Entomol. 75:31-35.

Hewitt, G.B., and N.E. Rees. 1974. Abundance of grasshoppers in relation to rangeland renovation practices. J. Range Manage. 27:156-160.

Hewitt, G.B., A.C. Wilton, and R.J. Lorenz. 1982. The suitability of legumes for rangeland interseeding and as grasshopper food plants. J. Range Manage. 35:653-656.

Hilbert, D.W., and J.A. Logan. 1981. A review of the population biology of the migratory grasshopper, *Melanoplus sanguinipes*. Colorado Agr. Exp. Sta. Bull. 577S.

Jepson, K.A. 1985. Response of grasshoppers to changes in vegetation structure and availability: a comparison of grazed and ungrazed sites in southeastern Arizona. Unpublished M.S. Thesis, Univ. Colorado.

Joern, A. 1979. Resource utilization and community structure in assemblages of arid grassland grasshoppers (Orthoptera: Acrididae). Trans. Amer. Entomol. Soc. 105:253-300.

Joern, A. 1982. Distributions, densities and relative abundances of grasshoppers (Orthoptera: Acrididae) in a Nebraska sandhills prairie. Prairie Nat. 14:37-45.

Joern, A. 1983. Small-scale displacements of grasshoppers (Orthoptera: Acrididae) within arid grasslands. J. Kansas Entomol. Soc. 56:131-139.

Joern, A. 1986. Experimental study of avian predation on coexisting grasshopper populations (Orthoptera: Acrididae) in a sandhills grassland. Oikos 46:243-249.

Joern, A., and K.P. Pruess. 1986. Temporal constancy in grasshopper assemblies (Orthoptera: Acrididae). Ecol. Entomol. 11:379-385.

Joern, A., and N.T. Rudd. 1982. Impact of predation by the robber fly *Proctacanthus milbertii* (Diptera: Asilidae) on grasshopper (Orthoptera: Acrididae) populations. Oecologia 55:42-46.

Johnson, D.L., and E. Pavlikova. 1986. Reduction of consumption by grasshoppers (Orthoptera: Acrididae) infected with *Nosema locustae* Canning (Microsporida: Nosematidae). J. Invertebr. Pathol. 48:232-238.

Kelley, G.D., and W.W. Middlekauff. 1961. Biological studies of *Dissosteira spurcata* Saussure with distributional notes on related California species (Orthoptera - Acrididae). Hilgardia 30:395-424.

King, J.W. 1973. Solar radiation changes and the weather. Nature 245:443-448.

King, J.W. 1975. Sun-weather relationships. Astron. Aeronautics 13:10-19.

Kirchner, T.B. 1977. The effects of resource enrichment on the diversity of plants and arthropods in a shortgrass prairie. Ecology 58:1334-1344.

Lavigne, R.J., and R.E. Pfadt. 1966. Parasites and predators of Wyoming rangeland grasshoppers. Univ. Wyoming Agr. Exp. Sta. Sci. Mono. 3.

MacCarthy, H.R. 1956. A ten-year study of the climatology of *Melanoplus mexicanus* (Sauss.) (Orthoptera: Acrididae) in Saskatchewan. Can. J. Agr. Sci. 36:445-462.

MacLeod, D.M., D. Tyrrell, and M.A. Welton. 1980. Isolation and growth of the grasshopper pathogen, *Entomophthora grylli*. J. Invertebr. Pathol. 36:85-89.

Mann, R., R.E. Pfadt, and J.J. Jacobs. 1986. A simulation model of grasshopper population dynamics and results for some alternative control strategies. Wyoming Agric. Exp. Sta. Sci. Mono. 51.

McNary, T.J., D.G. Milchunas, J.W. Leetham, W.K. Lauenroth, and J.L. Dodd. 1981. Effect of controlled low levels of SO_2 on grasshopper densities on a northern mixed-grass prairie. J. Econ. Entomol. 74:91-93.

Mitchener, A.V. 1953. A history of grasshopper outbreaks and their control in Manitoba, 1799-1953. Proc. Entomol. Soc. Ontario 84:27-35.

Mongkolkiti, S. and R.M. Hosford, Jr. 1971. Biological control of the grasshopper *Hesperotettix viridis pratensis* by the nematode *Mermis nigrescens*. J. Nematol. 3:356-363.

Moore, H.W. 1948. Variations in fall embryological development in three grasshopper species. Can. Entomol. 80-83-88.

Mukerji, M.K. 1987. Parasitism by *Scelio calopteni* Riley (Hymenoptera: Scelionidae) in eggs of the two dominant melanopline species (Orthoptera: Acrididae) in Saskatchewan. Can. Entomol. 119:147-151.

Mukerji, M.K., and S.H. Gage. 1978. A model for estimating hatch and mortality of grasshopper egg populations based on soil moisture and heat. Ann. Entomol. Soc. Am. 71:183-190.

Mulkern, G.B., K.P. Pruess, H. Knutson, A.F. Hagen, J.B. Campbell, and J.C. Lambley. 1969. Food habits and preferences of grassland grasshoppers of the north central Great Plains. North Dakota State Univ. Agr. Exp. Sta. Bull. 481.

Munro, J.A. 1949. Grasshopper outbreaks in North Dakota 1808-1948. North Dakota Hist. 16:143-164.

Nelson, D.R., W.D. Valovage, and R.D. Frye. 1982. Infection of grasshoppers with *Entomophaga* (=*Entomophthora*) *grylli* by infection of germinating resting spores. J. Invertebr. Pathol. 39:416-418.

Nerney, N.J. 1961. Effects of seasonal rainfall on range condition and grasshopper population, San Carlos Apache Indian reservation, Arizona. J. Econ. Entomol. 54:382-385.

Newton, R.C. 1956. Digger wasps, *Tachysphex* spp., as predators of a range grasshopper in Idaho. J. Econ. Entomol. 49:615-619.

O'Meilia, M.E., F.L. Knopf, and J.C. Lewis. 1982. Some consequences of competition between prairie dogs and beef cattle. J. Range Manage. 35:580-585.

Onsager, J.A. 1983. Relationships between survival rate, density, population trends, and forage destruction by instars on grasshoppers (Orthoptera: Acrididae). Environ. Entomol. 12:1099-1102.

Onsager, J.A., J.E. Henry, R.N. Foster, and R.T. Staten. 1980. Acceptance of wheat bran bait by species of rangeland grasshoppers. J. Econ. Entomol. 73:548-551.

Parker, J.R. 1930. Some effects of temperature and moisture upon *Melanoplus mexicanus mexicanus* Saussure and *Camnula pellucida* Scudder (Orthoptera). Univ. Montana Agr. Exp. Sta. Bull. 223.

Parker, J.R., and C. Wakeland. 1957. Grasshopper egg pods destroyed by larvae of bee flies, blister beetles, and ground beetles. USDA Tech. Bull. 1165.

Parker, J.R., R.C. Newton, and R.L. Shotwell. 1955. Observations on mass flights and other activities of the migratory grasshopper. USDA Tech. Bull. 1109.

Pfadt, R.E. 1949. Food plants as factors in the ecology of the lesser migratory grasshopper *Melanoplus mexicanus* (Sauss.). Univ. Wyoming Agr. Exp. Sta. Bull. 290.

Pfadt, R.E. 1977. Some aspects of the ecology of grasshopper populations inhabiting the shortgrass plains, p. 73-79. *In:* H.M. Kulman and H.C. Chiang (eds.) Insect ecology -papers presented in the A.C. Hodson lectures. Minnesota Agr. Exp. Sta. Bull. 310.

Pfadt, R.E., and R.J. Lavigne. 1982. Food habits of grasshoppers inhabiting the Pawnee Site. Univ. Wyoming Agr. Exp. Sci. Mono. SM-42.

Pickford, R. 1964. Life history and behavior of *Scelio calopteni* Riley (Hymenoptera: Scelionidae), a parasite of grasshopper eggs. Can. Entomol. 96:1167-1172.

Pickford, R. 1976. Embryonic growth and hatchability of eggs of the two-striped grasshopper, *Melanoplus bivittatus* (Orthoptera: Acrididae), in relation to date of oviposition and weather. Can. Entomol. 108:621-626.

Pickford, R., and P.W. Riegert. 1964. The fungous disease caused by *Entomophthora grylli* Fres., and its effects on grasshopper populations in Saskatchewan in 1963. Can. Entomol. 96:1158-1166.

Prescott, H.W. 1960. Suppression of grasshoppers by nemestrinid parasites. Ann. Entomol. Soc. Am. 53:513-521.

Randell, R.L., and M.K. Mukerji. 1974. A technique for estimating hatching of natural egg populations of *Melanoplus sanguinipes* (Orthoptera: Acrididae). Can. Entomol. 106:801-812.

Rees, N.E. 1973. Arthropod and nematode parasites, parasitoids, and predators of Acrididae in America North of Mexico. USDA, ARS Tech. Bull. 1460.

Rees, N.E. 1986. Effects of dipterous parasites on production and viability of *Melanoplus sanguinipes* eggs. (Orthoptera: Acrididae). Environ. Entomol. 15:205-206.

Rees, N.E., and J.A. Onsager. 1985. Parasitism and survival among rangeland grasshoppers in response to suppression of robber fly (Diptera: Asilidae) predators. Environ. Entomol. 14:20-23.

Salt, R.W. 1952. Some aspects of moisture absorbtion and loss on eggs of *Melanoplus bivittatus* (Say). Can. J. Zool. 30:55-82.

Sayer, H.J. 1962. The desert locust and tropical convergence. Nature 194:330-336.

Schlebecker, J.T. 1953. Grasshoppers in American agricultural history. Agric. Hist. 27:85-93.

Severin, H.C. 1944. The grasshopper mite *Eutrombidium trigonum* (Hermann) an important enemy of grasshoppers. South Dakota Agr. Exp. Sta. Tech. Bull. 3.

Shotwell, R.L. 1941. Life histories and habits of some grasshoppers of economic importance on the Great Plains. USDA Tech. Bull. 774.

Smith, C.C. 1940. The effect of overgrazing and erosion upon the biota of the mixed-grass prairie of Oklahoma. Ecology 21:381-397.

Smith, D.S., and F.E. Northcott. 1951. The effects on the grasshopper, *Melanoplus mexicanus* (Sauss.) (Orthoptera: Acrididae) of varying the nitrogen content in its food plant. Can. J. Zool. 29:297-304.

Smith, D.S. 1960. Effects of changing the phosphorus content of the food plant on the migratory grasshopper, *Melanoplus biluratus* (Walker) (Orthoptera: Acrididae). Can. Entomol. 92:103-107.

Smith, L.B. 1969. Possible effects of changes in the environment on grasshopper populations. Manitoba Entomol. 3:51-55.

Smith, R.C. 1954. An analysis of 100 years of grasshopper populations in Kansas (1854 to 1954). Trans. Kansas Acad. Sci. 57:397-433.

Smith, R.W. 1958. Parasites of nymphal and adult grasshoppers (Orthoptera: Acrididae) in western Canada. Can. J. Zool. 36:217-262.

Smith, R.W. 1964. A field population of *Melanoplus sanguinipes* (Fab.) (Orthoptera: Acrididae) and its parasites. Can. J. Zool. 43:179-201.

Streett, D.A., and J.E. Henry. 1984. Epizootiology of a microsporidium in field populations of *Aulocara elliotti* and *Psoloessa delicatula* (Insecta: Orthoptera). Can. Entomol. 116: 1439-1440.

Swinton, A.H. 1883. Data obtained from solar physics and earthquake commotions applied to elucidate locust multiplication and migration. USDA, U.S. Entomol. Commission Rpt. 3:65-85.

Uvarov, B. 1966. Grasshoppers and locusts: a handbook of general acridology. Vol. 1. Cambridge Univ. Press.

Visscher, S.N., R. Lund, and W. Whitmore. 1979. Host plant growth temperatures and insect rearing temperatures influence reproduction and longevity in the grasshopper, *Aulocara elliotti* (Orthoptera: Acrididae). Environ. Entomol. 8:253-258.

Wakeland, C. 1958. The high plains grasshopper: a compilation of facts about its occurrence and control. USDA Tech. Bull. 1167.

White, D.R. 1977. The solar output and its variation. Colo. Assoc. Univ. Press, Boulder, CO.

White, T.C.R. 1976. Weather, food, and plagues of locusts. Oecologia 22:119-134.

Wiens, J. 1977. Model estimates of energy flow in North American grassland bird communities. Oecologia 31:135-151.

Wiens, J., and M.I. Dyer. 1975. Rangeland avifaunas: their composition, energetics, and role in the ecosystem, p. 146-182. *In:* D.R. Smith (ed.) Management of forest and range habitats for non-game birds. USDA, Forest Service Gen. Tech. Rpt. WO-1.

12.   A HISTORICAL LOOK AT RANGELAND GRASSHOPPERS AND THE VALUE OF GRASSHOPPER CONTROL PROGRAMS

Robert E. Pfadt
Department of Plant, Soil, and Insect Sciences
University of Wyoming, Laramie, Wyoming 82071

Don M. Hardy
U.S. Department of Agriculture
Cheyenne, Wyoming 82003

In settling the western prairies, pioneers faced many adverse conditions: disease, hostile Indians, isolation, droughts, floods, prairie fires, and grasshoppers.   Historian Harold E. Briggs (1934) concluded that periodic attacks of grasshoppers were probably the most disastrous of the obstacles that beset the pioneer farmers of Dakota Territory.   During their first years, settlers broke only a few acres with oxen and planted these to vegetables, corn, and wheat.   The crops were raised mainly for home consumption, as transportation and local markets were lacking.   Destruction of crops and forage by grasshopper swarms left families without food for themselves and feed for their livestock.   General Alfred Sully, encamped between the Missouri and Yellowstone rivers during the summer of 1864 reported, "The only thing spoken about here is the grasshopper.   They are awful.   They actually have eaten holes in my wagon covers and in the tarpaulins that cover my stores."

The source of invading swarms of grasshoppers was little known until government entomologists, especially members of the U.S. Entomological Commission, began to study the problem in 1869.   The investigators found that great numbers of *Melanoplus spretus* (Walsh) developed in a "permanent region" located chiefly east of the Rocky Mountains.   After fledging, the insects migrated primarily southeastward, and wherever the swarms descended to rest they destroyed crops, forage, and other vegetation.

Although reviews of early grasshopper outbreaks have emphasized crop destruction (Munro 1949, Mitchener 1953, Smith 1954, Riegert 1968, 1980), it is evident that damage of prairie grasses likewise occurred.   Records indicate that the migrating hordes of *M. spretus* fed on prairie grasses as well as garden vegetables and field crops. Explorer and naturalist Henry Youle Hind (1859) described a swarm that alighted west of the present town of Souris, Manitoba in 1859. He wrote, "Those portions of the prairie which had been visited by the grasshoppers wore a curious appearance; the grass was cut uniformly to one inch from the ground, and the whole surface was covered with the small, round, exuviae (feces) of these destructive invaders."

Judging from the grasshoppers collected and named by early entomologists such as Thomas Say, Cyrus Thomas, and Samuel Scudder, and from the composition of grasshopper assemblages in mixedgrass prairie in more recent times, we may conjecture that other species also inhabited the prairie in destructive numbers in the 1800s. Maximilian (1843) noted that in the area near Bismarck, North Dakota in 1833, "Numerous insects of various species abound here, to the great annoyance of the inhabitants, such as mosquitoes, and innumerable grasshoppers, which quickly devour the plants in the prairie." In 1853 near Fort Totten, Isaac Stevens, while on an exploration trip for the Pacific Railroad reported, "The grass at the best is very poor and the great abundance of grasshoppers had made sad havoc with what had grown here." Destruction of grass in western Dakota Territory was noted also by General Alfred Sully in August 1864 (Lounsberry 1919). He wrote, "The country at the Little Missouri was covered with myriads of grasshoppers which had entirely destroyed the grass." Accompanying General Custer on the ill-fated Little Big Horn Expedition, Mark Kellogg, correspondent for the Bismarck Tribune, observed large numbers of grasshoppers infesting the prairie of western Dakota. His entry for May 25, 1876 read, "Past three days discovered the grass contained millions of infinitesimal-sized locust; too small to hurt grass now."

The species of grasshoppers composing outbreak populations in the northern mixedgrass prairie in the 1800s probably varied with time and locality; only the migrating swarms were of one species, the Rocky Mountain grasshopper, *M. spretus*. Since the last specimens of *M. spretus* were collected in 1902 and the species is now considered to be very scarce or perhaps extinct (Brooks 1958, Gurney and Brooks 1959), other species have been responsible for the damage of range grasses during later outbreaks. An early report incriminating another species of grasshopper was that of Bruner (1891), who found that *Dissosteira longipennis* (Thomas) infested 400 mi^2 of range in eastern Colorado and had caused severe damage to native grasses.

Making observations of a general grasshopper outbreak in Montana in the drought years of 1901-1903, Cooley (1904) found that in the eastern part of the state assemblages of species, two or three being dominant, were responsible for damage to range grasses. He noted the common grasshoppers were *Aulocara elliotti* (Thomas), *Cordillacris occipitalis* (Thomas), and *Melanoplus sanguinipes* (F.). Grasshoppers infested an extensive territory and in some localities they were so abundant that only bare ground was left. From this extreme there was every gradation down to no injury.

Bruner (1902) reported on this same outbreak which extended as far south as eastern Colorado and western Kansas. In 1901 he observed grasshopper damage to range grasses and a scarcity of forage, notwithstanding the fact that rainfall had been good. In the mixedgrass prairie around Guernsey, Wyoming he recorded the prevailing species as *A. elliotti*, *A. femoratum* (Scudder), *M. sanguinipes*, *M. occidentalis* (Thomas), and *M. packardii* Scudder. In the vicinity of Douglas and Casper, Wyoming he found the most abundant species to be *A. elliotti*, *A. femoratum*, *Metator pardalinus* (Saussure), *M. occidentalis*, *M. infantilis* Scudder, and *M. packardii*.

Gillette (1904), who actively collected grasshoppers and other insects in Colorado from 1891 to 1903, noted that on the eastern plains three species were "very common", viz, *Amphitornus coloradus* (Thomas), *Phlibostroma quadrimaculatum* (Thomas), and *Trachyrhachys kiowa* (Thomas) and two species "very abundant", viz, *A. elliotti* and *A. femoratum*. He also noted that *M. sanguinipes* was one of the most destructive grasshoppers to native range pastures. Except for the fact that *Ageneotettix deorum* (Scudder) would be currently ranked as very abundant, dominance among rangeland grasshoppers in recent times is no different from that described by Cooley, Bruner, and Gillette (Pfadt 1977, 1984).

DAMAGE BY RANGELAND GRASSHOPPERS

Historical accounts and current reports attest to the fact that grasshoppers infesting rangeland in the western states are a serious hazard facing ranchers whose livestock, principally cattle and sheep, require ample forage for efficient growth and successful reproduction. At densities of 25 to $50/yd^2$ these insects consume nearly all of the grass in a mixedgrass prairie by the middle of August (Figure 12.1). Observations show that between 95 and 100% of all green leaves are eaten. Only dead leaves and crowns remain and even these are partly consumed.

In Wyoming, the mixedgrass prairie which lies in the eastern third of the state and comprises 13,967,000 acres is the grazing land most susceptible to damage of rangeland grasshoppers. During outbreaks this region makes up from 82 to 98% of the grasshopper-infested land, while smaller areas of infestation occur in the bunchgrass steppe of central and western Wyoming.

FIG. 12.1 Mixedgrass prairie at Guernsey, Wyoming. Top, no economic damage to vegetation by grasshoppers, density $1/yd^2$, August 1983; bottom, heavy damage to vegetation by grasshoppers, density $35/yd^2$, August 1975.

Attempts to quantify the damage by rangeland grasshoppers have been partially successful (Morton 1939, Pfadt 1949, Anderson 1961, and others). Estimates indicate that grasshoppers annually destroy 21 to 23% of available range vegetation, which represents a loss of about $393 million/year (Hewitt and Onsager 1983).

Calculating the total damage of rangeland grasshoppers for the whole region affected and the average loss per acre provides information on the seriousness of the entire problem. These figures, however, do not reveal the catastrophe for individual ranchers whose properties may be engulfed in a seriously infested area. Even during the worst outbreaks, extensive areas are infested while others are free of injurious densities of grasshoppers. Hence, average figures of damage do not reflect the dire straits in which particular ranchers may find themselves.

Damage to range forage plants by grasshoppers is readily observable, but obtaining reliable quantitative data on the extent of damage is difficult. Reasons for the difficulty are several; the principal causes being the mobility of grasshoppers, variability of rangeland, deviations of weather, and the effects of caging upon both range plants and grasshoppers. Nevertheless, the importance of quantifying the damage of grasshoppers has prompted many field and laboratory studies which have been reviewed thoroughly by Bullen (1966) and Hewitt (1977). See also Chapter 6.

CONTROL OF RANGELAND GRASSHOPPERS

Early attempts at control of rangeland grasshoppers were discouraged by the enormous size of infested territories. Concerning the 1934-1936 outbreak, Morton (1937) wrote, "No widespread attempts have been made to control grasshoppers on rangeland because millions of acres were infested, and because the cost of bait and its application was too high in comparison with the rental value of range." Yet the economic impact of rangeland grasshoppers, sometimes infesting the entire holdings of individual ranchers, was ruinous. Because of the loss of carrying capacity, many ranchers were forced to sell their livestock at sacrifice prices. Wyoming ranchers were keenly aware of problems connected with both drought and grasshoppers. In 1936 the Wyoming Stock Growers Association at their annual meeting requested that the U.S. Department of Agriculture and the Wyoming Agricultural Experiment Station investigate the problem of rangeland grasshoppers in a resolution that reads:

RESOLUTION NO. 12. Calling on the agricultural branch of our State University, private foundations, and the U.S. Department of Agriculture, to set aside or appropriate sufficient funds to enable their entomologists to evolve such a method of control of grasshoppers, Mormon crickets, and other harmful insects which prey on the products of agriculture (Cow Country 64(1), 1936).

After the wishes of stock growers were made known, the U.S. Department of Agriculture approved a project for the study of rangeland grasshoppers in the spring of 1936 and designated Fred A. Morton as project leader. Mr. Morton was stationed at the Grasshopper Laboratory of the Bureau of Entomology and Plant Quarantine, Bozeman, Montana and conducted field research in Colorado, Montana, Nebraska, North Dakota, South Dakota, and Wyoming. The Wyoming Agricultural Experiment Station also responded to this request for research by initiating a project February 8, 1940 entitled "Control of Wyoming Range Grasshoppers." The title is somewhat misleading as the objectives of the project were: (1) to determine the important species of grasshoppers infesting Wyoming rangeland and to study their life histories and habits, and (2) to determine the effect of environmental factors upon development of rangeland grasshoppers for the purpose of developing methods of control. The leader of the project was C.H. Gilbert, apiculturist of the Wyoming station, who hired a field assistant, Robert E. Pfadt, to conduct the research. A field station was set up in an insectary constructed on the State Fair Grounds, Douglas, Wyoming, and the project was begun in the summer of 1940 with a study of the food habits of rangeland grasshoppers.

Within a decade, sufficient research conducted cooperatively between the USDA and Wyoming Agricultural Experiment Station was accomplished to allow the federal Grasshopper Control Division and the State Department of Agriculture to propose control of an outbreak of grasshoppers in the late 1940s.

Organized control of rangeland grasshoppers began in Wyoming in 1949 when a cooperative program was conducted by several parties: the Division of Grasshopper Control of USDA, the State Department of Agriculture, the Wyoming Agricultural Experiment Station, the Wyoming Cooperative Extension Service, several counties, and many ranchers. Surveys made in the summer of 1948 revealed nearly 3.5 million acres of Wyoming rangeland infested with damaging numbers of grasshoppers. Treatment of range was begun in Hot Springs County, Wyoming, where the Douglas DC-3 of the Bureau of Entomology and Plant Quarantine made initial baiting runs on June 6, 1949 spreading bran bait at 5 lb/acre. In this area 40,000 acres were treated, some with toxaphene bait and some with chlordane bait. Cowan (1949) reported uniformly good results with the exception of one small area where the grasshopper population averaged 100-150/yd^2. Although the average reduction was around 70%, the number of grasshoppers remaining was high enough to severely damage the grass. Because Cowan and others suspected that the original application of 5 lbs of bait/acre was insufficient to provide control, the area was later rebaited.

After the baiting of 1.5 million acres of infested Wyoming rangeland, a survey of adult grasshoppers indicated a residual infestation of 2.8 million acres. The following year, nearly 2.5 million acres were baited which brought the infested acres down to 1.1 million. Because certain species (e.g., *A. coloradus*) apparently did not feed on the bait and were unaffected by the treatment and research had shown that aldrin spray (2 oz aldrin in 1 gal diesel oil per acre) was effective in killing all species of grasshoppers, the program switched to spraying in 1951. The number of sprayed acres amounted to 442,610 in 1951, 69,024 in 1952, and only 9,261 in 1953, at the end of the campaign.

DURATION OF OUTBREAKS

An important factor to consider in the economics of control is the duration of an outbreak of rangeland grasshoppers. In the 1934-1936 (peak years) outbreak against which no organized control was conducted, damaging infestations lasted about six or seven years. Corkins (1937) reported that the rise in grasshopper numbers began in 1930 and populations reached their peaks in 1934 to 1936. He evaluated the outbreak as the longest and most severe in Wyoming's agricultural history and estimated that in 1934 six to eight million acres of rangeland in eastern Wyoming were completely denuded of vegetation. Morton (1939), who observed the same outbreak, reported that rangeland grasshoppers decreased during 1937 and 1938. During the past 40 years four major and two minor outbreaks have occurred lasting from four to six years. Wyoming ranchers have reacted to upsurges of grasshopper numbers by participating in the cooperative control program, but response has varied considerably. Although forage in treated acres has been protected and migration of grasshoppers prevented, the duration of outbreaks as measured by number of economically infested acres has not always been greatly affected.

For an individual rancher, however, the important questions are how long a damaging infestation by a particular population will continue on his rangeland and how long control will last after treating such a population early in its outbreak stage. Although state and federal workers have assumed that insecticidal treatment of an area for rangeland grasshoppers in Wyoming lasts for more than the year of application, this concept has been questioned by Blickenstaff et al. (1974), who reported that only two of nine selected studies showed protection from grasshoppers lasting beyond the year of application. Their statistical examination of

1,208,000 acres treated in Wyoming between 1952 and 1958 indicated, however, that an average of only 3.8% was retreated each season in the five years following the initial treatment. The two sets of data appear to be in conflict. Yet if we examine closely the nine studies selected by Blickenstaff et al. (1974) there is no inconsistency. One study was done in Wyoming in the Big Horn National Forest where a two-year life cycle of grasshoppers explained the lack of control the year after treatment. Two studies were made by Shotwell (1953, 1960) on populations of crop-infesting grasshoppers, *Melanoplus* spp., in Missouri and Kansas and are not comparable with rangeland populations which consist of other species treated in whole blocks. In three studies, treatments were made in the year before a crash of the check populations and cannot, for statistical reasons, be used to show control beyond the year of treatment. Two studies were located on the San Carlos Apache Indian Reservation, Arizona, where dense populations of *M. sanguinipes*, *M. cuneatus*, and *A. elliotti* regularly migrate soon after the grasshoppers become adult (Nerney 1961). Indeed, McAnelly and Rankin (1986) have found that in this region *M. sanguinipes* continues migratory behavior through adult life, the males promiscuously, the females between ovipositions. Since treatments were restricted to only 18,000 acres in 1952 and 1600 acres in 1965 out of a tract of 424,877 acres near the center of the reservation, the redistribution of grasshoppers that occurred was a predictable outcome. This leaves only one study with any value for determining duration of rangeland grasshopper control. It is the unpublished report of E. J. Hinman, which showed that near Fort Benton, Montana, insecticidal treatment of rangeland grasshoppers in 1950 and 1951, in blocks ranging from 640 to 8700 acres, gave at least three to four years of protection from grasshopper damage.

Results of our study of the cooperative rangeland grasshopper control program in eastern Wyoming agree with Hinman's study in Montana (Pfadt 1977). Nine sites were chosen for observation, seven were located within larger areas receiving treatment of ultra low volume (ULV) malathion at 8 fl oz per acre applied by airplane, and two were untreated check sites. Protection of forage was considered to have ended when populations of rangeland grasshoppers resurged from the treatment at densities of $8/yd^2$ or greater. Table 12.1 shows that in the seven treated sites, protection of forage lasted from three to six years.

Unfortunately, after reading the article of Blickenstaff et al. (1974), Hewitt and Onsager (1983) concluded that future benefits could not be depended upon, and based their detailed economic analysis of rangeland grasshopper control on forage protection for only the year of treatment. These authors failed to review all of the evidence presented by Blickenstaff et al. (1974), and they

TABLE 12.1
Long-term control of rangeland grasshoppers in Wyoming treated with malathion at 8 fl oz/acre.

| Location | Treatment year, pre-treat | No. grasshoppers/yd^2 | | | | | | Seasons protection |
		Posttreat 1 yr	Posttreat 2 yr	Posttreat 3 yr	Posttreat 4 yr	Posttreat 5 yr	
Ft. Laramie-1	11.4	0.3	2.0	3.1	11.4		4
Douglas	22.1	0.5	2.6	4.3	15.5		4
Guernsey-1	13.9	3.7	5.1	8.7	15.3	48.6	3
Hartville-1	16.8	4.1	1.6	3.0	2.6	6.7	6
Kaspiere	17.2	4.1	0.4	0.5	1.2	2.1	6
Lingle-2	14.9	6.2	1.5	4.1	5.2	24.5	5
Lingle-3	13.7	0.5	0.8	1.2	2.0	6.8	6
Check Hartville-2	13.7	10.4	24.8	15.2[a]			
Check Glendo	6.1	2.9	5.6	4.0	3.3	10.4	

[a] This site was included in the cooperative grasshopper control program and lost as a check site beginning in 1973.

ignored the Wyoming data. We conclude, contrary to Hewitt and Onsager (1983), that all of the reliable evidence indicate that control of rangeland grasshoppers lasts an average of five seasons and that costs can be amortized over this period.

Little is known on how long particular populations of rangeland grasshoppers remain at economic or outbreak densities. Data bearing on this important question are no doubt in the files of several researchers but have yet to be summarized and published. Our check site 5 mi. south of Guernsey, Wyoming provides some indication on duration of an untreated outbreak population. The site is a quarter section of state land that was set aside and removed from any insecticidal treatment to serve as a check for treated areas. The site is a flat to gently sloping area with sandy loam soil (Otero), a cover of mixed grasses and forbs, and contained a high density of grasshoppers at the start of sampling in 1975. On July 9, 1975 when summer species were in the late nymphal and young adult stages, the density was 34.5 grasshoppers/yd^2. A high density was present in 1974 but was not sampled. From 1975 to 1981 densities ranged from 15 to 65/yd^2, but in 1982 the density was only 1.6/yd^2, as a natural collapse of the population had occurred for unknown reasons. Thus, an economic infestation lasted for a period of at least eight years, from 1974 to 1981.

MANAGEMENT OF RANGELAND GRASSHOPPER POPULATIONS

Ranchers do not rush into a control program when densities of grasshoppers reach the nominal economic injury level of 8/yd^2. Most often they wait and hope nature takes care of their problem, but as a grasshopper population continues to grow and causes obvious loss of forage, ranchers are motivated to take action. Table 12.1 shows that in the seven study sites contained within large control blocks the densities ranged from 11.4 to 22.1/yd^2 and averaged 15.7/yd^2 in the year of treatment.

Almost half of 40 respondents in a recent survey of ranchers indicated that they had to borrow the money to finance their part of rangeland grasshopper control (Mann et al. 1987). Others were able to take money out of operating expenses, out of savings, or out of ranch income. Even so, the cooperative rangeland grasshopper control program has been popular among Wyoming ranchers because of its effectiveness in protecting forage, and no year goes by without some infested acreages being treated for economic numbers of rangeland grasshoppers.

Since the start of rangeland grasshopper control in 1949, improvements have been made in procedures, methods, and materials. Development of effective ultra low volume insecticides and the availability of large aircraft to dispense them made practical the treatment of several million acres within an optimal time interval of a few weeks. The major objective of the rangeland grasshopper control program has remained the same over these years. The goal is to reduce grasshopper populations to noneconomic levels by cooperative programs so that timely control prevents damage to native grasslands and prevents migration into surrounding grasslands and croplands. Although the objectives have often been accomplished successfully, the management of rangeland grasshopper populations has not been achieved. Outbreaks occur periodically and are followed by control efforts that are sometimes inadequate. Students of grasshopper biology have theorized that outbreaks of rangeland grasshoppers have their origin in certain favorable habitats. When population densities in these centers become great enough, grasshoppers migrate to surrounding grassland thereby initiating a general widespread outbreak. Insufficient research has been done to prove this hypothesis, and even if it were proven correct, the procedure of waiting for ranchers to request a control program does not fit into the management of populations. A new plan of action is needed for possible integrated pest management of rangeland grasshoppers that requires a permanent team of workers to annually survey the grasslands, giving special attention to the favorable habitats and treating these locations whenever densities rise above the economic injury level.

SUMMARY

Historical records reveal that grasshoppers have been serious pests of western grasslands as long ago as the 1800s. They still continue their depredations in present times causing severe problems for the livestock industry by consuming and clipping native grasses. In Wyoming, government-organized programs to control grasshoppers infesting rangeland started in 1949. Since then the area treated annually has ranged from several thousand to nearly three million acres. Insecticidal control of economic populations of rangeland grasshoppers lasts an average of five years. Although the cooperative control program has succeeded in protecting forage and preventing migration, it has yet to achieve the management of rangeland grasshopper populations.

REFERENCES

Anderson, N.L. 1961. Seasonal losses in rangeland vegetation due to grasshoppers. J. Econ. Entomol. 54:369-378.

Blickenstaff, C.C., F.E. Skoog, and R.J. Daum. 1974. Long-term control of grasshoppers. J. Econ. Entomol. 67:268-274.

Briggs, H.E. 1934. Grasshopper plagues and early Dakota agriculture, 1864-1876. Agr. Hist. 8:51-63.

Brooks, A.R. 1958. Acridoidea of southern Alberta, Saskatchewan, and Manitoba (Orthoptera). Can. Entomol. 90, suppl. 9.

Bruner, L. 1891. Destructive locusts of North America,together with notes on the occurrences in 1891. Insect Life 4 (1 and 2):18-24.

Bruner, L. 1902. Grasshopper notes for 1901. USDA Div. Entomol. Bull. 38 new ser. pp. 39-49.

Bullen, F. T. 1966. Locusts and grasshoppers as pests of crops and pasture - a preliminary economic approach. J. Appl. Ecol. 3:147-168.

Cooley, R.A. 1904. Grasshoppers. Montana Agr. Exp. Sta. Bull. 51:232-242.

Corkins, C.L. 1937. Grasshopper control in Wyoming. Wyoming State Dept. Agric. Office State Entomol. Circular No. 1.

Cowan, F.T. 1949. Grasshopper and Mormon cricket investigations. USDA ARS Bur. Entomol. Pl. Quar. Bozeman, Montana, Third Quarterly Report.

Gillette, C.P. 1904. Some of the more important insects of 1903 and an annotated list of Colorado Orthoptera. Colorado Agr. Exp. Sta. Tech. Series No. 6.

Gurney, A.B., and A.R. Brooks. 1959. Grasshoppers of the *mexicanus* group, genus *Melanoplus* (Orthoptera: Acrididae). Proc. U.S. Nat. Mus. 110:1-93.

Hewitt, G.B. 1977. Review of forage losses caused by rangeland grasshoppers. USDA, ARS Misc. Publ. 1348.

Hewitt, G.B., and J.A. Onsager. 1983. Control of grasshoppers on rangeland in the United States - a perspective. J. Range Manage. 36:202-207.

Hind, H.Y. 1859. A preliminary and general report on the Assiniboine and Saskatchewan exploring expedition. Queens Printer, Toronto.

Lounsberry, C.A. 1919. Early history of North Dakota. Liberty Press, Washington, D.C.

Mann, R., R.E. Pfadt, J.J. Jacobs. 1987. A survey of rancher attitudes and responses to changing range condition, grasshopper infestation and grasshopper control. (in press).

Maximilian, Prince of Wied. 1843. Travels in the interior of North America (1832-1834). Ackerman and Co., London.

McAnelly, M.L., and M.A. Rankin. 1986. Migration in the grasshopper *Melanoplus sanguinipes* (Fab.). I. The capacity for flight in non-swarming populations. Biol. Bull. 170:368-377. II. Interactions between flight and reproduction. Ibid. 170:378-392.

Mitchener, A.V. 1953. A history of grasshopper outbreaks and their control in Manitoba, 1799-1953. Ann. Report Entomol. Soc. Ontario 84:27-35.

Morton, F.A. 1937. Summary of 1936 rangeland grasshopper studies. USDA Bur. Entomol. Pl. Quar. Lab., Bozeman, Montana, Mimeo. Report.

Morton, F.A. 1939. Review of research program on range grasshoppers. USDA Bur. Entomol. Pl. Quar. Lab., Bozeman, Montana, Mimeo. Report.

Munro, J.A. 1949. Grasshopper outbreaks in North Dakota 1808-1948. North Dakota Hist. 16:143-164.

Nerney, N.J. 1961. Effects of seasonal rainfall on range condition and grasshopper population, San Carlos Apache Indian Reservation. J. Econ. Entomol. 54:382-385.

Pfadt, R.E. 1949. Range grasshoppers as an economic factor in the production of livestock. Wyoming Range Manage. Issue No. 7.

Pfadt, R.E. 1977. Some aspects of the ecology of grasshopper populations inhabiting the shortgrass plains. Minnesota Agr. Exp. Sta. Tech. Bull. 310:73-79.

Pfadt, R.E. 1984. Species richness, density, and diversity of grasshoppers (Orthoptera: Acrididae) in a habitat of the mixedgrass prairie. Can. Entomol. 116:703-709.

Riegert, P.W. 1968. A history of grasshopper abundance surveys and forecasts of outbreaks in Saskatchewan. Mem. Entomol. Soc. Can. 52.

Riegert, P.W. 1980. From arsenic to DDT: a history of entomology in western Canada. Univ. Toronto Press, Toronto.

Shotwell, R.L. 1953. The use of sprays to control grasshoppers in fall-seeded wheat in western Kansas. USDA Bur. Entomol. Pl. Quar. E-771.

Shotwell, R.L. 1960. Grasshopper control with sprays. Southwestern Missouri tests, 1955-1957. Missouri Agr. Exp. Sta. Res. Bull. 751.

Smith, R.C. 1954. An analysis of 100 years of grasshopper populations in Kansas (1854 to 1954). Trans. Kansas Acad. Sci. 57:397-433.

13. INTEGRATED MANAGEMENT OF RANGELAND GRASSHOPPERS

Jerome A. Onsager
Rangeland Insect Laboratory, Agricultural Research Service
U.S. Department of Agriculture, Bozeman, Montana 59717

In recent review articles, I have discussed several aspects of grasshopper management on range, including a rational basis for control (Onsager 1985), between-season and within-season variation in grasshopper population dynamics (Onsager 1986a), current registered tactics for grasshopper suppression (Onsager 1986b), and factors that influence selection of control tactics (Onsager 1986c). In this paper it will be necessary to repeat some general information from previous papers, but I will attempt to emphasize considerations that apply specifically to the shortgrass prairie ecosystem.

Rangeland in the United States is infested by a heterogeneous complex of grasshoppers that includes about 200 species. It is not unusual to encounter 30-40 species within an area of about 10 hectares during the course of a single season. In the shortgrass prairie ecosystem, no more than 15 species are responsible for most destruction of range forage. At any given location, however, only 3-5 species will usually comprise at least 75-95% of the total grasshopper population. Therefore, in most cases, there will be a definite interval in time when the preponderance of individuals of predominant species are vulnerable to a control measure.

There are at least three major philosophies of grasshopper control on rangeland. One can strive to treat economic infestations of mid-instar nymphs before serious forage destruction occurs in order to maximize protection of the current season's forage crop. One can treat economic infestations of sexually-immature adults in order to salvage a portion of the current crop and prevent infestation of subsequent crops. One can apply prophylactic treatments to incipient infestations when subsequent increases appear imminent. This discussion will attempt to identify appropriate tactics for each of the three philosophies. A fourth goal could be to treat infestations on range in order to

196

prevent migration to high-value cropland. While that goal certainly is within the scope of IPM, it will not be discussed in depth due to lack of quantitative data on success of such tactics. At the risk of over-simplification, I will suggest that migration problems can be prevented by diligent application of the first three philosophies. If populations are maintained within the total carrying capacity of their habitat, the local grasshopper species will have little inherent tendency to migrate.

Newton et al. (1954) reported the seasonal history of about 50 species of grasshoppers in Montana and Wyoming. The different life histories tend to partition total forage resources over time, and also tend to assure that essentially all forage resources are exposed to at least one or more species.

The rate and degree of forage destruction by a grasshopper infestation is a rather complex function of density, stage of development, and species composition. Based on observed distributions of different kinds of grasshoppers in a typical infestation of mixed grass prairie, a theoretical "average" grasshopper weighed 81.6 mg (dry weight) in the adult stage, and consumed 9, 22, and 53 mg of forage/day in the 4th instar, 5th instar, and adult stages, respectively (Onsager 1984). While the daily rate of forage destruction per grasshopper increases by an average factor of 2.42 with each stage of development, total daily consumption is moderated by mortality among grasshoppers over time. Under most circumstances, therefore, the rate of total daily forage destruction by a population of grasshoppers will tend to increase as the population develops, and will become maximum when most of the population reaches the adult stage (Capinera et al. 1983, Onsager 1984).

The hatch of common economic species of grasshopper can occur over varying periods of time. In an intensive study of the population dynamics of 6 important grasshopper species over a 3-year period (Onsager and Hewitt 1982), the frequency distribution for 1st instar nymphs of each species approximated a normal distribution. The normal distribution of 1st instar nymphs over 3-4 weeks of time gave rise to similar distributions of older nymphs over approximately the same intervals of time. However, each successive instar was represented by lower mean densities because of mortality. When the distributions of successive instars of given species were plotted over time, a typical exponential density decay curve was apparent.

Figure 13.1 illustrates relative seasonal abundance of a hypothetical normal distribution of 1,000 hatchlings, corrected for a constant mortality rate of 5%/day. The mean day of hatch and its standard deviation is 11±3 days, and each stage requires 7 days for development. The cumulative curve indicates potential population in the absence of mortality, and the dotted line represents actual population over time.

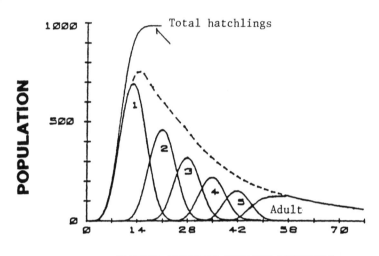

DAYS AFTER FIRST HATCH

FIG. 13.1   Seasonal abundance of nymphal and adult stages of a hypothetical population of 1000 grasshopper hatchlings (see text for survival and development parameters).

One purpose of Figure 13.1 is to illustrate extremes of variation that can occur within a single stage of a single species within a given season. For example, the density of lst instar nymphs varies approximately 700-fold within a 10-day period. Each nymphal stage is present over an interval of about 21 days. On given days, as many as four different stages are represented. If Figure 13.1 was repeated for about 30 different species whose seasonal curves originate (i.e., that begin to hatch) over a continuum of about 70 days, whose eggs may hatch over intervals as long as 60 days, whose individual mortality rates may differ drastically, and whose relative density may vary from trace to perhaps around 12 total adults/m², one could begin to visualize the astounding complexity that exists each season within a typical grassland community.

A second purpose of Figure 13.1 is that it closely depicts the observed seasonal development of *Aulocara elliotti* (Thomas) (Onsager and Hewitt 1982), an important pest of rangeland. Its biology differed markedly from that of another major pest, *Melanoplus sanguinipes* (F.). In the same study, eggs of *M. sanguinipes* hatched over an interval of at least 50 days, which caused both 1st instar nymphs and adults to be present over an interval of at least 11 days. Consequently, the different stages of *A. elliotti* tend to appear in the general population as a conspicuous pulse or wave over a relatively narrow time frame. In contrast, successive stages of *M. sanguinipes* occupy a wider time frame than *A. elliotti*, and an identical total population of *M. sanguinipes* will exist as lower densities that are spread over a longer period of time than *A. elliotti* (Onsager 1986a). Such differences in seasonal phenology may require drastic adjustment in management tactics for suppression of infestations. If the abscissa of Figure 13.1 was expressed as degree-days rather than calendar days (as in Kemp, Chapter 24) and the ordinate was grasshopper density rather than population, the figure would illustrate at least four important characteristics of a grasshopper problem. First, the height of the total curve and the area under the curve for 4th instar and later stages are indicators of the potential seriousness of the problem. Second, the anticipated stage of development can indicate an opportune time to sample in anticipation of a potential suppressive tactic. Third, the observed stage of development when threatening infestation is detected may limit the number of viable tactics that are available for suppression. Fourth, both the origin and the width of the base for stage-specific distribution curves (i.e., the mean and standard deviation associated with average stage-specific presence) affect the length of the "action window" during which each suppressive tactic is likely to achieve optimum efficacy. The importance of each of these characteristics will now be discussed separately.

Extremely severe infestations may require immediate and drastic non-selective control measures. Moderate to low infestations, however, may be amenable to reduction by less traumatic selective measures. It is important to understand that immediate effects of treatments do not necessarily have to be dramatic to bring about dramatic adjustments in population density. A relatively subtle increase in the daily mortality rate can have dramatic consequences over the course of a season (Onsager 1986b,c).

The key to providing a choice among treatment options is to sample early in the season when all options are still available. I believe that sampling should occur when the major economic species are predominantly 3rd instar nymphs. Field experiments (Onsager 1978) and modeling trials (Hardman and Mukerji 1982, Onsager 1984) agree that treatments provide maximum prevention of forage destruction if applied when most grasshoppers in an infestation are in the 3rd or 4th nymphal instar. According to Hewitt (1979), the 3rd instar tends to become important in northern states for 3 major reasons: (1) it begins to consume significant amounts of forage per day, (2) its appearance coincides with maturation of several important cool-season forage grasses, so most of the forage that is consumed will not be replaced by regrowth, and (3) the probability of catastrophic mortality becomes relatively low. The second criteria above may be of little importance in the southern shortgrass prairie, where the most important forage species are warm-season grasses.

The Rangeland Insect Laboratory has investigated 5 approved techniques for control of grasshoppers, broadcast spray application of 3 chemical insecticides (carbaryl, acephate, and malathion), and broadcast applications of 2 baits (carbaryl and a pathogen, *Nosema locustae* Canning). Some of the advantages and disadvantages of each have been discussed elsewhere (Onsager 1986c). Because the five techniques have widely divergent capabilities, it follows that no single technique can meet all grasshopper management needs on the shortgrass prairie. If, for example, the objective is to treat early in the season to prevent current-season forage destruction, then malathion is a risky treatment choice. If undetected mid-season infestations reach epidemic proportions and require immediate reductions, then carbaryl may not be cost competitive. In environmentally sensitive areas, *Nosema locustae* may not be a viable option if the problem is not diagnosed sufficiently early.

Action windows can indicate not only when given control tactics become efficacious, but also when the same tactics become futile. In the remainder of this discussion, I will estimate action windows for key grasshopper management activities.

Figure 13.2 is an extension of Figure 13.1, and assumes that adults live an average of 21 days, adults require 10 days to attain sexual maturity, mature females produce an egg-pod every 5 days, and each pod contains 10 eggs. It is intended to help summarize the importance of instar frequency distributions and their impact on action windows. By calculating the interval between the last percentile of hatchlings in Figure 13.1 (about day 17) and the first percentile of egg production in Figure 13.2 (about day 51), the total estimated control window is about 34 days. The earliest treatments will theoretically save the most forage from destruction, but all treatments within that window are theoretically capable of preventing infestation of subsequent forage crops.

In cropland situations where relatively few grasshopper species are involved, sampling to detect economic infestations should begin at about peak second instar. At that time, the grasshoppers will be concentrated along crop margins in fence rows, roadsides, grass waterways, and similar breeding sites. Early treatment of margins will prevent subsequent movement of grasshoppers over entire crop acreages. If additional insects hatch after treatment, a second application is a viable option in the case of cropland. On rangeland, however, economics dictate that only one treatment be applied, so it behooves the manager to gain maximum benefit from that treatment. As mentioned earlier, sampling for protection of range forage should occur near peak 3rd instar.

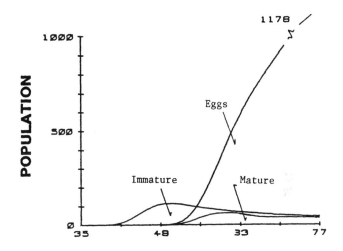

DAYS AFTER FIRST HATCH

FIG. 13.2   Seasonal abundance of immature adults, reproductive adults, and eggs for the hypothetical grasshopper population of Figure 13.1.

Bait treatments with either carbaryl or *Nosema locustae* should be applied during the 3rd and 4th instar stages for three important reasons. First, the early instars of several species actively forage for dry plant material on the soil surface, so early instars are most vulnerable to a dry bait carrier. Second, early instars are susceptible to lower dosages than later stages. Third, early bait treatments provide greater time for accumulation of subtle direct or indirect effects of these selective treatments.

Carbaryl spray treatments can begin against early 3rd instar nymphs (Onsager 1978). The key factors are a formulation that can tolerate cool temperatures and early-season precipitation, and that persists with sufficient potency to eliminate late-hatching nymphs for 2 to 3 weeks after treatment. Carbaryl spray treatments usually fall into disfavor at either about peak 4th instar or peak 5th instar stages. At those times, equivalent control can be achieved with lower treatment costs by switching to acephate or malathion, respectively.

Malathion spray treatments can begin after there is little probability of additional hatching, and when afternoon air

202

temperatures are likely to reach about 70° to 75°F. If an objective of treatment is to prevent infestation during the following season, then malathion treatments cannot be applied after significant oviposition has taken place.

Acephate spray treatments have not been characterized as well as have carbaryl or malathion, but are considered to fall between the latter treatments with respect to both persistence of residues and windows of optimum utility.

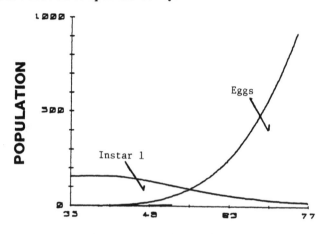

DAYS AFTER FIRST HATCH

FIG. 13.3   Seasonal abundance of 1st instar nymphs and of eggs produced by mature females (see text for survival and development parameters).

Figures 13.1 and 13.2 depict a relatively simple control problem comprised of a single species that tends to develop over time in a well-synchronized pulse or wave. A complex of several species or a preponderance of a species like *M. sanguinipes* that hatches over a long interval of time would pose a completely different problem. Figure 13.3 depicts the latter situation, and assumes a standard deviation of 15 days for average day of hatch, egg production of 20 eggs per pod, and all other parameters as for Figures 13.1 and 13.2. For simplicity, Figure 13.3 shows only the distributions of 1st instar nymphs and of eggs that are produced by mature females of the current season. In contrast to the 34-day control window between last hatch and first egg production that was estimated for Figures 13.1 and 13.2, Figure 13.3 indicates that the first percentile of eggs are produced on day 51, which is 22 days *before* the last percentile of hatchlings on day 73. Such an infestation could not possibly be controlled efficaciously with a short-lived chemical like ULV malathion.

Early applications would miss too many unhatched eggs to maximize prevention of current-season damage, and later applications would neither prevent current-season damage nor assure prevention of reinfestation the next season. A perfectly timed carbaryl spray treatment applied about day 50 could theoretically achieve both objectives. In my opinion, the most viable control tactic would be to inoculate with *Nosema locustae* between about days 28 to 49. Initial infections would cause some direct mortality, reduce vigor of infected survivors, and generate inoculum that would not only be transmitted during the current season through cannibalism but also persist to infect the next generation.

It should now be clear that integrated management of grasshoppers on rangeland will require much more sophisticated techniques for sampling and for selection of treatments than have been common in the past. This is especially true for the southern portion of the shortgrass prairie, where it appears that grasshopper hatching occurs over a long interval as shown in Figure 13.3. In my opinion, a useful strategy for improving both procedures is to intensify the Sentinel Site monitoring concept being developed primarily by Knight (1986) combined with the phenology model of Kemp and Onsager (1986). The combination would not only determine distribution and relative abundance of key stages of important species, but also would provide degree-day requirements for use in predicting action windows during subsequent seasons.

REFERENCES

Capinera, J.L., W.J. Parton, and J.K. Detling. 1983. Application of a grassland simulation model to grasshopper pest management on the North American shortgrass prairie, p. 335-344. *In:* W.K. Lauenroth, G.V. Skogerboe, and M. Flug, (eds.) Analysis of ecological systems: state-of-the-art in ecological modelling. Elsevier, New York.

Hardman, J.M., and M.K. Mukerji. 1982. A model simulating the population dynamics of the grasshoppers (Acrididae) *Melanoplus sanguinipes* (Fabr.), *M. packardii* Scudder, and *Camnula pellucida* (Scudder). Res. Pop. Ecol. 24:276-301.

Hewitt, G.B. 1979. Hatching and development of rangeland grasshoppers in relation to forage growth, temperature, and precipitation. Environ. Entomol. 8:24-29.

Kemp, W.P., and J.A. Onsager. 1986. Rangeland grasshoppers: modeling phenology of natural populations of six species. Environ. Entomol. 15:924-930.

Knight, S. 1986. The sentinel monitoring system for grasshoppers in Nebraska. Proc. Grasshopper Symposium, March 13-14, 1986, Bismarck, ND (in press).

Newton, R.C., C.O. Esselbaugh, G.T. York, and H.W. Prescott. 1954. Seasonal development of range grasshoppers as related to control. USDA/ARS, Bureau of Entomology and Plant Quarantine E-873.

Onsager, J.A. 1978. Efficacy of carbaryl applied to different life stages of rangeland grasshoppers. J. Econ. Entomol. 71:269-273.

Onsager, J.A. 1984. A method for estimating economic injury levels for control of rangeland grasshoppers with malathion and carbaryl. J. Range Manage. 37:200-203.

Onsager. J.A. 1985. An ecological basis for prudent control of grasshoppers in the western United States. Proc. Triennial Mtg., Pan Am. Acrid. Soc., 5-10 July 1981, Maracay, Venezuela 3:97-104.

Onsager, J.A. 1986a. Stability and diversity of grasshopper species in a grassland community due to temporal heterogeneity. Proc. Triennial Mtg., Pan Am. Acrid. Soc., 28 July - 2 August 1985, Saskatoon, Saskatchewan, Canada. 4:101-109.

Onsager, J.A. 1986b. Current tactics for suppression of grasshoppers on range, p. 60-66. In: J.A. Onsager (ed.) IPM on rangeland: state of the art in the sagebrush ecosystem. USDA, ARS 50.

Onsager, J.A. 1986c. Factors affecting population dynamics of grasshoppers and control options. Proc. Grasshopper Symposium, March 13-14, 1986, Bismarck, ND (in press).

Onsager, J.A., and G.B. Hewitt. 1982. Rangeland grasshoppers: average longevity and daily rate of mortality among six species in nature. Environ. Entomol. 11:127-133.

14. FUTURE PROSPECTS FOR MICROBIAL CONTROL OF GRASSHOPPERS

D. A. Streett
Rangeland Insect Laboratory, Agricultural Research Service
U.S. Department of Agriculture, Bozeman, Montana 59717

Major stimuli for increased interest in insect pest management (IPM) programs have been the development of resistance to insecticides in pest species, and a growing public concern over the environmental effects of toxic chemicals. The relatively high economic threshold of insects on rangeland offers an ideal opportunity for the implementation of IPM programs. A component of an IPM program with a major potential role in controlling insect pest species is the use of entomopathogens as microbial control agents. Pathogenic microbial agents which normally do not exhibit "rapid kill" can be successful for either short-term and long-term control strategies of grasshoppers on rangeland.

Microbial control agents can be divided into those used for short-term control, which commonly have some attributes similar to chemical insecticides, and those used for long-term control, which result in the establishment of the entomopathogen in a pest population and long-term regulatory activity. The ideal long-term microbial control agent should be moderately pathogenic, survive in the environment, be capable of transmission in the host population, be able to reduce host fecundity, and should maintain the host population levels at a stable state. *Nosema locustae*, a grasshopper pathogen registered in the United States as a microbial insecticide, meets all of these criteria and is a successful example of a long-term microbial control agent. A more virulent entomopathogen capable of short-term control is desirable to augment *N. locustae* in some IPM programs.

An evaluation of any entomopathogen as a candidate microbial control agent must consider epizootiology, microbial control, and commercialization. Epizootiology and microbial control are interrelated fields with respect to virulence and pathogenicity of the pathogen, transmission, persistence, and timing of application. Comprehensive reviews by Tanada (1963)

205

and Burges (1981) are recommended for discussion of these criteria. The other major area for evaluation of an entomopathogen is commercialization, and includes such considerations as efficacy, mass production and formulation, and safety and registration. These factors will be addressed in this review only when they represent major advantages or disadvantages for a specific entomopathogen.

This review will explore the present status of microbial control agents of grasshoppers and evaluate entomopathogens of current interest from the major groups of pathogenic microorganisms; i.e., fungi, protozoa, rickettsia, viruses, and bacteria. An excellent review by Henry (1977) provides an earlier assessment of microbial agents for controlling grasshopper populations.

FUNGI

Insect pathogenic fungi are a diverse group of microorganisms with some potential as microbial control agents. The mode of infection by fungi is normally through the host cuticle into the hemocoel rather than by ingestion. Conidia of entomopathogenic fungi attach to the insect cuticle, and under suitable conditions will germinate and penetrate through the host cuticle into the hemocoel. Growth of the fungus ensues with death resulting either from mechanical forces or toxin production. Most fungi pathogenic to grasshoppers are either in the Order Entomophthorales (specifically *Entomophaga*), or the Deuteromycetes (Fungi Imperfecti).

Several deuteromycetes have shown some promise as candidate microbial control agents including *Metarhizium anisopliae, Beauveria bassiana*, and *Verticillium lecanii*. Certainly these fungi have not received much attention as potential microbial control agents, probably because of the rather spectacular reduction of grasshopper populations observed during epizootics of *Entomophaga grylli*. However, this should not preclude them from consideration as candidate microbial control agents in an IPM program.

Present knowledge on these fungi is limited with respect to their potential for controlling grasshopper populations. Nonetheless, several *M. anisopliae* strains have been isolated from orthopteran hosts (Veen 1968, Reinganum et al. 1981). Indeed, the *M. anisopliae* strains infect grasshoppers *per os* (Veen 1966, Veen 1968), and thus could be used in a wheat bran bait application program as described by Onsager et al. (1980). The fungus *V. lecanii* commonly infects scale insects, but has shown some potential for controlling grasshopper populations (Harper and

Huang 1986). A mean mortality reported for grasshoppers sprayed with *V. lecanii* spores was 22% and ranged from 8 to 33%. Although these results are not conclusive, they indicate that *V. lecanii* may have some potential in an IPM program.

The ability to cultivate these fungi easily is a major asset toward eventual commercialization (Ferron 1981, Hall 1981). Indeed, some of these fungi are now commercially formulated and available. However, due to the high relative humidity required for spore germination and development, fungi in general do not lend themselves to commercial development as grasshopper microbial control agents. The additional major constraints of these fungi as microbial control agents are the abiotic (temperature, relative humidity, UV inactivation) and biotic (host-pathogen density) factors.

Entomophaga grylli is a fungal pathogen important in the natural control of grasshopper populations. Several epizootics by this fungal disease have been responsible for dramatic reductions in grasshopper populations (Pickford and Riegert 1964, Henry and Onsager 1982). The primary factor responsible for inducing an epizootic is high relative humidity associated with above-average rainfall, and as with other fungi, abiotic and biotic factors play a major role.

The biological cycle of *E. grylli* is characterized by a mycelial vegetative phase and a reproductive phase which is subdivided into two types of spore formation. Conidia are produced externally, and are important in the horizontal transmission of the fungus. A resting spore is produced internally, and is important for survival of the fungus during adverse environmental conditions.

Soper et al. (1983) recently differentiated *E. grylli* into two pathotypes on the basis of isoenzyme analysis and certain biological characteristics. Pathotype 1 occurs primarily in *Camnula* spp. and during the reproductive phase forms both conidia and resting spores. In contrast, pathotype 2 is found primarily in *Melanoplus* spp., and lacks a typical conidial cycle. Resting spores are presumably responsible for the horizontal transmission of this pathotype.

Recent evidence, though, has demonstrated the formation of atypical conidia produced directly from hyphal bodies (Humber and Ramoska 1986). These atypical conidia are referred to as "cryptoconidia" and their formation as "cryptoconidiogenesis." Cryptoconidia are infective to susceptible host species and can support horizontal transmission of the fungus in grasshoppers (Humber and Ramoska 1986). *E. grylli* pathotype 1 will also undergo cryptoconidiogenesis and this reproductive strategy may presumably be an important factor in the establishment of an epizootic in grasshopper populations.

The major constraints on the development of *E. grylli* as a microbial control agent are an inability to produce the fungus on artificial media, difficulties with *in vivo* production, and storage problems. Protoplast formation and multiplication have been successful in Grace's tissue culture medium with a grasshopper extract (MacLeod et al. 1980). However, development of *E. grylli* conidia or resting spores on artificial media has not been successful. Nevertheless, *E. grylli* protoplasts and resting spores are infectious in grasshoppers and can be used for production of conidia and resting spores, respectively (MacLeod et al. 1980, Nelson et al. 1982). The major liability with this *in vivo* production system is the labor intensive infection of grasshopper by injection. Nelson et al. (1982) also noted a decrease in resting spore germination after increasing storage time with about 5% germination observed after 3 months storage. The difficulties with mass production and storage of *E. grylli* must be rectified before it can be considered as a potential microbial control agent. Finally, research is needed on the selection or isolation of new *E. grylli* strains for improved efficacy, stability, and dispersal in lower relative humidity. Somatic hybridization and protoplast fusion are two techniques which may prove useful for this objective.

The rather complex abiotic and biotic factors required for initiation of an *E. grylli* epizootic make it an unlikely candidate for commercial development. Certainly the epizootiology of *E. grylli* in grasshopper populations and the effect of cryptoconidiogenesis on horizontal transmission will require additional research. The mass production and application of *E. grylli* in some type of inundative release program will be unlikely in the near future. Nevertheless, it may be feasible under favorable environmental conditions to use *E. grylli* in an inoculative release program.

PROTOZOA

Protozoa have received the most attention as potential microbial control agents of grasshoppers. Among the protozoa, microsporidia appear to offer the greatest potential for controlling grasshopper populations.

Microsporidia are intracellular parasites characterized by a unicellular spore containing a polar tube and sporoplasm. Transmission normally occurs when spores contaminating a food source are ingested by a susceptible host. Spore germination follows, forcing the polar tube to evert and inject the sporoplasm into a host cell. The microsporidium replicates either in a specific tissue or group of tissues, and eventually forms spores.

The infected host frequently will either die or show a loss of vigor and fecundity.

Three microsporidium species, *N. locustae, Nosema cuneatum*, and *Nosema acridophagus* have generated the most interest as potential microbial control agents of grasshoppers (Henry 1977). *N. locustae* has undergone extensive development as a microbial control agent for controlling grasshopper populations. Comprehensive reviews by Henry (1978, 1982) and Henry and Oma (1981) discuss the development and successful use of this entomopathogen to control grasshoppers. Rather than reiterate the conclusions of these excellent reviews, this chapter will focus on the microsporidium species *N. acridophagus* since it is more virulent than *N. cuneatum* and would complement *N. locustae* in an IPM program as a short-term control agent.

N. acridophagus was originally isolated from *Schistocerca americana*, but will infect several *Melanoplus* spp. (Henry 1967). The microsporidium infects the midgut, gastric caecae, gonads, adipose tissue, pericardial cells, and tissues associated with the nervous system. Mass production of *N. acridophagus* spores in sufficient quantities for field application has been difficult due to the virulence of this parasite in grasshoppers (Henry and Oma 1974). A study by Henry et al. (1979) has shown that *N. acridophagus* can be produced in the corn earworm, *Heliothis zea*, and at higher spore production levels than those in the grasshopper *M. sanguinipes*. In addition, *H. zea* can be reared easily for *in vivo* production, and the microsporidium develops faster in the corn earworm which ultimately increases spore production.

The desirable attributes of *N. acridophagus* for microbial control are its greater virulence and host range. However, while *N. acridophagus* is the most virulent microsporidium infecting grasshoppers, it is doubtful this protozoan will be used as a short-term microbial control agent. Mass production of *N. acridophagus* in *H. zea* will probably not be economical even for a "cottage-type" industry.

RICKETTSIA

Rickettsia are gram-negative rod-shaped microorganisms that commonly multiply inside host cells by binary fission. The rickettsia isolated from grasshoppers belong to the genus *Rickettsiella* with *R. schistocercae* placed in synonymy with the only recognized species *R. grylli* (Weiss et al. 1984). An ultrastructural study of a *Rickettsiella* sp. tentatively identified as *R. grylli* by Henry et al. (1986) provides a comprehensive review of grasshopper rickettsia and a description of the developmental cycle.

R. grylli is easily transmitted to grasshoppers by contaminated food, and reproduction occurs in the cytoplasm of adipose cells. Infected grasshoppers are lethargic just prior to morbidity, with mortality occurring about 14 days postinoculation (Henry et al. 1986).

Although *R. grylli* will presumably cause severe epizootics resulting in high mortality (Vago and Meynadier 1965, Bergoin and Vago 1966), it probably will not be considered as a potential microbial control agent due to safety considerations. *Rickettsiella melolonthae* and *R. grylli* have both been shown to infect mammals when inoculated either by the intraperitoneal or nasal route (Ignoffo 1973). Consequently, although *R. grylli* may be a significant factor in natural suppression of grasshopper populations, it probably will not be developed as a microbial control agent.

VIRUSES

Many of the viruses isolated from insects are pathogenic. Some of the more well-known epizootics reported in populations of insects have been caused by these viruses. Thus far, virtually all the viruses reported from Orthoptera are either entomopoxviruses or picornaviruses. This is remarkable considering the diversity of viruses found in other insect groups.

The picornaviruses reported from Orthoptera are the cricket paralysis virus, isolated from *Teleogryllus commodus* (Reinganum et al. 1970) and a crystalline array virus in *Melanoplus bivittatus* (Jutila et al. 1970). A comparison of pathogenicity and ultrastructure has shown that these two viruses are distinct (Henry 1977), but additional biochemical and biophysical characteristics need to be examined. There is some evidence that antibodies to cricket paralysis virus might be present in vertebrates (Longworth and Scotti 1978). These studies on a related virus suggest that the crystalline array virus presumably might represent a safety risk to vertebrates. Therefore, the question of crystalline array virus safety to vertebrates should be determined by direct experimentation.

The crystalline array virus is highly virulent, and replicates in the cytoplasm of muscle, tracheal matrix, and pericardial cells. The virus was originally detected in *M. bivittatus* when a group of field-collected eggs hatched in the laboratory and showed an unusually high mortality. Grasshoppers inoculated *per os* with a homogenate prepared from an infected *M. bivittatus* nymph showed almost 80% cumulative mortality after one week. These studies suggest the virus might be an excellent short-term microbial control agent of grasshoppers.

Entomopoxviruses are a group of insect viruses in the family Poxviridae with a linear double-stranded DNA genome. Virus particles are large, brick-shaped or oval, with globular surface units that give them a mulberry-like appearance (Matthews 1979). The virus particles may be occluded within a crystalline protein occlusion body referred to as a "spheroid," which is irregularly shaped and ranges in size from about 2-10 μm in diameter. Several hundred virus particles may be occluded in a single spheroid which offers some protection to the virus in the environment.

In nature, the route of infection for grasshopper entomopoxviruses is *per os*. After ingestion, the spheroid dissolves in the gut lumen, releasing virus particles which enter midgut cells. The virus eventually enters adipose cells where replication occurs in the cytoplasm, with virus particles either occluded within spheroids or released from the infected cells for secondary infection in other adipose cells. The grasshopper eventually dies, releasing the spheroids into the environment for infection in a susceptible host.

The only known hosts for entomopoxviruses are found in four orders of insects: Coleoptera, Diptera, Orthoptera, and Lepidoptera. Within the Orthoptera, six entomopoxviruses have been isolated from grasshoppers (Streett et al. 1986). The grasshopper entomopoxviruses are relatively host specific although some grasshopper entomopoxviruses will infect species from different subfamilies in the Acrididae (Oma and Henry 1986).

Entomopoxviruses are normally named after the host species of the original isolation. The *Melanoplus sanguinipes* entomopoxvirus was one of the first and most extensively studied viruses in this group (Henry and Jutila 1966). Laboratory studies have demonstrated the virulence of *M. sanguinipes* entomopoxvirus in *M. sanguinipes* and small-scale field studies have also been reported with grasshopper densities reduced to about 3 to 5 grasshoppers per sq. yard (Henry 1977). Several other economically important grasshopper species are also susceptible to grasshopper entomopoxviruses. In particular, the *Oedaleus senegalensis* entomopoxvirus and *Arphia conspersa* entomopoxvirus show promise for microbial control. Concern regarding the apparent relatedness of entomopoxviruses to vertebrate poxviruses has provided the impetus for comparative molecular biology studies.

Grasshopper entomopoxvirus DNA has been compared with DNA from the vertebrate poxvirus, vaccinia, using site-specific restriction endonucleases and Southern blot hybridization to detect DNA homology or relatedness (Langridge et al. 1983, Langridge

1984, Streett et al. 1986). These studies confirmed that the grasshopper entomopoxviruses isolated from different host species are separate although related viruses, and show no relatedness to vaccinia virus DNA. Some limited safety tests have also been performed with entomopoxviruses on mice and rats. Virus particles of *Amsacta moorei* entomopoxvirus were administered intracerebrally or intraperitoneally to mice, and *Choristoneura fumiferana* entomopoxvirus spheroids were fed to mice and rats. No adverse effects were observed in these vertebrate safety tests. Furthermore, wild mammals and laboratory mice did not exhibit any adverse effects after exposure to field concentrations of *C. fumiferana* entomopoxvirus (Ignoffo 1973). The molecular studies on the relatedness of entomopoxviruses to vertebrate poxviruses and the limited safety tests, although not conclusive, do suggest that entomopoxviruses are safe microbial control agents. However, these viruses will need to be thoroughly evaluated for safety prior to their use in an IPM program.

A narrow host range is the most serious constraint to the utilization of these viruses as microbial control agents. The host range of grasshopper entomopoxviruses is limited to a single species or only a few species of grasshoppers. Fortunately, most grasshopper species susceptible to entomopoxvirus infection are economically important species and in many cases are available for virus production. Grasshopper entomopoxviruses are presently produced *in vivo* in sufficient quantities for small-scale field applications. Additional studies will be necessary to increase the efficiency of *in vivo* mass production of grasshopper entomopoxviruses.

BACTERIA

Bacteria are ubiquitous microorganisms which lack cell organelles and reproduce by binary fission. A large bacterial flora is found in the alimentary tract of healthy insects. Many of these bacteria are reported as pathogenic, but most of those are actually facultative pathogens. The major constraint in identification of bacterial pathogens is the diagnosis of a septicemic disease before extensive damage to the midgut releases the gut flora into the hemocoel (Bucher 1981).

The few reports of bacterial epizootics for grasshoppers in the literature have not been corroborated (Bucher 1959). In an early investigation by D'Herelle (1911) a bacterial pathogen, *Coccobacillus acridiorum*, was isolated during an epizootic from the grasshopper *Schistocerca pallens*. Subsequent field trials by D'Herelle (1914 a, b) with *C. acridiorum* for controlling grasshopper populations were successful, but other researchers were unable to corroborate his successful results in the field (Sergent and L'Heritier 1914, Bequet 1915, Sergent 1916).

Bucher (1959), during a survey for bacterial pathogens of grasshoppers, identified as the most virulent entomopathogen a strain of *Serratia marcescens*. Subsequent studies demonstrated that *S. marcescens* did not actively invade the hemocoel and was pathogenic only when it was able to accidentally enter the hemocoel. Bacterial entomopathogens with the most potential for microbial control have either been obligate pathogens or crystalliferous sporeformers.

The most successful of these crystalliferous sporeformers has been the gram-positive bacterium, *Bacillus thuringiensis* (B.t.). This spore-forming bacterium, forms a proteinaceous parasporal crystal in the sporangium during sporulation. The parasporal crystal contains an endotoxin (δ-endotoxin) which upon ingestion by a susceptible insect produces septicemia and death. Almost 30 varieties of *B.t.* have been characterized by serological and biochemical tests. *B.t.* varieties may also differ in crystal type, potency, and host spectrum of insecticidal activity. Unfortunately, isolation of a *B.t.* variety with significant activity against grasshoppers has not been successful. *B.t.* spores do not persist in the field and infected insects rarely produce spores. Consequently, *B.t.* will be difficult to detect in a survey of grasshoppers for pathogenic microorganisms. Although this is somewhat discouraging, the recent discovery of a *B.t.* variety for controlling Diptera, and a *B.t.* variety effective against Coleoptera, emphasizes the need for a more extensive survey for *B.t.* varieties pathogenic to grasshoppers.

CONCLUSIONS

The focus of this review has been to evaluate potential entomopathogens for controlling grasshopper populations. The criteria for a realistic assessment of these entomopathogens are based on epizootiology, microbial control potential, and eventual commercialization. In some cases, it is difficult to evaluate grasshopper pathogens on the basis of current knowledge because of the paucity of information on some major groups of pathogenic microorganisms. Consequently, recommendations for potential entomopathogens of grasshoppers with emphasis on short-term control are categorized into two groups: 1) those entomopathogens currently available for microbial control, and 2) pathogens requiring further investigation.

Presently, the viruses show great promise for microbial control of grasshopper populations. Grasshopper entomopoxviruses and the crystalline array virus are both highly virulent and show

214

potential as short-term control agents. A major obstacle in the development of these viruses as microbial control agents has been insufficient vertebrate safety tests. Current evidence suggests that these viruses are safe, but more research is necessary before field trials.

In a survey for grasshopper entomopathogens the deuteromycetes and bacteria are the two pathogen groups that require the most emphasis. Selection of fungus varieties which are less dependent on abiotic factors and show a greater virulence to grasshoppers will merit more emphasis. Technical constraints which include economical mass production and formulation still exist and will need to be addressed before commercial development occurs.

A more extensive survey should be undertaken for crystalliferous spore-forming bacteria with activity against grasshoppers. It may be feasible to isolate *B.t.* varieties toxic to grasshoppers from soil samples. A more traditional variety screening system for sporeforming bacteria, in conjunction with a bioassay, may be more promising for isolation of *B.t.* varieties effective against grasshoppers. However, instead of screening new entomopathogenic isolates from nature, successful development of short-term microbial control agents for grasshoppers may depend on recombinant DNA technology. These microorganisms will either be viruses or bacteria, as these organisms have proven the most amenable to genetic manipulation. Genetically-engineered microorganisms for grasshoppers may be available within the next decade. Environmental Protection Agency approval for these genetically engineered microorganisms will be the major limitation toward eventual commercialization.

ACKNOWLEDGMENTS

The author is indebted to Dr. M.R. McGuire and E.A. Oma for reviewing this manuscript and for their helpful comments.

REFERENCES

Bequet, M. 1915. Deuxieme campagne les sauterelles *Stauronotus macroceanus* (Thun.) en Algérie au moyen du *Coccobacillus acridiorum* D'Herelle. Ann. Inst. Past. 29:520-536.
Bergoin, M., and C. Vago. 1966. Passage de *Rickettsiella grylli* VAGO AND MORTOGA (Wolbachiae) dans le tube digestif d'Insectes prédateurs et détritivores. Arch. Inst. Past. Tunis. 43:65-75.

Bucher, G.E. 1959. Bacteria of grasshoppers of western Canada: III. Frequency of occurrence, pathogenicity. J. Insect Pathol. 1:391-405.

Bucher, G.E. 1981. Identification of bacteria found in insects, p. 7-33. *In:* H.D. Burges (ed.) Microbial control of pests and plant diseases 1970-1980. Academic Press, New York.

Burges, H.D. 1981. Strategy for the microbial control of pests in 1980 and beyond, p. 797-836. *In:* H.D. Burges (ed.) Microbial control of pests and plant diseases 1970-1980. Academic Press, New York.

D'Herelle, M.F. 1911. Sur une épizootie de nature bactèrienne sèvissant sur les sauterelles au Mexique. C.R. Acad. Sci. 152:1413-1415.

D'Herelle, F. 1914a. Le coccobacille des sauterelles. Ann. Inst. Past. 28:280-328.

D'Herelle, F. 1914b. Le coccobacille des sauterelles. Ann. Inst. Past. 28:387-407.

Ferron, P. 1981. Pest control by the fungi *Beauveria* and *Metarhizium*, p. 465-482. *In:* H.D. Burges (ed.) Microbial control of pests and plant diseases 1970-1980. Academic Press, New York.

Hall, R.A. 1981. The fungus *Verticillium lecanii* as a microbial insecticide against aphids and scales, p. 483-498. *In:* H.D. Burges (ed.) Microbial control of pests and plant diseases 1970-1980. Academic Press, New York.

Harper, A.M., and H.C. Huang. 1986. Evaluation of the entomophagous fungus *Verticillium lecanii* (Moniliales: Moniliaceae) as a control agent for insects. Environ. Entomol. 15:281-284.

Henry, J.E., and J.W. Jutila. 1966. The isolation of a polyhedrosis virus from a grasshopper. J. Invertebr. Pathol. 8:417-418.

Henry, J.E. 1967. *Nosema acridophagus* sp. n., a microsporidian isolated from grasshoppers. J. Invertebr. Pathol. 9:331-341.

Henry, J.E., and E.A. Oma. 1974. Effects of infections by *Nosema locustae* Canning, *Nosema acridophagus* Henry, and *Nosema cuneatum* Henry (Microsporida: Nosematidae) in *Melanoplus bivittatus* (Say) (Orthoptera: Acrididae). Acrida 3:223-231.

Henry, J.E. 1977. Development of microbial agents for the control of Acrididae. Rev. Soc. Entomol. Argentina 36:125-134.

Henry, J.E. 1978. Microbial control of grasshoppers with *Nosema locustae* Canning. Misc. Pub. Entomol. Soc. Am. 11:85-95.

Henry, J.E., E.A. Oma, J.A. Onsager, and S.W. Oldacre. 1979. Infection of the corn earworm, *Heliothis zea*, with *Nosema acridophagus* and *Nosema cuneatum* from grasshoppers: Relative virulence and production of spores. J. Invertebr. Pathol. 34:125-132.

216

Henry, J.E., and E.A. Oma. 1981. Pest control by *Nosema locustae*, a pathogen of grasshoppers and crickets, p. 573-586. *In:* H.D. Burges (ed.) Microbial control of pests and plant diseases 1970- 1980. Academic Press, New York.

Henry, J.E. 1982. Production and commercialization of microbials: *Nosema locustae* and other protozoa, p. 103-106. *In:* Invertebrate pathology and microbial control. Proc. IIIrd Int. Colloq. Invertebr. Pathol. Brighton, United Kingdom.

Henry, J.E., and J.A. Onsager. 1982. Large-scale test of control of grasshoppers on rangeland with *Nosema locustae*. J. Econ. Entomol. 75:31-35.

Henry, J.A., D.A. Streett, E.A. Oma, and R.H. Goodwin. 1986. Ultrastructure of an isolate of *Rickettsiella* from the African grasshopper *Zonocerus variegatus*. J. Invertebr. Pathol. 47:203-213.

Humber, R.A. and W.A. Ramoska. 1986. Variations in Entomophthoralean life cycles: practical implications, p. 190-193. *In:* R.A. Samson, J.M. Vlak, and D. Peters (eds.) Fundamental and applied aspects of invertebrate pathology. Wageningen, The Netherlands.

Ignoffo, C. 1973. Effects of entomopathogens on vertebrates. Ann. N.Y. Acad. Sci. 217:141-172.

Jutila, J.W., J.E. Henry, R.L. Anacker, and W.R. Brown. 1970. Some properties of a crystalline-array virus (CAV) isolated from the grasshopper *Melanoplus bivattatus* (Say) (Orthoptera: Acrididae). J. Invertebr. Pathol. 15:225-231.

Langridge, W.H.R., E.A. Oma, and J.E. Henry. 1983. Characterization of the DNA and structural proteins of entomopoxviruses from *Melanoplus sanguinipes*, *Arphia conspersa*, and *Phoetaliotes nebrascensis* (Orthoptera). J. Invertebr. Pathol. 42:327-333.

Langridge, W.H.R. 1984. Detection of DNA base sequence homology between entomopoxviruses isolated from Lepidoptera and Orthoptera. J. Invertebr. Pathol. 43:41-46.

Longworth, J.F., and P.D. Scotti. 1978. Identification of nonoccluded viruses of invertebrates, p. 75-85. *In:* M.D. Summers and C.Y. Kawanishi (eds.) Viral pesticides: present knowledge and potential effects on public and environmental health. U.S. Environmental Protection Agency.

MacLeod, D.M., D. Tyrrell, and M.A. Welton. 1980. Isolation and growth of the grasshopper pathogen, *Entomophthora grylli*. J. Invertebr. Pathol. 36:85-89.

Matthews, R.E.F. 1979. Classification and nomenclature of viruses. Intervirology 12:160-164.

Nelson, D.R., W.D. Valovage, and R.D. Frye. 1982. Infection of grasshoppers with *Entomophaga* (=*Entomophthora*) *grylli* by injection of germinating resting spores. J. Invertebr. Pathol. 39:416-418.

Oma, E.A., and J.E. Henry. 1986. Host relationships of entomopoxviruses from grasshoppers. Proc. Grasshopper Symp., Bismarck, ND. (in press).

Onsager, J.A., J.E. Henry, R.N. Foster, and R.T. Staten. 1980. Acceptance of wheat bran bait by species of rangeland grasshoppers. J. Econ. Entomol. 73:548-551.

Pickford, R., and P.W. Riegert. 1964. The fungous disease caused by *Entomophthora grylli* Fres., and its effects on grasshopper populations in Saskatchewan in 1963. Can. Entomol. 96:1158-1166.

Reinganum, C., G.T. O'Loughlin, and T.W. Hogan. 1970. A nonoccluded virus of the field crickets *Teleogryllus occeanicus* and *T. commodus* (Orthoptera: Gryllidae). J. Invertebr. Pathol. 16:214-220.

Reinganum, C., S.J. Gagen, S.B. Sexton, and H.P. Vellacott. 1981. A survey for pathogens of the black field cricket, *Teleogryllus commodus*, in the western district of Victoria, Australia. J. Invertebr. Pathol. 38:153-160.

Sergent, E., and A. L'Heritier. 1914. Essai de destruction des sauterelles en Algerie par le *Coccobacillus acridiorum* de D'Herelle. Ann. Inst. Past. 28:408-419.

Sergent, E. 1916. Campagne d'experimentation de la methode bioloque contre les *Schistocerca peregrina* dans la vallee de la hauts Tafna, comune miste de Sebdou (department d'Oran). Existence d'une epizootie autochtone vaccinante (Mai-Juin-Juilet 1915). Ann. Inst. Past. 30:209-224.

Soper, R.S., B. May, and B. Martinell. 1983. *Entomophaga grylli* enzyme polymorphism as a technique for pathotype identification. Environ. Entomol. 12:720-723.

Streett, D.A., E.A. Oma, and J.E. Henry. 1986. Characterization of the DNA from Orthopteran entomopoxviruses, p. 408. *In:* R. Samson, J. Vlak, and D. Peters (eds.) Fundamental and applied aspects of invertebrate pathology. Wageningen, The Netherlands.

Tanada, Y. 1963. Epizootiology of infectious diseases, p. 423-475. *In:* E. Steinhaus (ed.) Insect pathology an advanced treatise. Vol. 2. Academic Press, New York.

Vago, C., and G. Meynadier. 1965. Une rickettsiose chez le criquet pèlerin (*Schistocerca gregaria* Forsk.). Entomophaga 10:307-310.

Veen, K.H. 1966. Oral infection of second-instar nymphs of *Schistocerca gregaria* by *Metarrhizium anisopliae*. J. Invertebr. Pathol. 8:254-256.

218

Veen, K. H.   1968.   Recherches sur la maladie, due à *Metarrhizium anisopliae* chez le criquet pelerin.   Madedelingen Landbouwhogeschool Wageningen 68-5:1-77.

Weiss, E., G.A. Dasch, and K. Chang.   1984.   Genus VIII *Rickettsiella* Phillip.   1956, 267.   p. 713-717.   *In:* N.R. Krieg (ed.) Bergey's manual of systematic bacteriology.   9th ed., Vol. 1. Williams and Wilkens, Baltimore.

15. SAMPLING RANGELAND GRASSHOPPERS

David C. Thompson
Department of Entomology
Colorado State University, Fort Collins, Colorado 80523

Grasshoppers are major agricultural pests in nearly every continent in the world, and are the most important invertebrate pest on rangeland in the western United States (Hewitt 1977). Grasshoppers commonly consume 21-23% of the available range forage (Hewitt and Onsager 1983). In some localities they consume essentially all available forage (Capinera and Sechrist 1982b). Rangeland, however, has a relatively low economic return per unit area; thus, intensive grasshopper monitoring and expensive control programs are usually unwarranted. Millions of hectares in the western United States are surveyed annually for grasshopper densities. Accurate and efficient sampling methods should be a high priority.

The best method for sampling rangeland grasshoppers would give accurate population estimates of nymphal and adult populations, be inexpensive to implement, and be easy to use. Capinera (Chapter 11) and Onsager (Chapter 13) discuss the importance of grasshopper density estimates and species composition estimates in making range management decisions. Most large scale grasshopper management programs do not account for the latter. Several techniques have been used to estimate grasshopper population numbers on a per unit area basis. Boundaries of sample areas, commonly one square foot or one square yard, can be visually delineated and grasshoppers flushed and counted from the defined area (USDA undated). This method, as recommended by the USDA, is quite inconsistent and leads to inaccurate estimation of population densities (Onsager 1977). Other techniques include physically defined sample areas (rings) which allow absolute delineation of sample boundaries (Richards and Waloff 1954, Onsager and Henry 1977), quick traps (Turnbull and Nicholls 1966), night cages (Anderson and Wright 1952), drop cages, and net samplers (Hills 1933, Smalley 1960). Unfortunately, it appears that ease of use and man hours necessary are inversely proportional to the accuracy and precision of the population estimate made from each sampling method.

Estimating species composition has been achieved in several ways. Joern and Pruess (1986) and Pfadt (1982, 1984) have identified grasshopper species as they jump from the observer in the field. This technique can only be used by the average surveyor if densities are low, if species richness within a community is small, and if the species present are morphologically distinct. Sweep net sampling is the most commonly used method to estimate grasshopper species composition. Shortcomings such as differential abilities of individual species and instars to avoid capture, influence of vegetation structure, and influence of environmental conditions on sampling effectiveness have long been recognized (see Southwood [1978] for a review). Capinera and Sechrist (1982a) describe a "flush-capture" technique which controls for much of the variance inherent in sweep net sampling. Ultimately, however, the various cage or net samplers used for density estimation give the most accurate species composition estimates. To facilitate identification, grasshoppers trapped within the cages can be collected individually or collected using some type of suction apparatus (e.g., D-Vac). Inaccurate estimation of the nymphal population remains the major bias between cage estimates and the other methods (Figure 15.1).

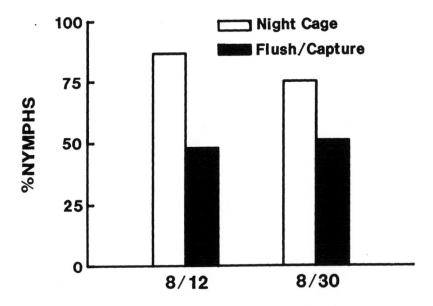

FIG. 15.1 Percentage of sample represented by nymphs (USDA Central Plains Experimental Range, Nunn, CO, pasture 8NC, August 1985).

Currently, the USDA is recommending establishment of "sentinel sites" -- semi-permanent sites where grasshopper densities and species composition are monitored several times each year. Sampling protocol requires the use of ring samples, but optimal sample unit size and number are unknown.

GRASSHOPPER DISPERSION

If it was possible to map the location of each grasshopper living in a specific area at a specific moment in time, a picture of the grasshopper distribution would be apparent. Knowledge of this pattern, sometimes referred to as the "dispersion" of the species, is necessary before any optimal sampling strategies can be explored (Karandinos 1976). Population ecologists generally agree upon three standard distribution types that can be used to describe natural populations: random, regular, and contagious.

In a random population there is an equal probability of an individual occupying any given point within the sampling area, and the presence of one individual does not influence the position of any nearby neighbors. A regular distribution typically occurs when individuals in a population are relatively crowded and move away from one another. Contagious distributions are the most common and also the most complex spatial patterns found in natural systems. There are always definite clumps or patches of individuals in a contagious distribution, but the final dispersion pattern varies considerably. Size of the clumps, distance between clumps, spatial distribution of the clumps, and spatial distribution of individuals within the clumps all interact to provide unique distributions.

Onsager (1977) found that grasshopper complexes in northern mixedgrass prairie were not significantly different from random. Prihar (1983), Anderson (1964), and Anderson and Wright (1952) stated that field observations indicate individual grasshopper species are not random; however, no numerical support for these observations was provided.

DISPERSION OF GRASSHOPPER COMPLEX

Most rangeland pest management programs consider characteristics of the entire grasshopper species complex rather than individual species. The distribution of the grasshopper complex on Colorado shortgrass prairie was defined to facilitate determination of optimal sampling strategies. Details are provided by Thompson (1987); key findings follow. Eight to twenty field sites were monitored each year from 1980-1986. Quantitative and qualitative estimates of grasshopper populations were obtained (Capinera and Sechrist 1982a). Four different ring sizes (0.10, 0.25, 0.58, and 1.0 m^2), fashioned of yellow garden hose, were arranged in transects of 40 to 100 rings, or in a pseudorandomized complete block grid design (100 of each sized ring). In 1980-1983, fields were sampled three times during the season; in 1984-1986, fields were sampled weekly.

The dispersion patterns of these grasshopper populations were analyzed using frequency distributions and various indexes of dispersion. Field samples were summarized into frequency distribution classes and, using chi-square "goodness-of-fit" tests (Cochran 1954, Snedecor and Cochran 1967), were compared to common theoretical distributions (Poisson and negative binomial). The fit of all actual frequency distributions to the Poisson and negative binomial distributions was inconclusive by itself; 18% of all fields fit only the negative binomial, while 10% fit only the Poisson. However, 67% fit both distributions. As these data demonstrate, caution must be observed when interpreting the fit of data to theoretical distributions. The quantitative expression of aggregation through a frequency series may reflect an actual dispersive tendency, or it may simply be a statistical artifact (Waters 1959).

Indexes of dispersion further define the spatial pattern exhibited by an organism (Myers 1978). Many indexes defining animal populations have been described; however, all are derived from various arrangements of the estimate (x) of the true mean (μ), the estimate (s^2) of the true variance (σ^2), and the sample size (Southwood 1978). All of the indexes measure the relative departure of a sample from randomness. The variance/mean ratio, Morisita's index (Morisita 1959, 1962, 1971), and Green's index (Green 1966) all produced identical results: 75% of the fields were random, 19% clumped, and 6% uniform.

Two other indexes, b in Taylor's power law (Taylor 1961, 1971) and the m*-m method (Iwao 1968), are based on regression techniques involving data from many different sites. The slope of the regression line obtained by regressing Lloyd's mean crowding index (Lloyd 1967) against mean density (m*-m), or regressing variance against mean on a log scale (b in power law), is characteristic of the distribution type. Neither index, when calculated from ring estimates, resulted in a slope significantly different from one. Significance is necessary to indicate a departure from randomness.

Quadrat size can have a strong influence on the perception of pattern within a population. Figure 15.2 shows how the perceptions of non-randomness in a grasshopper population change as quadrat size increases; note that randomness decreases and clumpedness increases with an increase in quadrat size. In a truly

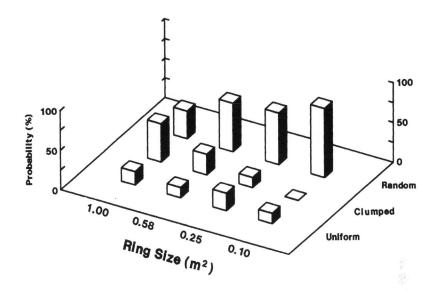

FIG. 15.2 Influence of ring size on the probability of a field sample exhibiting a particular distribution type (random, clumped, or uniform).

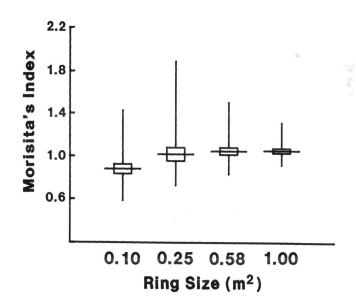

FIG. 15.3 Average Morisita's index (n=23) resulting from four different ring sizes. Means, standard errors, and ranges are given by horizontal line, square, and vertical line, respectively.

random population, which can be described by the Poisson distribution, different quadrat sizes should yield identical distribution estimates. Morisita (1959) investigated changes in I_δ with changes in quadrat size. The gradual increase of Morisita's index, averaged over 23 fields and using four different quadrat sizes (Figure 15.3), suggests that grasshopper complexes may be slightly clumped and that these clumps are larger than one square meter. Intuitively, this is not very surprising; it is commonly theorized that grasshoppers are more abundant where food resources (Joern 1986), egg laying sites (Davis and Wadley 1949), and environmental conditions (Hewitt 1977, Anderson et al. 1979) are favorable. Yet, is knowledge of the average clump size within one field or even on a larger scale (townships, sections, or ranges), really important? The latter information would be useful but its acquisition is not realistic. From a research or management viewpoint, clump sizes larger than a few square meters are unmeasurable. Van Horn (1972) sampled with 2 m^2 quadrats (drop cages), but sample units larger than this are not feasible. Apparent clumpedness within the grasshopper complex may be undefinable because of the small quadrat size relative to the clump size (Elliott 1977).

Most research and large-scale surveys use sampling units which are less than 1 m^2 in area, resulting in grasshopper populations which are functionally random. This assumption of randomness will simplify optimization of a sampling strategy for grasshopper complexes.

DISPERSION OF INDIVIDUAL SPECIES

The dispersion of individual species of grasshoppers within the complex can vary considerably. Information on the spatial distribution of specific grasshopper species is minimal. From a management perspective, if the most damaging species within a population are clumped, a much more intensive survey may be warranted.

Night cages were used to monitor the species composition and distribution patterns of individual grasshopper species at the Central Plains Experimental Range, Nunn, Colorado. Figure 15.4 presents the results at this site. *Phlibostroma quadrimaculatum* (Thomas), *Opeia obscura* (Thomas), and *Arphia pseudonietana* (Thomas) comprised over 95% of the population on this date. The bars represent a standardized dispersion index in which values start out at zero (solid line); those which are positive become increasingly more clumped as the index increases. The one negative value indicates that the species is tending more toward a uniform distribution. A species is not considered to be different

from random until it exceeds the dashed line (a value of 100).
Note that only *Phlibostroma quadrimaculatum* is significantly
clumped and that all others and the complex as a whole can be
considered random.

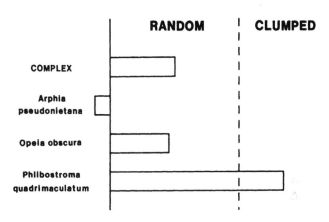

FIG. 15.4 Dispersion indexes for individual species and grasshopper
complex (CPER 8NC, 25 July 1985).

OPTIMAL SAMPLING PLAN

A standard technique to estimate sample number in a random
population uses the equation:

$$\text{sample number} = \frac{t^2}{x \cdot D^2}$$

where x is the estimate of the population mean, t is Student's t
value (1.96 for $P = 0.05$) and 'D' is the "reliability" of estimation
(normally 0.10). The number of samples necessary for a reliable
estimate increases as population mean decreases and as reliability
approaches zero.

What exactly is reliability? Reliability is subdivided into two
parts: precision and accuracy. Precision is the repeatability of a
sample strategy. In other words, if several samples are taken from
the same field using the same technique, how much will the
results differ? Accuracy is how close an estimate is to the true
population mean. These concepts were quantified for rangeland
grasshopper sampling.

Precision was estimated by determining the number of
samples necessary to stabilize the variance of the mean in each
field. Random subsamples of individual ring counts were drawn
from the raw data. Five separate running means were calculated
and averaged for each field (Table 15.1). The largest rings were
most precise because, on average, fewer rings were needed to

TABLE 15.1
Estimated sample size and respective densities (number/m^2) using various ring sizes (CPER 22NE, August 1984).

Ring Size	Number of Rings	Density Estimate
0.10 m^2	20.6	7.90
0.25 m^2	12.8	6.52
0.50 m^2	10.6	5.14
1.00 m^2	8.0	4.49

estimate the same population. The number of rings and corresponding means from all sites and ring sizes were plotted (Figure 15.5). The number of rings necessary for a precise estimate decreases as density increases for all ring sizes. Using forty 0.10 m^2 rings, populations of five or more grasshoppers per m^2 can be precisely estimated; with forty 0.25 m^2 rings, populations of 1.8 or more grasshoppers per m^2 can be precisely estimated and so on.

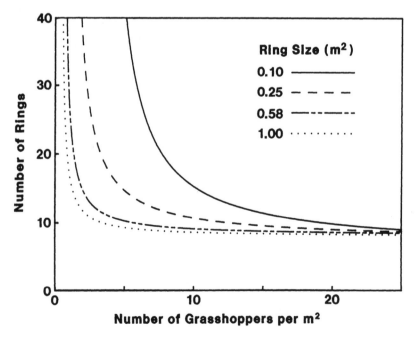

FIG. 15.5   Influence of ring size on sampling precision.

Precision increased as ring size increased. However, the density estimates from the same population sampled at the same time go down as ring size increases (Table 15.1). This represents a decrease in accuracy. Fifty night cages (0.25 m^2) incorporated into a pseudorandom grid of various sized rings provided density estimates which were considered 100% accurate and the best estimate of the actual population mean obtainable. Figure 15.6 compares accuracy among the four ring sizes and night cages as grasshopper density increases. The gray area delineates the critical area about the mean estimate obtained using night cages. It represents the area within which differences between mean estimates are found to be non-significant using two-sample t-tests (Snedecor and Cochran 1967). The farther estimates depart from the critical area, the more unreliable they become. At low densities, the difference in accuracy between sampling methods is minimal; however, as density increases the difference in accuracy becomes more pronounced. The night cage counts consistently provided higher estimates of grasshopper densities than even the smallest ring size, indicating that the ring counts are underestimating the true density.

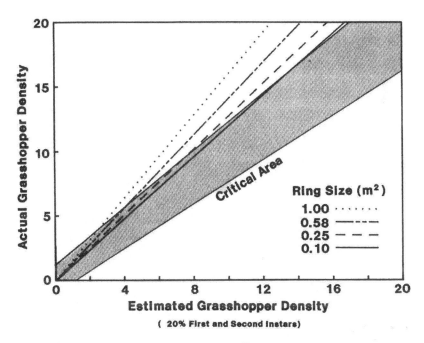

FIG. 15.6 Influence of ring size on sampling accuracy.

Error associated with the accuracy of ring samples appears to be due, in part, to the fact that many early instar grasshoppers are being overlooked. A significant correlation (r=.635) exists between percent of first instar grasshoppers in the population and the difference between night cage and ring estimates of that population. Although data spanning the entire range of nymphal percentages and grasshopper densities is unknown, Figure 15.7 graphically depicts the range of densities under which specific combinations of ring sizes and nymphal populations should provide estimates within the critical area about the actual population mean.

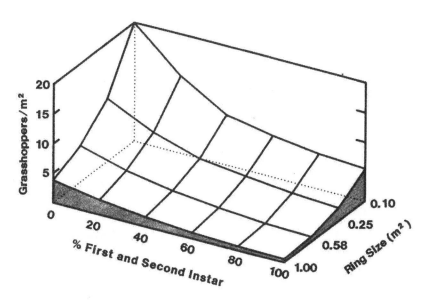

FIG. 15.7 Maximum densities which can be estimated with various combinations of nymphal percentages and ring size.

Ambient temperature also influences ring estimates. At densities of 1 to 5 grasshoppers/m^2, three temperature regimes (<15.5°C, 15.5-26.7°C, and >26.7°C) produced statistically different population estimates. Continued definition of errors inherent in grasshopper density estimation techniques, such as the influence of wind speed and cloud cover, will increase the value of sampling methods in range management decisions.

By combining accuracy and precision, the range of densities at which different ring sizes are most appropriate can be found. The lower bound for each ring size is determined by precision; the upper bound is based on accuracy (Table 15.2). For example, 0.10 m^2 rings cannot accurately estimate grasshopper populations over 13/m^2 (Figure 15.6), and twenty 0.10 m^2 rings cannot precisely estimate populations less than 7.7/m^2 (Figure 15.5). The acceptable accuracy, number of rings, or size of rings can be varied to obtain a sampling program which meets the specific objectives of individual projects.

TABLE 15.2
Range of grasshopper densities (number/m^2) between which various combinations of ring size and number are most appropriate.

Ring Size (m^2)	Number of Rings		
	Twenty	Forty	Sixty
0.10	7.7 to 13.0	5.0 to 13.0	3.7 to 13.0
0.25	3.2 to 7.0	1.8 to 7.0	1.1 to 7.0
0.50	1.2 to 4.0	0.7 to 4.0	0.5 to 4.0
1.00	0.8 to 2.5	0.4 to 2.5	0.3 to 2.5

MANAGEMENT CONSIDERATIONS

In many instances, grasshopper densities are well above the maximum values presented in Table 15.2 and in Figure 15.7. Grasshopper densities will be underestimated when numbers exceed these maximum values. This unavoidable consequence of the ring sampling technique should be considered. Typically, control tactics are only contemplated when populations are above these maximum levels. The loss in accuracy will have little impact on most final management decisions.

Samples taken when evaluating control programs or testing new insecticides and/or formulations commonly involve a pre-treatment count and several post-treatment counts. Mortality varies considerably; however, values above 90% are common (Johnson et al. 1986, Ewen et al. 1984, Onsager 1978). Large reductions in grasshopper abundance directly influence sampling strategies. Trumble (1985) suggests different sampling procedures for spider mites on strawberries before and after insecticide application. A similar strategy could be utilized with grasshoppers. Differentiating between insecticide-treated and untreated plots may be statistically easy because the large differences render tests less sensitive to variance within a site.

If, however, interest includes differentiating between treatments (e.g., *Nosema* vs. controls, or different insecticides) a dynamic sampling scheme should be used. This could involve pre-counts with 60 0.10 m² rings, when densities are greater than 10/m², and subsequent post-counts with 40 1.00 m² rings, when densities are less than 2.5/m². The goal of dynamic sampling is to maximize reliability and minimize variance at specific sites and over various grasshopper densities.

CONCLUSIONS

By comparing results of frequency distributions and several indexes of dispersion it appears that grasshopper complexes on shortgrass rangeland can be considered functionally random. Density levels of 0 to 25 grasshoppers/m² showed only a slight tendency toward aggregation at the higher levels; this may be an artifact caused by the decreased reliability in sampling at those levels. Research with night cages has shown that individual grasshopper species may have different distribution patterns, and that ring sampling underestimates populations with a large percentage of early instar nymphs.

An optimal sampling strategy should be dynamic, varying with the density of grasshoppers and, most importantly, with the objectives of an individual project. If densities are unknown prior to sampling, forty 0.25 m² rings will give good population estimates over most grasshopper densities of concern. However, objectives should dictate sampling strategies. As with any sampling strategy, the amount of time and labor that can be put forth is invariably limited by money or other resources.

ACKNOWLEDGMENTS

This work was supported by the Colorado Agricultural Experiment Station and the Cooperative National Plant Pest Survey and Detection Program. I would especially like to thank Dr. John L. Capinera for his support, encouragement, and guidance during the course of this research.

REFERENCES

Anderson, N.L., and J.C. Wright. 1952. Grasshopper investigations on Montana rangelands. Montana Agr. Exp. Sta. Bull. 486.
Anderson, N.L. 1964. Some relationships between grasshoppers and vegetation. Ann. Entomol. Soc. Am. 57:736-742.

Anderson, R.V., C.R. Tracy, and Z. Abramsky. 1979. Habitat selection in two species of short-horned grasshoppers: the role of thermal and hydric stress. Oecologia 38:359-374.

Capinera, J.L., and T.S. Sechrist. 1982a. Grasshopper (Acrididae) - host plant associations: response of grasshopper populations to cattle grazing intensity. Can. Entomol. 114:1055-1062.

Capinera, J.L., and T.S. Sechrist. 1982b. Grasshoppers (Acrididae) of Colorado: identification, biology and management. Colorado State Univ. Agr. Exp. Sta. Bull. 584S.

Cochran, W.G. 1954. Some methods for strengthening the common chi-square tests. Biometrics 4:417-51.

Davis, E.G., and F.M. Wadley. 1949. Grasshopper egg-pod distribution in the northern great plains and its relation to egg-survey methods. USDA Circ. 816.

Elliott, J.M. 1977. Some methods for statistical analysis of samples of benthic invertebrates. Freshwater Biol. Assoc. Ambleside, Great Britain.

Ewen, A.B., M.K. Mukerji, and C.F. Hinks. 1984. Effect of temperature on the toxicity of cypermethrin to nymphs of the migratory grasshopper, *Melanoplus sanguinipes* (Orthoptera: Acrididae). Can. Entomol. 116:1153-1156.

Green, R.H. 1966. Measurement of non-randomness in spatial distributions. Res. Popul. Ecol. 8:1-7.

Hewitt, G.B. 1977. Review of forage losses caused by rangeland grasshoppers. USDA Misc. Publ. 1348.

Hewitt, G.B., and J.A. Onsager. 1983. Control of grasshoppers on rangeland in the United States - a perspective. J. Range Manage. 36:202-207.

Hills, O.A. 1933. A new method for collecting samples of insect populations. J. Econ. Entomol. 26:906-910.

Iwao, S. 1968. A new regression method for analyzing the aggregation pattern of animal populations. Res. Popul. Ecol. 10:1-20.

Joern, A. 1986. Resource partitioning by grasshopper species from grassland communities, p.75-100. *In:* D. Nickle (ed.) Proceedings 4th triennial meeting, Pan American Acridological Society.

Joern, A., and K.P. Pruess. 1986. Temporal constancy in grasshopper assemblies (Orthoptera: Acrididae). Ecol. Entomol. 11:379-385.

Johnson, D.L., B.D. Hill, C.F. Hinks, and G.B. Schaalje. 1986. Aerial application of the pyrethroid deltamethrin for grasshopper (Orthoptera: Acrididae) control. J. Econ. Entomol. 79:181-188.

Karandinos, M.G. 1976. Optimum sample size and comments on some published formulae. Bull. Entomol. Soc. Am. 22:417-421.

Lloyd, M. 1967. 'Mean crowding'. J. Anim. Ecol. 36:1-30.

232

Morisita, M. 1959. Measuring of the dispersion of individuals and analysis of the distributional patterns. Mem. Fac. Sci. Kyusha Univ. Ser. E. 2:215-35.

Morisita, M. 1962. I_δ-index, a measure of dispersion of individuals. Res. Popul. Ecol. 4:1-7.

Morisita, M. 1971. Composition of the I_δ index. Res. Popul. Ecol. 13:1-27.

Myers, J.H. 1978. Selecting a measure of dispersion. Environ. Entomol. 7:619-621.

Onsager, J.A. 1977. Comparison of five methods for estimating density of rangeland grasshoppers. J. Econ. Entomol. 70:187-190.

Onsager, J.A. 1978. Efficacy of carbaryl applied to different life stages of rangeland grasshoppers. J. Econ. Entomol. 71:269-273.

Onsager, J.A., and J.E. Henry. 1977. A method for estimating the density of rangeland grasshoppers (Orthoptera, Acrididae) in experimental plots. Acrida 6:231-237.

Pfadt, R.F. 1982. Density and diversity of grasshoppers (Orthoptera: Acrididae) in an outbreak on Arizona rangeland. Environ. Entomol. 11:690-694.

Pfadt, R.F. 1984. Species richness, density, and diversity of grasshoppers (Orthoptera: Acrididae) in a habitat of the mixed grass prairie. Can. Entomol. 116:703-709.

Prihar, D.R. 1983. Abundance and damage of grasshoppers (Acridoidea) in grazingland vegetation in the Indian desert. Z. Angew. Entomol. 96:3-9.

Richards, O.W., and N. Waloff. 1954. Studies on the biology and population dynamics of British grasshoppers. Anti-locust Bull. 17.

Smalley, A.E. 1960. Energy flow of a salt marsh grasshopper population. Ecology 41:672-677.

Snedecor, G.W., and W.G. Cochran. 1967. Statistical methods, 6th ed. Iowa State Univ. Press, Ames.

Southwood, T.R.E. 1978. Ecological methods with particular reference to the study of insect populations, 2nd ed. Chapman and Hall, London.

Taylor, L.R. 1961. Aggregation, variance and the mean. Nature 189:732-35.

Taylor, L.R. 1971. Aggregation as a species characteristic. Stat. Ecol. 1:357-77.

Thompson, D.C. 1987. Distributional properties and optimal sampling of shortgrass rangeland grasshoppers. Unpublished Ph.D. Dissertation, Colorado State Univ.

Trumble, J.T. 1985. Implications of changes in arthropod distribution following chemical application. Res. Popul. Ecol. 27:277-285.

Turnbull, A.L., and C.F. Nicholls. 1966. A "quick trap" for area sampling of arthropods in grassland communities. J. Econ. Entomol. 59:1100-1104.

USDA. undated. Grasshopper survey, a species field guide. USDA, APHIS, PPQ.

Van Horn, D.H. 1972. Grasshopper population numbers and biomass dynamics on the Pawnee site from fall of 1968 through 1970. USIBP Technical Report 148.

Waters, W.E. 1959. A quantitative measure of aggregation in insects. J. Econ. Entomol. 52:1180-1184.

16.   ECOLOGY OF THE RANGE CATERPILLAR, Hemileuca oliviae COCKERELL

Peter B. Shaw, David B. Richman, John C. Owens, and Ellis W. Huddleston
Department of Entomology, Plant Pathology, and Weed Science
New Mexico State University, Las Cruces, New Mexico 88003

The range caterpillar, *Hemileuca oliviae* Cockerell (Lepidoptera: Saturniidae), is a major pest on rangeland grasses in New Mexico.   Larvae consume grass which would otherwise be available for sheep and cattle.

The range caterpillar was the object of considerable research early in this century, when it was first noted as a pest in New Mexico.   Ainslie (1910) presented a detailed study of the range caterpillar.   Subsequent work through the early 1930's is reviewed by Watts and Everett (1976).   Problems with the range caterpillar and research on the insect subsided after the 1930's.   In the 1960's range caterpillar outbreaks were again observed in New Mexico and in the 1970's intensive research was initiated.

Ecology has been defined as "the scientific study of the interactions that determine the distribution and abundance of organisms" (Krebs 1972).   The interactions determining the abundance and distribution of the range caterpillar could be categorized in a number of different ways.   For the purposes of this paper we recognize three broad categories of factors with which range caterpillars interact: the abiotic environment, plants, and natural enemies.

In addition to describing ecological interactions determining the distribution of the range caterpillar, we will summarize the results of recent research on distribution.

ABIOTIC ENVIRONMENT

Abiotic factors can affect the abundance and distribution of the range caterpillar on a broad scale. Hansen et al. (1982b) noted one case in which a severe rain storm killed 48.6% of the first instar larvae in an area.   In the same study the authors also noted that only 21.6% of the eggs hatched at one site, which may have been due to the poor nutrition of the females as larvae. The poor nutrition of the larvae could have been due to a drought at that same site during the previous summer. Thus, a low rate of egg hatch could be attributed to an abiotic condition--low rainfall. Fritz et al. (1987) described a case in which the geographic pattern of rainfall apparently drove a local population extinct. Fritz et al. (1987) also suggested that range grass fires caused by lightning can reduce range caterpillar populations.

The abiotic environment also can act through subtle effects on the behavior and physiology of the range caterpillar. Temperature affects the insect in several ways. To avoid elevated temperatures range caterpillar exhibits three behavioral responses (Capinera et al. 1980): 1) keeping the substrate between incident sunlight and the ventral surface of the caterpillar; 2) seeking elevated, vertical positions; and 3) inactivity at mid-day. The authors further observed that feeding by the range caterpillar occurs when temperatures are moderate, especially in the morning and early evening. Hansen et al. (1984b) stated that older larvae fed more often during these periods, but that the relationship was not statistically significant. Capinera et al. (1980) also found that blackening a larva caused an elevation in its body temperature, and suggested that the lighter color of later instars reduces the absorption of solar energy.

Temperature has also been shown to affect adult behavior, in that the occurrence of calling behavior in adult females appears to depend on ambient temperature and not to be an endogenous circadian rhythm (Smith and Turner 1979).

Water relationships for the egg and early instars of the range caterpillar have been examined in several studies. Everett and Turner (1979) found that larval eclosion was highly sensitive to humidity, with higher humidity increasing the rate of eclosion. They postulated the existence of "very sensitive humidity sensors" in the larvae. In a related study, Capinera et al. (1981) concluded that, although higher vapor densities do stimulate more rapid eclosion, "relatively low levels of humidity can stimulate eclosion". They felt that the characteristics of the egg chorion could explain the results observed, without need to postulate the existence of "humidity sensors". Compared to other species' eggs, Capinera et al. (1981) found that range caterpillar eggs were quite resistant to desiccation. They suggested that this resistance to desiccation was due to the presence of a thick waxy layer on the eggs and the lack of aeropyles. Capinera (1980) also investigated the relationship between aggregation and osmoregulation, finding that there was no significant difference in water loss between isolated larvae and larvae in groups of 10.

Dispersal by older range caterpillar larvae appears to be

affected by temperature, wind speed, humidity, and other factors in a complex manner (Hansen 1984a,b).

PLANTS

Feeding

One of the most interesting aspects of range caterpillar ecological relationships is its feeding preference. Range caterpillars feed almost exclusively on grasses (Hansen et al. 1984b), and much experimental evidence (Capinera 1978, Capinera et al. 1983b, Shaw et al. 1987a) suggests a preference by the range caterpillar for C_4 over C_3 grasses. As Capinera (1978) points out, this finding appears to contradict Caswell et al.'s (1973) hypothesis that C_3 plants provide a better source of food for herbivores than C_4 plants, owing to the rather indigestible nature of C_4 plants with their numerous bundle sheath cells. Based on the information available in 1978, the preference of the range caterpillar for C_4 plants did indeed present a paradox.

A study by Riley et al. (1984) appears to shed some light on the problem. Riley et al. (1984) studied the energetics of the range caterpillar and found high maintenance costs associated with eating blue grama grass, *Bouteloua gracilis* (H.B.K.) Lag. They suggested that the high maintenance costs may have been due to the rather indigestible nature of blue grama (a C_4 plant) and, further, that high maintenance costs were associated with high consumption rates by the range caterpillar. Range caterpillar appears to compensate for the poor nutritive quality of blue grama by eating large amounts; large quantities of frass are also produced. Riley et al. (1984) reported that the average daily production of litter and frass by the range caterpillar in the last three instars was nearly 6 times that reported for various grasshopper species. The idea that the range caterpillar's high consumption rate compensates for the poor nutrient quality of C_4 grasses it eats is supported by the fact that Riley et al. (1984) reported much higher consumption rates than Schowalter et al. (1977), who fed the range caterpillar on an artificial diet, which presumably was more nutritious than the blue grama.

As suggested by Riley et al. (1984), the high consumption rate of the range caterpillar has implications for the rangeland ecosystem. The production of such prodigious amounts of frass and litter would mean that range caterpillars could play a major role in nutrient cycling in the range grassland. Such a role would correspond well with the suggestion by Caswell et al. (1973) that detritus should assume great importance in communities dominated by a C_4 species.

The possibility that the range caterpillar could compensate for the relatively poor nutritive quality of C_4 plants with high consumption rates means that there is no inherent contradiction between Capinera's (1978) finding of a preference for C_4 grasses and the suggestion of Caswell et al. (1973) that C_4 plants are of a poorer nutritive quality for insects. The question really is: why should range caterpillar eat C_4 plants if this activity requires greater maintenance costs and, hence, greater consumption?

At least two evolutionary hypotheses can be offered to explain the preference of range caterpillar for C_4 plants. First, if C_3 plants are of higher nutritive value to most insects than C_4 plants, then C_3 plants would be subject to intense herbivory by insects and be a scarce resource. As suggested by Capinera et al. (1983b), any insect that could successfully utilize C_4 plants would have an abundant resource to exploit. Thus, competition over evolutionary time could explain the preference of the range caterpillar for C_4 plants.

A second evolutionary hypothesis for the preference of range caterpillar for C_4 plants relates to the probable center of origin of the two species. Plants that photosynthesize by the C_4-dicarboxylic acid pathway are of tropical or subtropical origin (Capinera et al. 1983b). Nearly all the described species of *Hemileuca* are known from Mexico (Ferguson 1972); thus, it would not be surprising for species in this genus to have evolved the ability to utilize C_4 plants and even to exhibit a preference for them. As noted in Fritz et al. (1986), a very common host of the range caterpillar, blue grama grass, belongs to a genus which has its greatest diversity in Mexico (Gould 1979).

Capinera et al. (1983b) studied proximate mechanisms involved in the preference of the range caterpillar for C_4 plants. They concluded that astringency, a measure of the tannin content of plants, explained the greatest amount of variance of any of the plant characteristics considered. They further noted that C_3 plants had significantly higher levels of astringency and total tannin than C_4 plants.

In their discussion of tannins and astringency, Capinera et al. (1983b) suggested that the low level of tannin in blue grama is "somewhat surprising", considering the fact that blue grama is an apparent resource (Rhoades and Cates 1976). As such, blue grama would presumably have been under selective pressure to evolve anti-herbivory defenses, such as high tannin levels. This assumes, however, that blue grama would have been subject to intense consumption by herbivores, such as the range caterpillar, where both evolved. If, as suggested, both species evolved in the area of Mexico, this may not have been true. Recent evidence suggests that the range caterpillar in Mexico is kept at relatively low levels by the action of parasites and other factors (Fritz et al. 1986, 1987).

Pupation and Oviposition

Experiments by Bellows et al. (1983b) showed that *Hemileuca oliviae* chooses pupation and oviposition sites on the basis of plant form. Pupation sites are most often plant species which are stiff and upright in form, such as *Aristida* spp., *Panicum* spp., *Stipa neomexicana* (Thurb.) Scribn., and *Gutierrezia sarothrae* (Pursh) B&R, and oviposition sites most often include species with a "strong, upright flowering stem", such as *Aristida* spp. and *Stipa neomexicana*. At one of their four sites, Fritz et al. (1987) observed a strong preference for pupation on *Aristida* spp. by range caterpillars. They also noticed a strong larval preference for *Aristida* spp. at the same site, but did not specify the stage of the larvae.

Plant Productivity and Growth Responses

Plant productivity has been cited as one of the most important factors affecting range caterpillar density. In a study which included a comparison of range caterpillar populations at two field sites, Bellows et al. (1982a) suggested that the higher mortality observed among fourth and fifth instar larvae at one of the sites could well have been due to lower grass productivity (as well as the harmful effects of lower ground cover). Beavis et al. (1981) also noted that range caterpillars have higher densities and are more developmentally advanced in swales than on upper slopes and ridges. They suggested that higher plant productivity in the swales could account for this difference (along with the possibility of greater ground cover affording protection from hot soil surfaces). Bellows et al. (1984) observed that adult female *H. oliviae* generally expend little effort on dispersal, and that this was probably due to the much greater importance of fluctuations in grass productivity from one year to the next than variations over broad geographic ranges (meaning that *H. oliviae* should devote its energy primarily to reproduction rather than dispersal).

The growth responses of plants could also affect a local population of range caterpillars over time, as suggested in a modeling study by Capinera et al. (1983a). The results of this study indicated that, for low to moderate densities of range caterpillars feeding on blue grama, compensatory plant growth would reduce the damage inflicted by the range caterpillar on the grass. In contrast, the model showed that a high level of herbivory by the range caterpillar resulted in greater reduction of , above-ground standing plants "than might be expected". In essence, the differing responses of plants to high vs. low and moderate levels of range caterpillars could affect the future food supply of the range caterpillar population.

Habitat

The local habitat of the rangeland, as characterized by the dominant grass species present, can affect the abundance and distribution of the range caterpillar. A study by Beavis et al. (1982) showed that the habitat of the rangeland can either be characterized as blue grama habitat or feathergrass habitat. Furthermore, they showed that range caterpillar densities were greater in blue grama habitats than in feathergrass habitats for all developmental stages but eggs, and that for the later stages of the range caterpillar, greater losses occurred in feathergrass than in blue grama habitats. Beavis et al. (1982) suggested that the differences observed could have been due to the presence of more satisfactory diet in the blue grama habitat, less exposure to high surface temperature due to higher ground cover in the blue grama habitat, or possibly to less exposure to predators in the blue grama habitat. The category of habitat is not neatly separated from other factors, however, because Mexican feathergrass is a C_3 plant and blue grama is a C_4 plant.

NATURAL ENEMIES

Several studies point to the substantial impact of insect and mammalian natural enemies on range caterpillar abundance. Bellows et al. (1982a) found that range caterpillar experienced much higher mortality in instars 1 to 3 than in later instars. They suggested that this high early mortality could have been due to insect predators. Shaw et al. (1987b) explored this hypothesis in field experiments; their work revealed a number of potential arthropod predators and definitely showed that ants, *Crematogaster punctulata* Emery, attacked first instar range caterpillars. The ant predation in Shaw et al.'s (1987b) experiments is interesting in light of Capinera et al.'s (1980) study showing that aggregation by early instars of the range caterpillar can reduce ant predation. The phenomenon of high mortality in early instars of the range caterpillar was incorporated into a pest management model by Bellows et al. (1983a). Their model indicated that early spraying would be most effective, but that "early" could be over a 40 day period. If we assume that predators are an important factor in causing the early mortality, then it may be beneficial to delay any spraying with a broad spectrum insecticide until the end of the 40 day period, so as not to harm arthropod predators. Insect predators also prey upon later instars of range caterpillar; Hansen et al (1982a) observed carabid beetles, *Pasimachus* sp., preying on fifth and sixth instars of the range caterpillar.

The activities of mammals appear to influence strongly the distribution and abundance of range caterpillar. Bellows et al. (1982c) gathered compelling evidence that predation by woodland rodents, especially the mouse *Peromyscus truei* (Schufelt), on range caterpillar eggs and pupae restricts the range caterpillar to grasslands. Also, the density of eggs can be substantially reduced through accidental destruction by sheep foraging during the winter. Bellows et al. (1982b) found that areas grazed by sheep experienced a 74.5% loss of eggs during the winter, while adjacent, protected areas only had a 38.3% loss. Fritz et al. (1987) surmised that overgrazing on Mexican rangeland sharply reduces range caterpillar populations. The accidental destruction of early instars of the range caterpillar by large herbivores, such as bison, suggests the selective advantage of one of the behavioral traits of early instars of the range caterpillar--aggregation. Capinera (1980) hypothesized that the aggregation of early instars could serve to make them more visible to large herbivores, and, hence, less likely to be eaten, since the larvae possess urticating spines. While mammals appear to play a major role in reducing range caterpillar numbers, there is little evidence lizards and birds exert significant influence (Bellows et al. 1982c).

The importance of the range caterpillar egg parasite, *Anastatus semiflavidus* Gahan, is problematic. Historically the parasitoid is credited with high levels of parasitization (Caffrey 1921, Wildermuth and Frankenfeld 1933). Watts and Everett (1976), on the other hand, reported never observing parasitization rates of eggs higher than two or three percent. In contrast, Hansen (1978) reported rates of parasitization of egg masses as high as 85% and rates of egg parasitization as high as 55%. The life history characteristics of *A. semiflavidus* have been studied in detail by Mendel et al. (1987). This study provided evidence that the parasitoid was better adapted to high temperatures than the range caterpillar. Mendel et al. (1987) also discussed the possible influence of low autumn temperatures on parasitoid activity.

Under certain conditions larval and pupal parasitoids can substantially reduce range caterpillar populations. Hansen et al. (1982b) noted that parasitization by tachinids can reach levels of 50% of sixth instar larvae. Table 16.1 lists known parasitoids of the range caterpillar.

TABLE 16.1
Known parasitoids of the range caterpillar in New Mexico and
Chihuahua, Mexico.   Sources: Arnaud (1978), Fritz et al. (1986),
Krombein et al. (1979), Watts and Everett (1976).

Stage	Parasitoid	New Mexico	Chihuahua
Egg			
	Eupelmidae		
	Anastatus semiflavidus Gahan	X	X
	Torymidae		
	Microdontomerus n. sp.*		X
	Eulophidae		
	Tetrastichus sp.*		X
Larva			
	Tachinidae		
	Exorista mella (Walker)**	X	
	Compsilura concinnata (Meigen)**	X	
	Lespesia archippivora (Riley)**	X	
	Euphorocera claripennis (Macquart)**	X	
	Leschenaultia fulvipes (Bigot)		X
	Chetogena sp.		X
	Braconidae		
	Apanteles electrae Viereck**	X	
	Ichneumonidae		
	Enicospilus sp.		X
Pupa			
	Chalcididae		
	Spilochalcis mariae (Riley)**		X
	S. phais Burks	X	
	Brachymeria ovata (Say)**	X	X
	Pteromalidae		
	Pteromalus hemileucae (Gahan)	X	
	Ichneumonidae		
	Coccygomimus sanguinipes (Viereck)**	X	
	Itoplectis viduata (Gravenhurst)**	X	
	I. conquisitor (Say)**	X	

*May be hyperparasitic
**Known to have a broad host range

GEOGRAPHICAL DISTRIBUTION

The Range Caterpillar in Chihuahua, Mexico

In 1980 E.W. Huddleston collected larvae of the range caterpillar near Rancho la Campana in Chihuahua, Mexico (Watts et al. 1982). Studies on the distribution of the range caterpillar in Mexico (Fritz et al. 1987), the parasitoids and predators of the Mexican population (Fritz et al. 1986), and a comparison of the Mexican and New Mexican range caterpillars based on biochemical and morphometric data (Dubach et al. 1987), indicate that the populations are conspecific, but have somewhat different natural enemies (Table 16.1). As in the northern populations, the Mexican population overwinters in the egg stage. The life cycle is, however, slightly shifted in time, with egg hatch, larval development, pupation, and adult emergence delayed 1 to 2 months.

Fritz et al. (1986, 1987) found no evidence that the range caterpillar had ever been a pest in Mexico. In Mexico the egg parasite *A. semiflavidus* destroyed more than 50% of the eggs observed in the field (Fritz et al. 1986). In addition, many eggs apparently were destroyed by cattle (Fritz et al. 1987). It thus appears that the range caterpillar is under "natural" biological control in Mexico.

Historical Distribution of the Range Caterpillar

At the present time the range caterpillar apparently is limited to three isolated populations: 1) the northern population, centered in Union, Colfax, Mora, and Harding counties in New Mexico and adjacent parts of Colorado, Oklahoma, and Texas; 2) the central population, centered in Lincoln, De Baca, and Guadalupe counties in New Mexico; and 3) the southern population, centered in Chihuahua, Mexico (see Fritz et al. 1987). A possible fourth population may have existed in the Davis Mountains of west Texas, based on specimens in the British Museum (Natural History) (E.W. Huddleston, per. comm.) (Figure 16.1).

This distribution pattern could have been due either to the formation of relict isolated populations through changing environmental conditions or to the accidental introduction of the species into new favorable habitats by human activity. In the first case, the pattern could have been caused by the reduction of grassland in relatively recent times. During the late Wisconsin glacial period (22,000-11,000 B.P.) much of the southwestern United States and northern Mexico was covered with forests. By the middle Holocene (8000 years B.P.) these forests were confined to higher elevations, and the present Chihuahuan Desert was probably

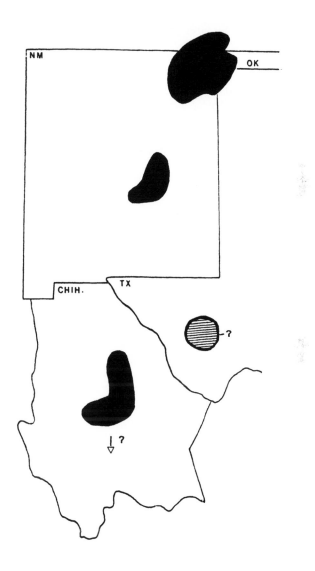

FIG. 16.1 Distribution of the range caterpillar, *Hemileuca oliviae*, in New Mexico, Texas, Oklahoma, Colorado, and Chihuahua. See text for explanation of west Texas area.

grassland (Van Devender and Spaulding 1979). As suggested earlier, the range caterpillar could have evolved in the grassland of Mexico. After the glacial retreat in the north of the continent and the spread of the grassland, *H. oliviae* could have expanded its range to the north. Continued drying of the Southwest and the ensuing retreat of the grassland could then have left isolated range caterpillar populations in their present distribution.

The second alternative states that the disjunct distribution is the result of some human agency, such as cutting hay that contained range caterpillar eggs, and then transporting it to another area previously unoccupied by the range caterpillar. W. Dick-Peddie (per. comm.) points out that there was probably little such transport of hay because of generally favorable forage conditions in New Mexico up to the 1870s. Even in such currently arid areas as the Jornada del Muerto, grass was plentiful in the 1850s (Herbel et al. 1972). Most likely the current distribution reflects a relict population pattern, although we cannot rule out human agency.

ACKNOWLEDGMENT

This is Scientific Paper No. 273 of the New Mexico State University Agricultural Experiment Station.

REFERENCES

Ainslie, C.N. 1910. The New Mexico range caterpillar. USDA Bur. Entomol. Bull. 85, Part V.

Arnaud, P.H., Jr. 1978. A host-parasite catalog of North American Tachinidae (Diptera) Misc. Publ. USDA 1319:1-860.

Beavis, W.D., J.C. Owens, M. Ortiz, T.S. Bellows, Jr., J.A. Ludwig, and E.W. Huddleston. 1981. Density and developmental stage of range caterpillar *Hemileuca oliviae* Cockerell, as affected by topographic position. J. Range Manage. 34:389-392.

Beavis, W.D., J.C. Owens, J.A. Ludwig, and E.W. Huddleston. 1982. Grassland communities of east-central New Mexico and density of the range caterpillar, *Hemileuca oliviae* (Lepidoptera: Saturniidae). Southwestern Nat. 27:335-343.

Bellows, T.S., Jr., M. Ortiz, J.C. Owens, and E.W. Huddleston. 1982a. A model for analyzing insect stage-frequency data when mortality varies with time. Res. Popul. Ecol. 24:142-156.

Bellows, T.S., Jr., J.C. Owens, and E.W. Huddleston. 1982b. Persistence of egg masses of the range caterpillar in grazed and ungrazed rangeland. J. Econ. Entomol. 75:574-576.

Bellows, T.S., Jr., J.C. Owens, and E.W. Huddleston. 1982c. Predation of range caterpillar, *Hemileuca oliviae* (Lepidoptera: Saturniidae) at various stages of development by different species of rodents in New Mexico during 1980. Environ. Entomol. 11:1211-1215.

Bellows, T.S., Jr., J.C. Owens, and E.W. Huddleston. 1983a. Model for simulating consumption and economic injury level for the range caterpillar (Lepidoptera: Saturniidae). J. Econ. Entomol. 76:1231-1238.

Bellows, T.S., Jr., J.C. Owens, and E.W. Huddleston. 1983b. Plant species utilization by different life stages of the range caterpillar, *Hemileuca oliviae* (Lepidoptera: Saturniidae). Environ. Entomol. 12:1315-1317.

Bellows, T.S., Jr., J.C. Owens, and E.W. Huddleston. 1984. Flight activity and dispersal of range caterpillar moths, *Hemileuca oliviae* (Lepidoptera: Saturniidae). Can. Entomol. 116:247-252.

Caffrey, D.J. 1921. Biology and economic importance of *Anastatus semiflavidus*, a recently described egg parasite of *Hemileuca oliviae*. J. Agr. Res. 21:373-384.

Capinera, J.L. 1978. Studies of host plant preference and suitability exhibited by early-instar range caterpillar larvae. Environ. Entomol. 7:738-740.

Capinera, J.L. 1980. A trail pheromone from silk produced by larvae of the range caterpillar *Hemileuca oliviae* (Lepidoptera: Saturniidae) and observations on aggregation behavior. J. Chem. Ecol. 6:655-664.

Capinera, J.L., L.F. Wiener, and P.R. Anamosa. 1980. Behavioral thermoregulation by late-instar range caterpillar larvae *Hemileuca oliviae* Cockerell. J. Kansas Entomol. Soc. 53:631-638.

Capinera, J.L., S.E. Naranjo, and M.J. Packard. 1981. Vapor density and water loss from eggs of the range caterpillar, *Hemileuca oliviae*. Environ. Entomol. 10:97-104.

Capinera, J.L., J.K. Detling, and W.J. Parton. 1983a. Assessment of range caterpillar (Lepidoptera: Saturniidae) effects with a grassland simulation model. J. Econ. Entomol. 76:1088-1094.

Capinera, J.L., A.R. Renaud, and N.E. Roehrig. 1983b. Chemical basis for host selection by *Hemileuca oliviae*: role of tannins in preference of C_4 grasses. J. Chem. Ecol. 9:1425-1437.

Caswell, H., F. Reed, S.N. Stephenson, and P.A. Werner. 1973. Photosynthetic pathways and selective herbivory: a hypothesis. Am. Nat. 107:465-480.

246

Dubach, J.M., D.B. Richman, and R.B. Turner. 1987. Genetic and morphological variation among geographical populations of the range caterpillar. Ann. Entomol. Soc. Am. (in press).

Everett, T.D., and R.B. Turner. 1979. Influence of temperature and humidity on embryogenesis and larval eclosion of the range caterpillar. Southwestern Entomol. 4:59-64.

Ferguson, D.C. 1972. Bombycoidea: Saturniidae. The moths of America north of Mexico. Fasc. 20, part 2:1-277.

Fritz, G.N., A.P. Frater, J.C. Owens, and E.W. Huddleston. 1987. Density and distribution of *Hemileuca oliviae* (Lepidoptera: Saturniidae) in Chihuahua, Mexico. Environ. Entomol. (in press).

Fritz, G.N., A.P. Frater, J.C. Owens, E.W. Huddleston, and D.B. Richman. 1986. Parasitoids and predators of *Hemileuca oliviae* (Lepidoptera: Saturniidae) in Chihuahua, Mexico. Ann. Entomol. Soc. Amer. 79:686-690.

Gould, F.W. 1979. The genus *Bouteloua* (Poaceae). Ann. Missouri Bot. Garden 66:348-416.

Hansen, J.D. 1978. The rate of egg parasitism of the range caterpillar by *Anastatus semiflavidus* from select areas of Union County. Clayton Livestock Research Center Progress Report No. 6.

Hansen, J.D., J.A. Ludwig, J.C. Owens, and E.W. Huddleston. 1982a. Movement of late instars of the range caterpillar, *Hemileuca oliviae* (Lepidoptera: Saturniidae). J. Ga. Entomol. Soc. 17:76-87.

Hansen, J.D., J.C. Owens, and E.W. Huddleston. 1982b. Life tables of the range caterpillar, *Hemileuca oliviae* (Lepidoptera: Saturniidae). Environ. Entomol. 11:355-360.

Hansen, J.D., J.A. Ludwig, J.C. Owens, and E.W. Huddleston. 1984a. Larval movement of the range caterpillar, *Hemileuca oliviae* (Lepidoptera: Saturniidae). Environ. Entomol. 13:415-420.

Hansen, J.D., J.A. Ludwig, J.C. Owens, and E.W. Huddleston. 1984b. Motility, feeding, and molting in larvae of the range caterpillar, *Hemileuca oliviae* (Lepidoptera: Saturniidae). Environ. Entomol. 13:45-51.

Herbel, C.H., F.N. Ares, and R.A. Wright. 1972. Drought effects on a semidesert grassland range. Ecology 53:1084-1093.

Krebs, C.J. 1972. Ecology. The experimental analysis of distribution and abundance. Harper and Row, New York.

Krombein, K.V., P.D. Hurd, Jr., D.R. Smith, and B.D. Burks. 1979. Catalog of Hymenoptera in America north of Mexico. Vol. 1. Symphyta and Apocrita (Parasitica). Smithsonian Inst. Press.

Mendel, M.J., P.B. Shaw, and J.C. Owens. 1987. Life history characteristics of *Anastatus semiflavidus* (Hymenoptera: Eupelmidae), an egg parasitoid of the range caterpillar, *Hemileuca oliviae* (Lepidoptera: Saturniidae), over a range of temperatures. (In prep.).

Rhoades, D.F., and R.G. Cates. 1976. A general theory of plant antiherbivore chemistry, p. 168-212. *In:* J.W. Wallace and R.L. Mansell (eds.) Biochemical interaction between plants and insects. Plenum Press, New York.

Riley, S.L., T.S. Bellows, Jr., J.C. Owens, and E.W. Huddleston. 1984. Consumption and utilization of blue grama grass, *Bouteloua gracilis*, by range caterpillar larvae, *Hemileuca oliviae* (Lepidoptera: Saturniidae). Environ. Entomol. 13:29-35.

Schowalter, T.D., W.G. Whitford, and R.B. Turner. 1977. Bioenergetics of the range caterpillar, *Hemileuca oliviae* (Ckll.). Oecologia 28:153-161.

Shaw, P.B., J.C. Owens, and E.W. Huddleston. 1987a. Host preference and its modification in the range caterpillar. (in prep.).

Shaw, P.B., J.C. Owens, E.W. Huddleston, and D.B. Richman. 1987b. Role of arthropod predators in the mortality of early instars of the range caterpillar, *Hemileuca oliviae* (Lepidoptera: Saturniidae). Environ. Entomol. (In press).

Smith, W.E., and R.B. Turner. 1979. Effects of age and temperature on the calling behavior of the female range caterpillar moth, *Hemileuca oliviae* (Lepidoptera: Saturniidae). Southwestern Entomol. 4:254-257.

Van Devender, T.R., and W.G. Spaulding. 1979. Development of vegetation and climate in the southwestern United States. Science 204:701-710.

Watts, J.G., and T.D. Everett. 1976. Biology and behavior of the range caterpillar. New Mexico St. Univ. Agr. Exp. Sta. Bull. 646.

Watts, J.G., E.W. Huddleston, and J.C. Owens. 1982. Rangeland entomology. Annu. Rev. Entomol. 27:283-311.

Wildermuth, V.L., and J.C. Frankenfeld. 1933. The New Mexico range caterpillar and its natural control. J. Econ. Entomol. 26:794-798.

17. DEVELOPMENT AND IMPLEMENTATION OF A RANGELAND IPM PROGRAM FOR RANGE CATERPILLAR

Ellis W. Huddleston
Department of Entomology, Plant Pathology, and Weed Science
New Mexico State University, Las Cruces, New Mexico 88003

Charles R. Ward
Agricultural Science Center at Alcalde
New Mexico State University, Las Cruces, New Mexico 88003

John C. Owens
College of Agriculture and Home Economics
New Mexico State University, Las Cruces, New Mexico 88003

The range caterpillar *Hemileuca oliviae* Cockerell is a serious pest in the areas where it occurs in the United States. Although apparently cyclic in nature, populations during outbreak years can reach densities that result in larval accumulations into "windrows." In "windrows" there may be 300 to 500 larvae per square meter in a moving band which is 3 to 10 meters in width. Every green blade and stem of grass is consumed by the larvae, presenting a stark contrast with the green grass in front of the windrow. The difference in color may be visible for a mile or more.

The range caterpillar infests an estimated 7.7 million ha (19 million acres) of rangeland in the United States in what appears to be two distinct populations, separated by 260 or more kilometers. The area separating the two populations is traversed in part by the Pecos and Canadian rivers. The southern extent of the northeastern New Mexico population appears to be that escarpment which separates the plains or steppe area north of it from rougher, brushy terrain to the south.

The first outbreak to be studied scientifically was in 1908 and 1909 (Ainslie 1910). The species had been described from a single specimen collected by Cockerell in 1897 and labeled Santa Fe, New Mexico. Ainslie (1910) reported interviews with ranchers who indicated that the outbreak he studied was first noted in 1904; the outbreak continued until 1916 (Wildermuth and Frankenfeld 1933). Based on reports from ranchers, there had been an earlier outbreak in the period 1885 to 1895 (Ainslie 1910). Another outbreak, lasting from 1926 until about 1932 was studied by Wildermuth and Frankenfeld (1933). Significant populations, associated with the present outbreak, were noted in 1960 in isolated areas in Lincoln and Colfax counties, NM.

In 1966, the first cooperative (USDA, APHIS-State of New Mexico-rancher) spray program was initiated using toxaphene on 47,380 ha (117,074 acres) in Colfax County. Table 17.1 is a summary of cooperative and rancher control programs in New Mexico (unpublished data, New Mexico Department of Agriculture, Las Cruces, NM.) In the 21-year period, 1966 to 1986, about 2 million ha (5 million acres) were treated at a cost in excess of $5.6 million.

The earlier literature implies that the previous outbreaks lasted 11, 13 and 6 years respectively and that the pest was virtually absent in the intervening years. The excessive length of the present outbreak, 21 years, poses several questions. First, because this is the first outbreak to be treated with pesticides, has there been an effect from pesticides on the duration of the outbreak? Second, have the annual grasshopper surveys by USDA, APHIS and the NM Department of Agriculture detected populations which would not have been reported in the past? Third, has the sequence of abiotic and biotic conditions been such that the present outbreak has lasted longer than any of the three previous outbreaks? Present implications tend to favor the latter two possibilities. Some parts of the central New Mexico population centered in Lincoln County, were treated with pesticides in 1967, 70, 71, 72, 73, 75 and 77; this population has been of extremely low density since the early 1980s. Greatly increased survey activities were initiated in the early 1960s, chemical control became available for the first time, cooperative programs made pesticide use economically feasible, and state and federal agencies encouraged pesticide use. Also, prior to 1976, the economic threshold had been considered to be 7.2 to $9.6/m^2$ $(6-8/yd^2)$ instead of the $2.4/m^2$ $(2/yd^2)$, or less, presently accepted.

BIOLOGY AND BEHAVIOR OF THE RANGE CATERPILLAR AS IT RELATES TO IPM

Eggs

The range caterpillar overwinters in the egg stage. Eggs are laid from late September through late November. Hatching is triggered by late spring and summer rainfall, generally from May until early July depending on the area and season. Hatch usually occurs earlier in the northern limits of the distribution because

TABLE 17.1
Cooperative control programs and rancher control programs in New Mexico (unpublished data, New Mexico Department of Agriculture).

Type of Program[1]	Year	Counties[2]	Pesticide[3]	Acres[4]	Total Cost	State and Private Cost
C	1966	CO	tox	117,074	$ 84,110	$ 56,074
C	1967	L	tox	14,084	8,042	5,362
		H,CO,U	tox	456,202	189,212	116,768
R	1970	CO,U,H,CH,L	tox	662,000	291,280(Rancher)	
R	1971	CO,U,L	tox	162,000	71,280(Rancher)	
R	1972	CO,U,H,CH,L	tox	200,000	88,000(Rancher)	
C	1973	CH,L	car	246,441	218,172	64,557
C	1975	L	car	51,240	64,306	38,052
C	1976	CO,U,H,M	tri	689,308	822,294	500,511
C	1977	U	tri	115,710	164,018	109,345
		H	tri	45,224	53,530	35,686
		CO,M	tri	96,795	137,206	89,354
		L	tri	200,752	286,919	82,920
C	1978	CO,M	tri	522,500	876,520	785,640
		H	tri	71,549	120,027	65,674
		U	tri	609,228	1,022,017	817,250
C	1979	CO,M	car	127,524	218,440	145,051
		U,H	car	42,450	69,130	44,078
C	1980	U,CO	car	52,316	117,711	78,474
		U	car	41,574	91,799	37,452
S/C		U,CO	car/tri	71,190	156,306	156,306
C	1982	CO,U,M,H	tri	21,370	60,249	40,168
C	1983	CO,U,M,H	tri	144,496	304,959	203,306
R	1984	CO,U,M	per	28,014	30,815(Rancher)	
				4,850(G)	5 (Rancher)	
R	1985	CO,U,H,M,S	per	68,347	71,764(Rancher)	
			per	28,240(G)	5 (Rancher)	
R	1986	CO,U,M,H	per	159,802	138,229(Rancher)	
				15,000(G)	5 (Rancher)	

1   C = Cooperative APHIS 1/3, State 1/3, Rancher 1/3, S/C = State 1/2, Rancher 1/2, R = Rancher total costs.
2   CH = Chaves, CO = Colfax, H = Harding, L = Lincoln, M = Mora, S = San Miguel and U = Union.
3   car = carbaryl, per = permethrin, tox = toxaphene and tri = trichlorfon.
4   Aerial application except as noted (G = ground).
5   Ground application costs not available.

the first significant rainfall is earlier. Watts and Everett (1976) found that embryogenesis is complete in early April, and after embryogenesis hatch can occur whenever there is a significant rainfall event. Because ranchers normally monitor rainfall at several locations on their property, IPM scouts can use telephone reports from individual ranchers to schedule survey trips.

Range caterpillars normally lay their full complement of eggs in a single, cylindrical mass of up to 150 eggs. Because egg masses are ca. 0.7 cm x 3 to 5 cm and are laid on dry grass culms or stems of other vegetation, egg surveys are rapid and more accurate than those for most insects. Egg surveys can be conducted from late October until hatch. While Ludwig et al. (1979) and Hansen et al. (1982b) used the plotless wandering quarter technique for sampling range caterpillar densities for research, belt transects 1 m x 100 m have been adequate and much more rapid for surveys for eggs and larvae in the IPM program.

Larvae

During the first three instars, range caterpillar larvae are highly gregarious and remain in clumps on stems of vegetation during most of the daylight hours. For IPM survey, early instar larvae are easily seen in the belt transects. Clumps can be picked and broken apart to estimate numbers. By the fourth instar, when caterpillars become solitary, individuals are large enough to be seen easily as they rest head down on grass and forb stems. Hansen et al. (1981) have shown that under normal field conditions, the larvae have six instars which can be separated by head capsule width and spine characteristics. Other workers have reported different numbers of instars depending on environment and diet (Ainslie 1910, Schowalter et al. 1976, Capinera 1984).

Early studies reported range caterpillars feeding on about 40 species of range and cultivated grass crops (Wildermuth and Caffrey 1916). Later studies found that grasses with the C^4 metabolic pathway were preferred and more suitable for growth than species possessing the C3 pathway (Capinera 1978). Thus, the food for range caterpillar larvae is, in most cases, the same grasses utilized by domestic herbivores, creating direct competition for limited resources. Additional losses occur when larvae cut off parts of the leaves without eating them. Grass which is not eaten by the caterpillars is rendered unpalatable to livestock by the irritating spines on the active larvae and by their cast skins which remain on the grass through the growing season and into the winter. Extensive feeding by range caterpillars increases the danger of soil erosion by wind and water, reduces penetration of rainfall, and reduces plant vigor (Huddleston et al. 1976).

Pupae

While egg hatch may vary from late April to late July, pupation appears to occur within a narrow time frame in early September. Watts and Everett (1976) suggested that photoperiod and perhaps the quality and quantity of available food influence the time of pupation. Larvae spin loose cocoons in woody half shrubs such as *Gutierrezia* sp. (broom snakeweed), strong forbs, or teepee-shaped enclosures of grass stems. Before molting, especially in the later instars, larvae create the teepee-shaped grass structures and leave the cast skin(s) at the apex. Because the cast skin(s) are irritating to livestock, this may be a behavioral modification to preserve suitable pupation sites from overgrazing by large herbivores. Pupae are easily sighted and unemerged pupae found during egg surveys provide information on parasitism and predation.

Adults

Adult emergence begins in September, peaks in October and usually ends in November. Adults have vestigal mouthparts and live only a few days; males appear to outlive females (Watts and Huddleston 1977). Adults emerge from the pupae about noon, usually mate just before dusk of the same day, and lay their eggs within the following 12 to 18 hours. Females die soon after they lay their eggs and appear to mate only once. Males live five to seven days and are capable of multiple matings.

EVOLUTION OF IPM TACTICS FOR RANGE CATERPILLAR MANAGEMENT

IPM tactics have been described as the building blocks upon which IPM strategies are developed. Research during the past decade has focused on the evaluation of selected tactics needed for development and implementation of effective strategies for range caterpillar management.

Economic Threshold (Economic Injury Level)

Based on field observations, the original economic threshold for range caterpillars was considered to be $7.2/m^2$ ($6/yd^2$). Huddleston et al. (1976) conducted the first research to define the economic threshold. Their data defined a food consumption curve for various densities of larvae from the fourth instar until pupation. Based on 1975 chemical control costs of $1.26 per acre and a forage value of $20.00 per ton ($8.00 per AUM), the economic threshold was ca 2.4 larvae/m^2 (2/yd^2).

They introduced the ranchers' threshold for cooperative range caterpillar control programs. Because the rancher was paying one-third the cost, his economic threshold was about 0.8 larvae/m^2 (0.7/yd^2). Similar food consumption values were derived independently by laboratory and modelling efforts (Bellows et al. 1983, Capinera et al. 1983, Riley et al. 1984).

Torell and Huddleston (see Chapter 26) have shown that forage value is a very important variable affecting the economic threshold for grasshoppers. Percent control and the numbers of years the population is reduced also have significant impacts on the economic threshold. These data also are applicable to range caterpillar management.

Population Dynamics

An insect with a single generation per year that overwinters in the egg stage will decline in numbers during the larval stage. Recognizing that effective pest management programs require quantitative knowledge of population dynamics and natural mortality factors, Hansen et al. (1982b) presented the first life tables constructed from field data collected in northeastern New Mexico. Bellows et al. (1982a) developed a model of insect stage frequency when mortality varies with time using data from central New Mexico. These studies show that mortality is greatest in the early instars and much less for the latter instars. For IPM program planning, a calculated reduction of about 80% from winter egg numbers to 4th instar densities has proven useful. Data from the Bellows et al. (1983) paper indicate that ca 95% of the forage consumption occurs in the 5th and 6th instars. These values indicate that chemical control can be delayed until late fourth instar with negligible forage losses. By delaying control operations, as long as prudent, opportunities for increased natural mortality are enhanced.

Reinfestation of Treated Areas

One IPM tactic is to treat only selected areas of a ranch where densities are above the economic threshold. This situation occurs quite often on large ranches with varying topography (Beavis et al. 1981). Hansen et al. (1982a) studied the hourly and daily movement of fifth and sixth instar larvae and found that most movement was southerly and in a straight line with the maximum distance recorded for a larva being 13.7 meters over a 24-hour period. They concluded that dispersal of larvae was not a mechanism for rapid spread of range caterpillar populations. Later studies on all six instars yielded similar results (Hansen et al. 1984). Neither study was conducted with densities at or near the "windrow" level. General observations and rancher reports indicate that the distance traveled by larvae at or near "windrow" densities increases, but actual distances have not been measured.

Bellows et al. (1984) concluded that most females oviposit within 10 m of the pupation site. These studies indicate that, within the density ranges normally encountered, little migration into treated areas can be expected in the year of treatment. These data support "hot spot" spraying as an IPM tactic.

Biological Control

Several species of insect parasitoids are known from New Mexico. The egg parasitoid *Anastatus semiflavidus* Gahan is the best studied and apparently is distributed throughout the range of range caterpillar. Caffrey (1921) reported parasitism ranging from 0 to 75% in the 1915 crop of eggs in New Mexico, and Wildermuth and Frankenfeld (1933) credit this parasitoid with ending the 1904-1916 outbreak. Since then, parasitism rates have rarely exceeded one or two percent except in isolated locations (Hansen et al. 1982b). A larval parasitoid, the tachinid fly *Exorista mella* (Walker), appears to be widely distributed in the U.S. The most common pupal parasitoid is the chalcid wasp, *Brachymeria ovata* (Say) (Watts and Everett 1976). Fritz et al. (1986) identified six new species of parasitoids, two of which may be hyperparasites, in a Mexico population.

Several predatory insects have been reported as feeding on range caterpillar. These include a carabid beetle and a camel cricket by Watts and Everett (1976), two species of camel crickets by Wildermuth and Caffrey (1916), and robber flies (Ainslie 1910). Several other insects, including ants and lady beetles, have been suggested as predators.

Introductions of new insects and augmentation of natural populations have been attempted, but have not been shown to be practical. Watts and Everett (1976) found no evidence of the establishment of one parasite and three predatory beetles which were introduced from 1913 to 1916. Extensive releases of *A. semiflavidus* were made in the 1930s but no report of the success of these releases was ever published. The larval parasite *E. mella* was cultured and released in New Mexico by Dr. R. E. Fye (USDA, ARS, Tucson, Ariz.) Results were not encouraging enough to continue the project. Three species of rodents were found to prey on range caterpillar eggs and pupae. One species, *Peromyscus truei* (Schufelt), was found to be an avid predator of range caterpillar pupae in woodlands. It appears that this species may be an important factor in restricting range caterpillars to grasslands (Bellows et al. 1982b).

Conservation of natural enemies as an IPM tactic is aided by the fact that high levels of egg and pupal parasitism are readily apparent to the scouts. Escape holes in eggs and pupae are easily seen with the naked eye.

Rangeland Management

Range caterpillar population densities appear to be lower on severely overgrazed rangeland. Bellows et al. (1982c) found that significantly more egg masses disappeared in areas grazed by sheep than in adjacent exclosures. Ainslie (1910) indicated that some pupae may be knocked from plants by the livestock.

Burning was tried by Ainslie (1910), but he reported that later in the season the number of caterpillars in the burned area was equal to the number in the unburned area. Small plots were burned in 1976 with the results that egg clusters in areas where the heat was most intense were burned beyond recognition; where heat was less intense, the egg shells burst like popcorn. Eggs in yucca over 20 cm from the ground hatched (G. L. Nielsen, unpublished report, New Mexico Department of Agriculture). Burning does not appear to be a practical IPM tactic because many areas do not produce sufficient fuel to carry a successful fire.

Pheromones and Males Sterility

The range caterpillar has been shown to have a strong pheromone system (unpublished data, E. Huddleston and R. Turner). Field observations have been supported by laboratory experiments that show that females begin "calling" when declining environmental temperature reaches 7°C. Efforts to characterize and synthesize the pheromone have not yet been successful.

Ward et al. (1986) demonstrated that reasonable levels of sterility can be produced in field-collected, irradiated pupae. While the male sterile technique appears feasible, a completely acceptable artificial diet has not been developed, so mass rearing is not economic.

Chemical Control

Evaluation of insecticides for range caterpillar control was initiated in the early 1960s. Toxaphene was rapidly adopted because of cost and reliability. This material was the center of controversy in 1970 and no cooperative programs were conducted; ranchers conducted independent spray programs using toxaphene in

1970, 71, 72 and 73. Two insecticides, trichlorfon in the Dylox 1.5 oil formulation and carbaryl in the Sevin-4-oil formulation, were used in the period from 1974 until 1984. Permethrin was registered for use on range caterpillars in 1984. Huddleston et al. (1987) have shown that permethrin is efficacious at extremely low dosages either by ground or aerial application. The simple, inexpensive ground application developed in these and other studies (Huddleston et al. 1986) provided the tools for an IPM tactic of treating small, expanding populations before significant spread.

Additional research has been conducted to find alternatives to the currently registered insecticides. The insect growth regulators diflubenzuron and triflumuron have been found to be highly effective. For certain environmentally sensitive applications, *Bacillus thuringiensis* has been shown to be effective (unpublished data of the authors).

EVOLUTION OF THE IPM STRATEGIES

A pilot integrated pest management program was initiated in 1978 as a cooperative program with New Mexico State University, NM Department of Agriculture, USDA-APHIS, NM Cooperative Extension Service, NM Agricultural Experiment Station, and area ranchers. An egg mass survey was initiated in April 1978 and areas above the economic threshold of 12 eggs/m^2 (10/yd^2) were resurveyed in a larval survey. A total of 616,000 ha (1,522,000 acres) was surveyed at a cost of $0.015/ha ($0.006/a). A single sample was taken near the center of each square mile (259 ha). Local residents were employed as field scouts and supervised by a scout leader under the direction of the Union County Extension Agent. Ranchers provided transportation on the ranch and accompanied the scouts.

In 1979, the pilot program was expanded into parts of Colfax and Mora counties. An aggregate total of 1,000,000 ha (2,430,720a) was sampled at a cost of $0.012/ha ($0.005/a). A second scout supervisor was added as well as additional scouts.

Phase II was initiated in 1980 in a 48 x 64 km (30 x 40 mile) block in Union and Colfax counties. Egg surveys were conducted on a square mile basis; however, larval densities were determined and control decisions were made on a one-fourth square mile (64.8 ha) basis. USDA, APHIS cooperated by reducing its criteria from 4,000 ha (10,000 a) blocks to areas above the economic threshold. The two scout supervisors were able to handle the scouting with assistance from NMDA and APHIS survey personnel. Small agricultural aircraft were used and scheduled as needed based on larval development.

As experience was gained, more and more responsibility was placed on the ranchers to survey their own land and request help from the scouts when they were in doubt. In 1984, with the registration of permethrin, ranchers could afford their own program and operated independently of state and federal financial support. Special financial support for scouting from Cooperative Extension ceased September 30, 1986.

The Range Caterpillar IPM program was the first rangeland IPM program developed in the nation. The research efforts which culminated in the development of the highly effective, low cost ground application technique, and the very low dosages of permethrin reduced control costs from almost $7.50/ha ($3.00/a) to a few cents per ha. The future of the range caterpillar IPM program will be in the hands of the ranchers and the County Extension Agents beginning in 1987.

ACKNOWLEDGMENTS

More than any single individual, David Graham, Union County Extension Agent, should be credited with the success of this program. He believed in the concept from the start and convinced the ranchers and others involved of the future of the program. The other County Extension Agents, especially Steve Fernandez, Colfax County, were essential to the success. The untiring efforts and enthusiasm of the two scout supervisors, Lavonne Childress and Mary Gail Baker, were critical to the success of the program. The dedication and inspiration of the many scientists and graduate students associated throughout the years with this project contributed immeasurably to the success of the IPM approach and we are deeply appreciative of their contributions. This research and the extension efforts were supported in part by USDA CSRS Grant 801-15-35, USDA CES Pest Management grants, and grants from WRPIAP, The Cotton Foundation, FMC Corporation, ICI Americas, Mobay Chemical Company, Thompson Hayward, and others.

The continuing efforts and support of United States Senator Peter Domenici is gratefully acknowledged by each and every person who has been involved in this IPM program. His work resulted in federal funding which made much of this program possible. Funding from the New Mexico legislature also was responsible for the success of this project.

REFERENCES

Ainslie, C.N. 1910. The New Mexico range caterpillar. USDA Bur. Entomol. Bull. 85, Part V:57-96.

Beavis, W.D., J.C. Owens, M. Ortiz, T.S. Bellows Jr., J.A. Ludwig, and E.W. Huddleston. 1981. Density and developmental stage of range caterpillar *Hemileuca oliviae* Cockerell, as affected by topographic position. J. Range Manage. 34:389-392.

Bellows, T.S., Jr., M. Ortiz, J.C. Owens, and E.W. Huddleston. 1982a. A model for the analysis of insect stage-frequency data when the mortality rate varies with time. Res. Popul. Ecol. 24:142-156.

Bellows, T.S., Jr., J.C. Owens, and E.W. Huddleston. 1982b. Predation of range caterpillar, *Hemileuca oliviae* (Lepidoptera: Saturniidae) at various stages of development by different species of rodents in New Mexico during 1980. Environ. Entomol. 11:1211-1216.

Bellows, T.S., Jr., J.C. Owens, and E.W. Huddleston. 1982c. Persistence of egg masses of the range caterpillar in grazed and ungrazed rangeland. J. Econ. Entomol. 75:574-576.

Bellows, T.S., Jr., J.C. Owens, and E.W. Huddleston. 1983. Model for simulating consumption and economic injury level for the range caterpillar (Lepidoptera: Saturniidae). J. Econ. Entomol. 76:1231-1238.

Bellows, T.S., Jr., J.C. Owens, and E.W. Huddleston. 1984. Flight activity and dispersal of range caterpillar moths, *Hemileuca oliviae* (Lepidoptera: Saturniidae). Can. Entomol. 116:247-252.

Caffrey, D.J. 1921. Biology and economic importance of *Anastatus semiflavidus,* a recently described egg parasite of *Hemileuca oliviae*. J. Agr. Res. 21:373-384, Pl. 68.

Capinera, J.L. 1978. Studies of host plant preference and suitability exhibited by early-instar range caterpillar larvae. Environ. Entomol. 7:738-40.

Capinera, J.L. 1984. Instar number in range caterpillar *Hemileuca oliviae* Cockerell (Lepidoptera: Saturniidae). J. Kansas Entomol. Soc. 57:344-347.

Capinera, J.L., J.K. Detling, and W.J. Parton. 1983. Assessment of range caterpillar (Lepidoptera: Saturniidae) effects with a grassland simulation model. J. Econ. Entomol. 76:1088-1094.

Fritz, G.N., A.P. Frater, J.C. Owens, E.W. Huddleston, and D.B. Richman. 1986. Parasitoids of *Hemileuca oliviae* (Lepidoptera: Saturniidae) in Chihuahua, Mexico. Ann. Entomol. Soc. Amer. 79:686-690.

Hansen, J.D., J.A. Ludwig, J.C. Owens, and E.W. Huddleston. 1982a. Movement of late instars of the range caterpillar, *Hemileuca oliviae* (Lepidoptera: Saturniidae). J. Georgia Entomol. Soc. 17:76-87.

Hansen, J.D., J.A. Ludwig, J.C. Owens, and E.W. Huddleston. 1984. Regression models of diurnal movement of instars of the range caterpillar, *Hemileuca oliviae* (Lepidoptera: Saturniidae). J. Georgia Entomol. Soc. 19:417-425.

Hansen, J.D., J.C. Owens, and E.W. Huddleston. 1981. Relation of head capsule width to instar development in larvae of the range caterpillar, *Hemileuca oliviae* Cockerell (Lepidoptera: Saturniidae). J. Kansas Entomol. Soc. 54:1-7.

Hansen, J.D., J.C. Owens, and E.W. Huddleston. 1982b. Life tables of the range caterpillar, *Hemileuca oliviae* (Lepidoptera: Saturniidae). Environ. Entomol. 11:355-360.

Huddleston, E.W., E.M. Dressel, and J.G. Watts. 1976. Economic threshold for range caterpillar larvae on blue grama pasture in Northeastern Lincoln County, New Mexico,in 1975. New Mexico Agr. Exp. Sta. Res. Rep. 314.

Huddleston, E.W., R. Sanderson, and J. Ross. 1986. Practical use of the PMS spray analyzer for pesticide application research, p. 134-141. *In:* L.D. Spicer and T.M. Kaneko (eds.) Pesticide formulation and application systems. Fifth Volume, ASTM STP, 915, Amer. Soc. for Testing and Materials, Philadelphia, PA.

Huddleston, E.W., R. Sanderson, J.B. Ross, and J.A. Henderson. 1987. Comparison of a wind-assisted dispersal technique and aerial application for range caterpillar (Lepidoptera: Saturniidae) management. J. Econ. Entomol. 80:226-229.

Ludwig, J.A., M. Ortiz, Jr., and J.C. Owens. 1979. An evaluation of the plotless wandering-quarter method for estimating the density of aggregated organisms. Bull. Ecol. Soc. Am. 60:138 (abstract).

Riley, S.C., T.S. Bellows, J.C. Owens, and E.W. Huddleston. 1984. Consumption and utilization of blue grama grass, *Bouteloua gracilis,* by range caterpillar larvae, *Hemileuca oliviae* (Lepidoptera: Saturniidae). Environ. Entomol. 13:45-51.

Schowalter, T.D., W.G. Whitford, and R. Turner. 1976. Growth characteristics of the range caterpillar, *Hemileuca oliviae* (Ckll.), on an artificial diet. Southwestern Entomol. 1:164-167.

Ward, C.R., D.B. Richman, E.W. Huddleston, J.C. Owens, R.T. Staten and M. Ortiz, Jr. 1986. Sterility and longevity of adult range caterpillars reared from irradiated field-collected pupae. J. Econ. Entomol. 79:87-90.

Watts, J.G., and T.D. Everett. 1976. Biology and behavior of the range caterpillar. New Mexico Agr. Exp. Sta. Bull. 646.

Watts, J.G., and E.W. Huddleston. 1977. Effect of age and number of matings on oviposition of the range caterpillar. Bull. New Mexico Acad. Sci. 17:31-32.

Wildermuth, V.L., and D.J. Caffrey. 1916. The New Mexico range caterpillar and its control. USDA Bull. 443.

Wildermuth, V.L., and J.C. Frankenfeld. 1933. The New Mexico range caterpillar and its natural control. J. Econ. Entomol. 26:794-798.

18. ECOLOGY AND MANAGEMENT OF HARVESTER ANTS IN THE SHORTGRASS PLAINS

Lee E. Rogers
Surface System Dynamics Group
Battelle Pacific Northwest Laboratories
P.O. Box 999, Richland, Washington 99352

All ants belong to the family Formicidae within the order Hymenoptera and all are social, living in colonies comprised of workers, reproductives, and immatures. Ants, however, are quite diverse in terms of their ecological and social adaptations. Some ants are entirely arboreal while others excavate deep nests in the soil. Some are entirely dependent on their slave workers for their care while others have become entirely parasitic and only occur within the nests of other ant species. Food specialization is highly diverse among the Formicidae. While most species select a variety of food materials, some are limited to the "honeydew" secretions from Homopterans. The harvester ants are species that rely primarily on seeds as their food source. The ecology and management of harvester ants in the shortgrass plains, particularly the western harvester ant, *Pogonomyrmex occidentalis* (Cresson), are the focus for this chapter.

HARVESTER ANT ECOLOGY

The collection and storage of seeds by ants was recorded by many of the ancient scholars, including Solomon, Hesiod, Aesop, Aelian, Plutarch, Orus Apollo, Plautus, Horace, Virgil, Ovid, and Pliny (Wilson 1972). These early writers all lived within the Mediterranean region where seed-harvesting ants are particularly abundant. During the 17th-19th centuries, workers from Europe, who were outside the range of harvesting ants, could find no evidence that ants harvest or store seeds. It was not until studies were extended to the dry and warm temperate zones that the existence of harvester ants was confirmed (Wilson 1972).

The harvesting habit probably represents the most important adaptation by ants to arid and semi-arid regions where the flora is relatively short lived. Several genera have adapted to eating seeds. In the new world these include *Pogonomyrmex*, *Veromessor*, *Solenopsis*, and *Pheidole*. The genus *Pogonomyrmex* is the preeminent group of harvester ants in North America (Cole 1968). As with other harvester ants, *Pogonomyrmex* sp. workers forage over the area surrounding the nest, collecting seeds and carrying them back to the nest for storage within underground chambers. Other foods such as insect prey are also collected and taken to the nest, but such collections are not stored over long periods. Seeds, therefore, are the primary food source for the colony through long periods of food scarcity.

The western harvester ant is particularly abundant within the shortgrass ecosystem, and is reported to have the most consistent impact on rangeland of any insect other than grasshoppers (Hewett et al. 1974). Therefore, control of the western harvester ant has been a primary concern to the farmers and ranchers inhabiting rangeland regions of the West. Consequently, this chapter will primarily consider the life cycle and habits of this species.

Nests of the Western Harvester Ant

The presence of western harvester ants is easily recognized within shortgrass rangeland. Their large mounds are conspicuous and often exceed 6 to 12 inches in height and 3 to 5 feet in diameter at the base (Lavigne 1966, 1969). The mound is generally located near the center of a large bare circle or disc that is kept free of plants. The cleared disc varies from a few feet to several yards in diameter.

The entrance to the nest is usually located on the south or southeastern portion of the mound. This entrance leads to a multitude of interconnecting chambers within the mound. Tunnels extend from beneath the mound to depths of ca. 2 m (Hewitt et al. 1974, Rogers 1972), allowing the ants to spend the winter months below frostline. Although a single tunnel may extend to these depths, nests that have been occupied for several years may have many tunnels. Rogers and Lavigne (1974) found an average of five tunnels extending to an average depth of 1.4 m in western harvester ant colonies located within the shortgrass region of northeastern Colorado.

Presently, an established method does not exist for estimating ant densities within a nest other than excavating the nest and counting the ants. Consequently, there are relatively few estimates concerning the number of ants that actually occupy a nest. Lavigne (1969) excavated 33 colonies in Wyoming and found the number to vary from 412 to 8796 ants per colony. Chew (1960) excavated a single colony in Arizona and found 8700 ants; he thought this number was average for that area based on the size of the mounds. Rogers (1972) excavated 11 colonies in northeastern Colorado and found an average of 2676 ants per colony.

A simpler and more common method to determine ant abundance is to count the number of ant nests within a given area. Costello (1944) determined the number of colonies present in different stages of secondary succession on abandoned farmland in Colorado. He found up to 10 colonies per ha in the Russian thistle stage, 7.5 to 28 colonies in the forb stage, 40 to 140 colonies in the *Aristida* stage, and up to 32.5 colonies in a shortgrass disclimax stage. Fautin (1946), while studying the northern desert shrub biome in western Utah, found western harvester ant colonies present in shad scale, horsebrush, greasewood, and sagebrush communities. The greatest colony density (10 to 20 colonies per ha) occurred in the sagebrush communities. Similar findings were reported in Cole's (1932) studies of harvester ant densities in the Tooele Valley of Utah and in the Craters of the Moon National Monument in Idaho. Rogers (1972) counted the number of western harvester ant nests within four differentially grazed pastures in northeastern Colorado. He found no significant difference in densities between the light and moderately grazed pastures, although these pastures contained over twice as many nests as the heavily grazed pastures.

Daily, Seasonal, and Annual Cycles

New colonies are established each year by mated queens. Winged ants, which comprise the reproductive males and females of the population, develop from the first eggs of the new season. Large numbers of these reproductives emerge simultaneously from several colonies, frequently after a thunderstorm, and fly to the most prominent nearby feature (i.e., a hilltop, tree, fence post, or building) where they form mating clusters. The females disperse following mating, sometimes over considerable distances (Race 1966). The males are not so fortunate. Any male lucky enough to mate does so at the sacrifice of his own life. The females shed their wings after dispersing and search rapidly over the soil surface for a suitable location to start a new nest. The new queen digs a chamber into the soil and lays her first eggs. Following hatching this first brood is fed entirely from the fat reserves of the queen. Upon reaching maturity this brood takes over all foraging, nest building, and brood-care activities, and the only duty of the queen is to lay eggs.

Daily aboveground activities of harvester ants are dependent on soil surface temperatures. The ants open the mounds when the soil surface reaches 24°C, but little aboveground activity will take place until the surface temperature reaches approximately 28°C.

At this temperature the ants start foraging and foraging activities continue at a brisk pace until the surface temperature reaches 47°C, which precipitates a drastic reduction in the number of foragers leaving the colony. After the sun has passed its zenith and the soil temperatures have cooled, foraging will resume and continue until evening. The ants generally stop foraging when the surface temperature falls to 32°C. They start to close the mound when the surface temperature reaches 29°C and the last ant will have retreated inside the colony for the night by the time the soil surface temperatures have fallen to 25°C. These relationships have been described mathematically (Rogers 1974).

Foraging Activities of Western Harvester Ants

Western harvester ants collect and store large numbers of seeds. Seeds are foraged from the area surrounding the nest and are carried into the mound, where the husks are removed and then taken out of the mound to a midden pile at the edge of the cleared disc. Stevens (1965) studied the foraging behavior of western harvester ants in Wyoming. He found that the forage taken over the entire season averaged 43 percent forb seed, 47 percent grass seed, and 11 percent insect parts and grasshopper fecal pellets. These ants also forage a variety of other materials. Although seeds comprised 39 percent of the food items collected in a study conducted at the Pawnee National Grasslands in eastern Colorado, litter comprised 24 percent of the food items collected (Rogers 1974). This ratio was not expected since western harvester ants are not known to use litter for food. Insect prey, insect body parts, and plant reproductive parts (other than seeds) each comprised about 10 percent of the total number of forage particles in the Colorado study. Smaller percentages of bits and pieces of fecal matter and mineral substances were also foraged (Rogers 1974).

IMPORTANCE OF WESTERN HARVESTER ANTS

Many soil animals are important to the formation of soil because their activities result in the pulverization, granulation, and transfer of much soil. Ants and earthworms are two of the most important groups of invertebrate animals that move soil vertically (Jacot 1936). Earthworms, however, are not abundant in the semi-arid rangelands. Talbot (1953) studied the importance of ants in soil formation in Michigan and estimated that *Lasius niger neoniger* Emery brought 85.5 g/m^2 of soil to the surface. Rogers (1972) has conducted the only study to determine the amount of soil moved by western harvester ants in the shortgrass region. He has estimated that the ants move 2.8 kg of soil over the life of each colony, which amounts to about 9 g/m^2 of soil moved on a moderately grazed rangeland.

Relatively few attempts have been made to assess quantitatively the amount of seed used by the ants or the impact of seed removal on the surrounding community. Lavigne and Fisser (1966) reported that western harvester ants, from an average of 25 colonies per ha, may actually remove up to 2.3 kg of grass seed and more than 4.5 kg of forb seed. Rogers (1974), using an indirect technique, estimated that western harvester ants removed about 2 percent of the seed biomass available to them. Trevis (1958) studied the forager ant species *Veromessor pergandei* (Mayr) and found that these ants harvested only 1 percent of the seeds produced in his Mojave Desert study area. He suggested that *Veromessor* may influence the abundance of certain plant species by selective foraging. Whitford (1978) studied the seed-foraging activities of three *Pogonomyrmex* spp. (*P. rugosus* Emery, *P. desertorum* Wheeler, and *P. californicus* Buckley) in the Chihuahuan Desert near Las Cruces, New Mexico. He found that even under intensive foraging pressure from these ants, more than a million seeds per acre escaped predation. Even so, he suggested that the *Pogonomyrmex* spp. may have an effect on the relative abundance of the plant species present. Although the effects of seed harvesting by western harvester ants cannot be conclusively stated, existing evidence indicates that the seed-harvesting activities of these ants do not have a substantial impact on the abundance of rangeland plants.

The ants' habit of removing vegetation from around the nest is well known (Headlee and Dean 1908, Cole 1932, Costello 1947, Killough and Hull 1951, Lavigne 1966, Lavigne and Fisser 1966, Race 1966), but it has not been determined if these clearings significantly reduce the food resources available for other consumer groups. The disc-clearing activities of harvester ants can remove substantial amounts of vegetation, as much as 1/7 of a hectare in central Wyoming (Lavigne and Fisser 1966). However, other studies by Wight and Nichols (1966) in the Big Horn Basin of Wyoming and Rogers and Lavigne (1974) in the Pawnee National Grassland of northeastern Colorado showed that increased vegetative production around the perimeter of the clearings largely compensated for the plants removed.

Western harvester ants have provided direct benefits to humans and other higher-order consumers. McCook (1883) noted that it was a custom of American Indians to put insect-infested furs and blankets near the ant mounds in order to have them cleaned. Wyoming farmers have collected gravel from ant mounds to feed to their chickens for gizzard grit (Lavigne 1966), and uranium prospectors have scanned harvester ant mounds with

Geiger counters since the ants may transport radioactive materials from substantial depths to the surface.

Western harvester ants are also a food source for higher-order consumers including horned toads (Knowlton 1938, Knowlton and Baldwin 1953), the sagebrush swift (Knowlton 1947, Knowlton 1953, Knowlton and Valcarce 1950), chipping sparrows (Knowlton and Harmston 1941), rock wrens (Knowlton and Harmston 1942), flickers (Knowlton and Stains 1943), and sage grouse (Knowlton and Thornley 1942). Giezentanner and Clark (1973) reported that male sage grouse use the ant mounds as strutting locations during the mating season.

MANAGING HARVESTER ANTS

The first attempts to control harvester ants probably occurred shortly after the first farmers and ranchers settled in harvester ant country. The first published reference of control of harvester ants is by Popenoe (1904), who indicated that regular cultivation of infested fields discouraged colony establishment and that large, well-established colonies could be destroyed by fumigating the nests with carbon bisulfide. Lavigne (1966) describes the early attempts of Wyoming ranchers to control harvester ants by such diverse methods as smothering mounds with cow manure, pouring crank-case oil down the entrances, submerging partially filled cans of oil in the ground so that worker ants would fall into the cans, burning over fields, and treating the mounds with nicotine sulphate.

Various cultural practices have proved to be effective in managing many insect populations (Johansen 1971); however, few options exist for the cultural management of western harvester ants. Rogers and Lavigne (1974) reported that intensive cattle grazing can result in reduced colony densities. They found that where cattle grazing had been regulated for over 30 years the number of harvester ant colonies was substantially reduced because of pressure from heavy grazing. Controlled burning has also proved effective against many insect pests (Metcalf 1962). However, the large areas around the ant nests that have been denuded of vegetation effectively prevent any substantial mortality associated with fire (Cole 1932).

Cultivation may be an effective method of controlling colony densities. However, ants will establish their nests in both cultivated and uncultivated areas, although continual cultivation of an area is detrimental to colony maintenance (Cole 1932, Severin 1955). I have observed colonies of the harvester ant *P. owyheei* Cole in eastern Washington gradually to die out over 2 years after the area was converted to irrigated agriculture. Cole (1932) reported that the *P. occidentalis* brood does not mature in soils of high moisture content, which may account for their inability to maintain the colony under irrigated conditions.

Much research was conducted during the 1950s and 1960s to determine the effectiveness of new insecticides in controlling harvester ants (Barnes and Nerney 1953, Crowell 1963a, Haws and Knowlton 1951, Knowlton 1959, List 1954, Lavigne 1966, Race 1964, Severin 1955). Much of this research consisted of screening tests using chemicals available in bait form for controlling fire ants.

The individual treatment of nests located near farmsteads or stockhandling areas is an effective means for controlling the presence of western harvester ants. Unfortunately, the extensive dispersal of newly mated queens each year and the observation that ants from smaller colonies may move to larger colonies where the ants have been killed probably makes any long-term control within limited areas impossible (Lavigne 1966, Crowell 1963b).

CONCLUSIONS

Most species of ants play an important role in the functioning of natural ecosystems. Unfortunately, so little is known about the ecology of even the most common species that it is difficult to fully appreciate their importance. However, as described in this chapter, we are beginning to perceive the range of their activities that influences ecological functioning and the human economy. Their indirect effects may be the most important of all. These include aeration and incorporation of organic matter into the soil during nest construction, dispersal of seeds (seeds may be lost by foragers or sprout during storage in the nest), predation on other insects, and vegetation removal. They also serve as a food base for a variety of other consumers. Any further elaboration on the fascinating topic of how harvester ants influence the human economy and the functioning of rangeland ecosystems will need to await the results of future studies.

ACKNOWLEDGMENT

This work was supported by U.S. Department of Energy contract DE-ACO6-76RCO 1830.

REFERENCES

Barnes, O.L., and N.J. Nerney. 1953. The red harvester ant and how to subdue it. USDA Farmers' Bull. 1668.

268

Chew, R.M. 1960. Note on colony size and activity in *Pogonomyrmex occidentalis* (Cresson). J. New York Entomol. Soc. 69:81-82.

Cole, A.C. 1932. The relation of the ant, *Pogonomyrmex occidentalis* Cr., to its habitat. Ohio J. Sci. 32:133-146.

Cole, A.C., Jr. 1968. *Pogonomyrmex* harvester ants. A study of the genus in North America. Univ. Tennessee Press, Knoxville, TN.

Costello, D.F. 1944. Natural revegetation of abandoned plowed land in the mixed prairie association of northeastern Colorado. Ecology 25:312-326.

Costello, D.F. 1947. Harvesters of the plains. Nature Mag. 40:146-149.

Crowell, H.H. 1963a. Control of the western harvester ant, *Pogonomyrmex occidentalis*, with poisoned baits. J. Econ. Entomol. 56:295-298.

Crowell, H.H. 1963b. New insecticide effective as harvester ant control. Oregon Agr. Progr., Winter 1963. 10:14.

Fautin, R.W. 1946. Biotic communities of the northern desert shrub biome in western Utah. Ecol. Monogr. 16:251-310.

Giezentanner, K.I., and W.H. Clark. 1976. The use of western harvester ant mounds as strutting locations by sage grouse. Condor 76:218-219.

Haws, L.D., and G.F. Knowlton. 1951. Heptachlor in western harvester ant control. Utah State Agr. Exp. Sta. Mimeo. Ser. 381.

Headlee, R.J., and G.A. Dean. 1908. The mound building prairie ant (*Pogonomyrmex occidentalis* Cresson). Kansas State Agr. Exp. Sta. Bull. 154: 165-80.

Hewitt, G.B., E.W. Huddleston, R.J. Lavigne, D.N. Ueckert, and J.G. Watts. 1974. Rangeland Entomology. Range Science Series No. 2. Society for Range Management, Denver, CO.

Jacot, A.P. 1936. Soil structure and biology. Ecology 17:359-79.

Johansen, C. 1971. Principles of insect control,p.171-190. *In:* R.E. Pfadt (ed.) Fundamentals of Applied Entomology, 2nd ed. The MacMillan Co., New York.

Killough, J.R., and A.C. Hull, Jr. 1951. Ants denude 90,000 acres in Big Horn Basin of Wyoming. Wyoming Stockman-Farmer 57:8.

Knowlton, G.F. 1938. Horned toads and ant control. J. Econ. Entomol. 31:128.

Knowlton, G.F. 1947. The sagebrush swift in pasture insect control. Herpetologica 4:25.

Knowlton, G.F. 1953. Some insect food of *Sceloporus g. graciosus* (B.-G.). Herpetologica 9:70.

Knowlton, G.F. 1959. Control of the harvester ant. Utah State Univ. Ext. Serv. Leaf. 48. (Revised, June 1963).

Knowlton, G.F., and B.A. Baldwin. 1953. Ants in horned toads. Herpetologica 9:70.

Knowlton, G.F., and F.C. Harmston. 1941. Insect food of the chipping sparrow. J. Econ. Entomol. 34:123-124.

Knowlton, G.F., and F.C. Harmston. 1942. Insect food of the rock wren. Great Basin Nat. 3:22.

Knowlton, G.F., and G.S. Stains. 1943. Flickers eat injurious insects. Can. Entomol. 75:118.

Knowlton, G.F., and H.F. Thornley. 1942. Insect food of the sage grouse. J. Econ. Entomol. 35:107-108.

Knowlton, G.F., and A.C. Valcarce. 1950. Insect food of the sage brush swift in Box Elder County, UT. Herpetologica 6:33-34.

Lavigne, R.J. 1966. Individual mound treatments for control of the western harvester ant, *Pogonomyrmex occidentalis*, in Wyoming. J. Econ. Entomol. 59:525-532.

Lavigne, R.J. 1969. Bionomics and nest structure of *Pogonomyrmex occidentalis* (Hymenoptera: Formicidae). Ann. Entomol. Soc. Amer. 62:1166-1175.

Lavigne, R.J., and H.G. Fisser. 1966. Controlling western harvester ants. Mountain States Regional Pub. 3.

List, G.M. 1954. Western harvester ant control tests. Colorado Agr. Exp. Sta. Tech. Bull. 55.

McCook, H.C. 1883. The occident ant of Dakota. Proc. Acad. Nat. Sci. Philadelphia 35:294-296.

Metcalf, R.L. 1962. Destructive and useful insects. McGraw-Hill, New York.

Popenoe, E.A. 1904. "*Pogonomyrmex occidentalis*". Can. Entomol. 36:360.

Race, S.R. 1964. Individual colony control of the western harvester ant, *Pogonomyrmex occidentalis*. J. Econ. Entomol. 57:860-864.

Race, S.R. 1966. Control of western harvester ants on rangeland. New Mexico Agr. Exp. Sta. Bull. 502.

Rogers, L.E. 1972. The ecological effects of the western harvester ant (*Pogonomyrmex occidentalis*) in the shortgrass plains ecosystem. Tech. Rep. 206. Colorado State Univ.

Rogers, L.E. 1974. Foraging activity of the western harvester ant in the shortgrass plains ecosystem. Environ. Entomol. 3:420-424.

Rogers, L.E., and R.J. Lavigne. 1974. Environmental effects of western harvester ants on the shortgrass plains ecosystem. Environ. Entomol. 3:994-997.

Severin, H.C. 1955. Harvester ants and their control. South Dakota Farm and Home Res. 6:36-37, 48-49.

Stevens, L.J. 1965. The food preference and foraging habits of the western harvester ant, *Pogonomyrmex occidentalis* (Cresson). Unpublished M. S. Thesis, Univ. Wyoming.

Talbot, M. 1953. Ants of an old field community on the Edwin S. George Reserve, Livingston County, Michigan. Lab. Vert. Biol. Contrib. 63:1-13.

Trevis, J.L. 1958. Interrelations between the harvester ant, *Veromessor pergandei* (Mayr), and some desert ephemerals. Ecology 39:695-704.

Whitford, W.G. 1978. Foraging in seed-harvester ants *Pogonomyrmex* spp. Ecology 59:185-189.

Wight, J.R., and J.T. Nichols. 1966. Effects of harvester ants on production of saltbush community. J. Range Manage. 19:68-71.

Wilson, E.O. 1972. The insect societies. Belknap Press of Harvard University Press, Cambridge, MA.

19. BELOWGROUND ARTHROPODS OF SEMIARID GRASSLANDS

David Evans Walter
Natural Resource Ecology Laboratory
and Department of Entomology
Colorado State University, Fort Collins, CO 80523

A great variety of arthropods inhabit the semiarid soils of the grama-buffalo grass prairies (shortgrass steppe). One approach to dealing with this variety has been to divide soil arthropods into faunal size classes. The mesofauna are those arthropods greater than 1 mm in length. The larger members of the mesofauna are sometimes called macroarthropods. However, most soil arthropods are very small, generally less than 1 mm in length, and collectively are referred to as microarthropods. The majority of these animals are members of the decomposer subsystem, consumers of the microbial decomposers of decaying plant and animal material, or predators of those consumers. The microarthropods of semiarid grasslands, unlike those of forest soils, are tolerant of dry conditions. They occur in great numbers throughout the soil profile, at least to the depth of the hardpan. Root-feeding arthropods, including some important arthropod pests of grasses, spend all or part of their life cycle belowground. A simplified food web emphasizing the position of belowground arthropods in the shortgrass steppe is presented in Figure 19.1. The substrate pool consists of the products of primary production (roots, exudates, algae, litter), and returns to the substrate pool (microbial and animal bodies, feces, microbial products, etc.). Returns to the substrate pool via death, defecation, etc., are not shown. A more detailed food web may be found in Hunt et al. (1987).

Because soil arthropods are cryptic in their habits, they are difficult to observe and to sample. In grasslands, most previous workers have relied on behavioral extraction techniques developed for mesic forest biome soils that are not efficient for dryland faunas. Diversity, abundance and biomass of soil microarthropods in the shortgrass steppe have been underestimated in previous studies (Walter, Kethley and Moore in press). The information in the following chapter is based on my own observations of live

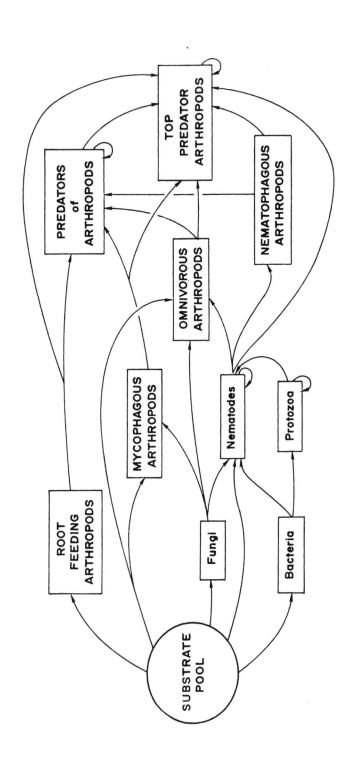

FIG. 19.1 A simplified belowground food web.

microarthropods extracted from native sod from areas of short- to midgrass steppe near Sidney, Nebraska; Cheyenne, Wyoming; and the Central Plains Experimental Range (CPER), near Nunn, Colorado. Slide mounted material from the above sites, and numerous other grassland collections from eastern Wyoming, Colorado, New Mexico, and western Nebraska were used for species determinations and examination of gut contents. In the following tables, feeding information has been based, by necessity, primarily on my observations. When available, literature information for congeneric species from other habitats has been included. A "?" indicates a potential food source that I have not been able to confirm from observation or the literature. A "+" indicates degree of feeding, either infrequent (+), moderate (++), or high (+++) rates of feeding. A high rate of feeding (+++) implies that an arthropod may be reared on that food source.

This chapter has three goals. The first is to survey some of the important taxa of soil arthropods found in the shortgrass steppe. The second is to review recent information on the abundance and trophic behavior of soil microarthropods in semiarid grasslands. The third goal is to suggest that, although soil microarthropods have long been recognized as being important mediators of decomposition and nutrient cycling, they are also directly impacting two other components of the shortgrass system: plants and nematodes.

SAMPLING TECHNIQUES FOR MICROARTHROPODS

In semiarid grasslands, soil microarthropod investigators have relied on behavioral techniques such as the Merchant-Crossley high-gradient Tullgren funnel which uses a light bulb to establish a temperature/humidity gradient through an intact soil core (Merchant and Crossley 1970). The soil animals are driven from the core by the heat and dropping humidity and fall through a screen into a funnel which leads to a vial of ethanol, where they are preserved. High-gradient extractors were designed for use in mesic, highly organic forest or meadow litter layers and surface soils where the dominant soil microarthropod fauna are oribatid mites and Collembola that are robust, active, and respond well to humidity gradients. In contrast, the microarthropod fauna of arid grasslands is dominated by small, soft-bodied prostigmatic mites, astigmatic mites, and fragile deep-soil collembolans (Leetham and Milchunas 1985, Walter, Kethley and Moore 1987) which are well adapted to dry conditions. At a variety of short- to midgrass steppe sites, Walter et al. have determined that high-gradient extraction underestimates microarthropod abundance by a factor of four, biomass by about one half, and species diversity by one half.

PREDATORY ARTHROPODS

A variety of surface dwelling insects and arachnids are peripheral predators on the belowground system. These include numerous species of carabid and staphylinid beetles, the larvae of asilid, dolichopodid, empidid, and cecidomyiid flies, anthocorid, cydnid, and *Geocoris* bugs, ants, spiders, and anystoid mites. Species lists and some trophic information for these groups have been developed for the CPER (formerly the Pawnee Site of the U.S. International Biological Program) near Nunn, Colorado (see Lloyd and Grow 1971, Bell 1971, Harris and Paur 1972, Kumar et al. 1975, Leetham 1977).

Diplurans in the families Japygidae and Campodeidae are mesofaunal predators that are not uncommon in some of the richer soils of the shortgrass steppe. Campodeids are often found in association with symphylans, but it is not clear if this represents similar microclimatic requirements or a predator-prey relationship. A similar association between campodeids and symphylans has been noted in Europe (Kuhnelt 1976). The symphylans in the shortgrass steppe (*Symphylellina* sp.) have nematodes and hypogasturid springtails in their guts. Pseudoscorpions such as *Parachermes nubilis* Hoff, centipedes, and a few small species of *Bembidion* and *Tachys* (Carabidae) are other common mesofaunal predators.

Predatory Mites

The predatory component of the belowground system is primarily the realm of the Acarina, especially the Mesostigmata (Table 19.1). The mite predators in the belowground system fall into three functional categories based on feeding behavior. These include top predators, specialized predators of other arthropods, and specialized nematode predators.

Some mesostigmatic mites restrict their feeding to a particular group of animals, for example *Veigaia* sp. prefers springtails and *Alliphis* nr. *halleri* (G. and R. Canestrini) feeds only on nematodes. Others find a variety of prey acceptable, for example, I have been able to culture *Cosmolaelaps vacua* (Michael) on nematodes (Rhabditida and Tylenchida) and on mites (*Tyrophagus* spp.). Mesostigmatic mites in the families Laelapidae, Ascidae, Rhodacaridae, and Phytoseiidae are able to feed at the top of the belowground food web by preying on bacterial-, fungal- and plant-feeding nematodes, as well as on arthropods such as mites, collembolans, and the eggs and young of insects. The most abundant group of top predators are hypoaspine laelapid mites in the genera *Hypoaspis* (*sensu latu*, including *Gaeolaelaps*), *Cosmolaelaps*, *Pseudoparasitus*, and *Ololaelaps*. Most species are between 500 and 1000 μm in length and are rapidly moving, voracious predators that can develop from egg to adult in less than two weeks if food is abundant and soil temperature is above 20°C.

TABLE 19.1
Predatory Mesostigmata from the shortgrass steppe. + = degree of feeding, ? = potential food, - = not fed upon.

	Feeding Categories		
Taxon	Fungi	Nematodes	Arthropods
Laelapidae			
Ololaelaps veneta		+++	+++
Pseudoparasitus n. sp.		+++	+++
Pseudoparasitus austriacus		+++	+++
Cosmolaelaps vacua	?	+++	+++
Hypoaspis cf. karawaiewi		+++	++
Hypoaspis nr. giffordi		+++	+++
Hypoaspis nr. similisetae		+++	+++
Ascidae			
Asca nesoica	?	+++	+++
Asca nr. piloja	?	+++	+++
Gamasellodes vermivorax	?	+++	+++
Arctoseius cetratus		+++	+++
Protogamasellus mica		+++	+++
Rhodacaridae			
Rhodacarus denticulatus		+++	+++
Rhodacarellus silesiacus		+++	+++
Rhodacarellus subterraneus		+++	+++
Phytoseiidae			
Amblyseius nr. brevispinus	?	+++	+++
Amblyseius spp.	?	+	+++
Veigaiidae			
Veigaia sp.		+	+++
Eviphididae			
Eviphis sp.		+++	-
Alliphis nr. halleri		+++	-
Crassicheles sp.		+++	-
Macrochelidae			
Macrocheles n. sp.		+++	+++

In the upper layers of the soil, ascid mites in the genus *Asca* and phytoseiid mites in the genus *Amblyseius* are often very abundant. Species in both genera will attack soft-bodied arthropods such as tydeid and tetranychid mites, and will also attack nematodes. *Amblyseius* species near *brevispinus* (Kennett) is unusual for a phytoseiid mite in that it may be easily cultured with only nematodes for a food source. Deeper in the soil profile, smaller predators such as the ascid mites *Arctoseius* and *Gamasellodes*, or the rhodacarid mites *Rhodacarus* and *Rhodacarellus* also are abundant. Developmental times for ascid, phytoseiid and rhodacarid mites in the shortgrass steppe are similar to the hypoaspine laelapid mites.

Phoretic mites in the families Eviphididae, Parasitidae, and Macrochelidae tend to be associated with cattle dung in the shortgrass steppe. *Macrocheles perglaber* Filipponi and Pegazzano and *M. dimidiatus* Berlese, predators of the eggs and early instar larvae of muscid flies, are restricted to cattle dung. However, *Macrocheles* n. sp., *Alliphis* nr. *halleri*, and species of *Eviphis* and *Crassicheles*, which are predators of nematodes, are able to invade the soil and are often collected in samples of native sod.

In contrast to the Mesostigmata, predaceous prostigmatic mites tend to prey only on other arthropods (Table 19.2). Members of the family Rhagidiidae are small, white, soft bodied predators which occur deep in the soil profile where they attack small mites and collembolans. Elongate purple paratydeid mites in the genus *Tanytydeus* are presumed to be predators, but like many grassland mites, their feeding habits are not known. I have observed related mites in the genus *Paratydeus* feeding on both fungi and nematodes.

The members of the superfamilies Bdelloidea and Raphignathoidea are the most common prostigmatic predators in the shortgrass steppe. The bdellid mite, *Spinibdella cronini* (Baker and Balock), is a very abundant large red predator of mites and collembolans which uses silk spun from the mouthparts to tie down prey items to the substrate before draining their body fluids. Another large red bdellid, *Cyta latirostris* (Hermann), preys on hard-bodied mites such as *Scutacarus* and *Bakerdania*. The smaller cunaxid mites in the genera *Armascirus*, *Pulaeus*, and *Pseudobonzia* occur lower in the soil profile. *Armascirus* sp. will attack both collembolans and small mites. *Pulaeus clarae* Heyer, a voracious predator of soft-bodied mites, spins a tightly woven silken cocoon in which it becomes quiescent and molts. The cocoon is effective protection from its cannibalistic brethren, and may also function to establish favorable microenvironmental conditions for molting.

TABLE 19.2

Predatory Prostigmata from the shortgrass steppe.   + = degree of feeding, ? = potential food, - = not fed upon.

Taxon	Feeding Categories		
	Fungi	Nematodes	Arthropods
Bimichaelidae			
Alycus roseus	-	+++	-
Rhagidiidae			
Coccorhagidia sp.			+++
Hammenia sp.			?
Bdelloidea			
Spinibdella cronini	-	-	+++
Cyta latirostris	-	-	+++
Armascirus sp.	-	-	+++
Pulaeus clarae	-	-	+++
Pseudobonzia sp.			?
Raphignathoidea			
Neophyllobius sp.			+++
Stigmaeus rayneri			+++
Neognathus sp.	?	-	eggs
Raphignathus gracilis	?	-	?
Cryptognathus barrasi	?	-	?
Cheyletidae			
Prosocheyla n. sp.		-	+++
Paratydeidae			
Tanytydeus n. sp.	?	?	?
Paratydeus sp.	+	+	?
Erythraeidae			
Balaustium sp.	?		+++

For most of the Raphignathoidea, feeding habits are not well known.   I have been able to rear *Stigmaeus raynei* Summers on the nymphs of oribatid mites.   However, another common stigmaeid in the shortgrass steppe, *Eustigmaeus segnis* (Koch) feeds on mosses (Gerson 1972), which are common in patches of bare soil in the prairie.   Camerobiid mites in the genus *Neophyllobius* are known to be predators of scale insects (Zaher and Gomaa 1979). Caligonellid mites in the genus *Neognathus* are the most common raphignathoid mites in the shortgrass steppe.   I have observed *Neognathus* feeding on the eggs of *Tyrophagus*, but have never seen them attack a moving animal. To date there is absolutely no information on the trophic behavior of cryptognathid mites, although they are not uncommon in grassland soils (McDaniel and Bolen 1979).

Omnivorous Microarthropods

The term omnivore, which once meant an animal with a very broad diet, has been used in the ecological literature to mean an animal which feeds at more than one trophic level (Pimm 1982). Many of the predatory mites in the belowground system will attack prey at several trophic levels, and in all probability, some of the mesostigmatic, raphignathoid, and paratydeid predators also derive some of their nourishment from fungi or other foods when preferred prey are scarce. In above-ground systems where predatory mite feeding behavior is better known, many predators of mites also feed on pollen, honeydew, or plant sap (see Tanigoshi 1982, Overmeer 1985). In addition, many mycophagous microarthropods will attack and consume nematodes (Muraoki and Ishibashi 1976, Rockett 1980, Santos and Whitford 1981).

Many of the mycophagous microarthropods of semiarid grasslands have been shown to feed at several trophic levels (Walter 1987, Walter et al. 1986). Among these animals is a subset which derive significant portions of their diets from both microbes and small animals. In this discussion, I am using the term omnivore to describe these animals (Table 19.3).

Species in the genus *Anotylus* are among the more common staphylinid beetles in samples of native sod from the shortgrass steppe. These animals consume rotting plant material, algae, fungi, nematodes, and (at least in the laboratory) the eggs and larvae of their own species. Among the other omnivores in Table 19.3, feeding on arthropods (including cannibalism) has been observed primarily against molting, injured, or dead individuals. The most common astigmatic mites in grassland soils are species of *Tyrophagus*. These mites will attack almost any organic material (animate or inanimate) with a high protein or lipid content (Walter et al. 1986), and have been observed feeding on the eggs of Collembola (J.C. Moore per.obs.) and corn rootworm (G.J. House per. obs.).

Three species of large (>500 μm) oribatid mites have been observed to avidly attack and consume nematodes even when more usual fare (fungi and algae) is available. *Pilogalumna* n. sp. is an oribatid mite with feeding habits nearly as indiscriminant as those of *Tyrophagus*. When nematodes are added to a algal-fungal diet, *Pilogalumna* n. sp. double the size of their egg clutches (to about 24 eggs). It is likely that other species of oribatid mites have the potential to feed on nematodes as well as on microbes. Unlike the oribatid mites, *Alicorhagia fragilis* Berlese, a member of the primitive mite group Endeostigmata, is primarily a predator of nematodes. Although *Alicorhagia* will consume fungi and algae, nematode-free cultures of *Alicorhagia* soon decline to extinction. With only nematodes as a food source, I have been able to maintain thriving cultures for more than 18 months (generation time is about 1 month). Species of *Alicorhagia* also spin silken cocoons before molting (Walter in press).

TABLE 19.3
Omnivorous microarthropods from the shortgrass steppe. + = degree of feeding, ? = potential food.

| Taxon | Feeding Categories | | | |
	Fungi	Algae	Nematodes	Arthropods
Oribatida				
Pilogalumna n.sp.	+++	+++	+++	*
Haplozetes sp.	+++	++	++	
Ceratozetes sp.1	+++	++	++	
Astigmata				
Tyrophagus similis	+++	+++	+++	*
Tyrophagus zachvatkini	+++	+++	+++	*
Endeostigmata				
Alicorhagia fragilis	++	+	+++	
Alicorhagia usilata	++		+++	
Collembola				
Hypogastura scotti	+++	+++	+++	
Coleoptera				
Anotylus sp.	+++	++	++	*

*dead or injured arthropods, or cannibalism

MICROPHYTOPHAGOUS MICROARTHROPODS

Microarthropods which consume small "plants", such as bacteria, fungi, algae, lichens, and mosses are called microphytophages by soil zoologists. Most of the diversity, abundance, and biomass of soil microarthropods properly belongs to the microphytophagous feeding category. Functionally, these animals may be categorized by their modes of feeding.

Engulfing microphytophages bite off and ingest pieces of microbial tissue, and often particles of substrate as well. Most oribatid mites are engulfing microphytophages. Some species, such as the cosmopolitan *Oppiella nova* (Oudemans), are extremely polyphagous. I have cultured *Oppiella nova* on powdered mushrooms, algae, and several species in each of the fungal genera *Penicillium*, *Paecilomyces*, *Aspergillus*, and *Cladosporium*. Other oribatid mites such as *Oribatula minuta* (Ewing) or *Zygoribatula fusca* (Ewing) have much narrower diets. In the shortgrass steppe, most oribatid mites (except damaeids, gymnodamaeids, and cosmochthonoids) will readily consume algae, and as noted above, some of the larger oribatid mites are nematophagous. Some oribatids, such as *Oppiella nova* and *Pilogalumna* n. sp., can complete their development in 3-5 weeks. However, most oribatid mites develop at much slower rates, and probably complete only 1-2 generations a year in the northern Great Plains.

I have collected species in 29 genera of oribatid mites from shortgrass steppe sites. In coniferous and deciduous forests soils, oribatid mites are the dominant taxocene, and there are often several sympatric species in each genus (Walter and Norton 1984). In semiarid grasslands, oribatid mites, while often abundant, are low in diversity (usually one species per genus per site) and are overshadowed by the more abundant Prostigmata. Prominent shortgrass steppe oribatid genera include *Aphelacarus*, *Beklemishevia*, *Brachychthonius*, *Cosmochthonius*, *Phyllozetes*, *Sphaerochthonius*, *Nothrus*, *Trhypochthonius*, *Joshuella*, *Fosseremus*, *Epidamaeus*, *Tectocepheus*, *Passalozetes*, *Banksinoma*, *Oppia*, *Oppiella*, *Microppia*, *Oribatula*, *Zygoribatula*, *Scheloribates*, *Haplozetes*, *Peloribates*, *Ceratozetes*, and *Pilogalumna*. Deep in the soil profile, taxa characteristic of more mesic (*Euphthiracarus*, *Epilohmannia*), and warmer areas (*Lohmannia*) may be found. The unusual oribatid mite family Pediculochelidae (*Paralycus* sp.) is sometimes common in shortgrass sod samples from Colorado. Taxa in the superfamilies Hypochthonoidea, Phthiracaroidea, Eremaeoidea, Liacaroidea, Oribatelloidea, and Pelopoidea are very rare or absent from shortgrass steppe sites, although they are common in nearby montane or relict conifer habitats.

Collembola or springtails are another important group of microphytophages. *Isotoma notabilis* Schaefer, *I. uniens* Christiansen and Bellinger, *Folsomia elongata* (MacGillivray), *Sminthurides pumilis* (Krausbauer), *Hypogastura scotti* (Yosii) and *Pseudosinella* sp. are especially abundant in the upper layers of the soil after rainfall. Deeper in the soil profile, isotomid (*Isotomodes productus* (Axelson), *Folsomides americanus* Denis, *Isotoma* spp.), hypogasturid (*Willemia vashita* Wray) and onychiurid (*Tullbergia granulata* Mills, *Onychiurus* sp.) springtails are often very abundant. *Tullbergia granulata*, a small slow-moving white collembolan, lacks a springing mechanism, but does possess a chemical defense. When attacked, *T. granulata* exudes small droplets of fluid along the sides of the abdomen which discourage some predators. Most collembolans will feed on algae as well as fungi, and some springtails have been shown to ingest the spores and hyphae of vesicular-arbuscular mycorrhizae (Moore et al. 1985). Psocoptera in the family Liposcelidae, aleocharine staphylinid and corylophid beetles, pauropods (Pauropodidae), and Protura (*Eosentomon* sp. - actually a fluid feeder) are common and locally abundant microphytophages in the shortgrass steppe.

An alternative to engulfing particulate microbial material is to puncture and drain the contents of a microbial cell. As mentioned above, Protura and some springtails are believed to feed on the fluids contained in fungal hyphae. The most abundant mite taxa in the shortgrass steppe are those which feed on the fluids of fungi and algae. These fluid-feeding microphytophages belong to three taxa of prostigmatic mites: Endeostigmata, Heterostigmata, and Eupodina.

The most primitive element of the mite order Acariformes, which includes the Prostigmata, Oribatida and Astigmata, is the supercohort Endeostigmata. The least derived members of the Endeostigmata are engulfers, either fungivores such as *Terpnacarus*, *Oehserchestes*, and *Micropsammus*, nematophages such as *Alycus roseus* (Koch) (Walter in press), or omnivores such as *Alicorhagia*. Many of the taxa in the Endeostigmata, however, do not ingest particulate matter (Walter in press). These include prominent dryland fauna such as the mycophagous-algivorous nanorchestid mites (*Nanorchestes*, *Speleorchestes*), and the exceedingly bizarre Nematalycidae. Nematalycid mites are elongate and worm-like, resembling nematodes more than mites. Adults in the genus *Gordialycus* may be more than 1 mm in length. Species of *Gordialycus*, *Psammolicus*, and *Cunliffea* occur in the shortgrass steppe, often at great depth in the soil.

The Heterostigmata also are fluid feeders, primarily on fungal hyphae. Pygmephorid mites in the genera *Siteroptes*, *Pediculaster*, and *Bakerdania* are common and often abundant inhabitants of the upper layers of prairie sod. A number of taxa of pyemotid mites occur in the shortgrass steppe and might be confused with pygmephorid mites. Species of *Pyemotes* are parasites (often fatal) of Homoptera, Coleoptera, Diptera and Hymenoptera (Krantz 1978). Scutacarid mites, especially species of *Scutacarus*, are another common group of mycophagous heterostigmatic mites in grasslands. Tarsonemid mites are also abundant mycophagous, and unlike pygmephorids and scutacarids, phytophagous members of the soil fauna. Species in the genus *Tarsonemus* will feed on algae as well as fungi (Walter 1987). A species of *Steneotarsonemus* may be found at least to the depth of 10 cm in association with native grasses. All known species of *Steneotarsonemus* are phytophagous (E. Lindquist pers. comm.), so presumably these animals are feeding on the crowns or roots of native grassses.

Eupodine mites are small, white, soft-bodied mites which often reach extremely high abundances in the shortgrass steppe. Members of the family Eupodidae (*Eupodes*, *Linopodes*) are abundant, generally mycophagous members of the soil fauna. Species of *Eupodes* often have green or orange gut contents when extracted from soil samples. In my laboratory, species of *Eupodes* readily consume a variety of soil algae which colors their guts green (or orange as the algae are digested). It is possible that eupodid mites are also opportunistically feeding on higher plants, since the closely related family Penthaleidae contains plant parasitic pests such as the winter grain mite, *Pethaleus major* (Duges). I have also observed a species of *Eupodes* to feed, at low rates, on nematodes, and have successfully cultured them on a fungus (*Cladosporium herbarum*). Tydeid mites in the genus *Tydeus* also readily consume algae and suck fluids from fungal hyphae. Santos and Whitford (1981) have observed microtydeid mites (a taxonomic nightmare, with at least five genera of tiny (<130 μm in length) mites collectively referred to as microtydeids) feeding on the eggs and active instars of nematodes and on fungi. I have observed microtydeids from the shortgrass steppe feeding on fungal hyphae, but have been unable to confirm their predation on nematodes. Tydeid mites in aboveground systems have been successfully reared on diets of fungi, pollen, honeydew, and the eggs of spider mites (Brickhill 1958, Knop and Hoy 1983). Ereynetid mites (*Ereynetes*) resemble tydeid mites and are often collected at low densities in the shortgrass steppe. Because of the structure of their mouthparts, *Ereynetes* spp. are presumed to be predatory (J. Kethley per. comm.). I have been able to maintain small cultures of *Ereynetes* sp. for several months with a variety of fungi and nematodes available as food sources, but have never observed feeding.

MACROPHYTOPHAGOUS ARTHROPODS

Arthropods which feed on higher plants are a prominent component of the shortgrass steppe system (Kumar et al. 1975). Lloyd and Kumar (1977) found 28 species of root-feeding soil mesofaunal (> 1 mm) arthropods on the Central Plains Experimental Range, with densities of several hundred per square meter in some years. Densities of root-feeders were greatest in plots ungrazed by cattle.

Herbivorous Microarthropods

When soil samples of the upper layers of native sod or dryland wheat fields are examined, there are often numerous phytophagous microarthropods collected (Crossley et al. 1975, Leetham and Milchunas 1985) (Table 19.4). Many of these, such as the spider mites (Tetranychidae), are feeding in the sheaths and on the aboveground portions of the grasses and herbs included in the soil core. Some herbivores, for example the brown wheat mite [*Petrobia latens* (Müller)], descend to the soil to lay their eggs, so adult females, eggs and larvae are briefly part of the soil fauna. For other species such as the false spider mite, *Pseudoleptus* nr. *palustria* Pritchard and Baker, or the previously mentioned tarsonemid mite, *Steneotarsonemus* sp., feeding sites are unknown, and may be the crowns or roots of native grasses.

Two unusual families in the Tetranychoidea are common inhabitants of grassland soils where they probably feed on the roots of plants. An undescribed species of *Linotetranus* (Linotetranidae) is found in association with the roots of native grasses. This mite is a close relative of the false spider mites, but it has lost all pigmentation, has an elongate narrow-sided body with transverse scissures which allow the body to bend and move through soil channels, and may be found to at least 60 cm depth in the soil (Leetham and Milchunas 1985). In contrast, species in the genus *Tuckerella* (Tuckerellidae), although they have lost all pigment, are covered with very ornate setae, and would not appear to be very well adapted to movement in the soil. In spite of their morpholgy, tuckerellids are common in prairie soils (McDaniel et al. 1975).

Herbivorous Macroarthropods

A variety of phytophagous arthropods larger than 1 mm may be found in the soil. Some of these animals, such as the chinch bugs (*Blissus* spp.), feed on aboveground portions of plants, and hide in the surface litter. Most, however, feed on the tissues or sap of plant roots (Table 19.5).

White grubs. The larvae of scarab beetles that feed on roots are collectively referred to as white grubs. Because the larvae and adults are large and distinctive, and the larval damage is often severe, white grubs have long been recognized as important pests of rangelands. At low levels of infestation, white grub damage may be confused with drought stress or low fertility (Ueckert 1979). At high densities, however, large patches of turf may be killed. Species in four genera of white grubs are important in the shortgrass steppe.

TABLE 19.4
Herbivorous microarthropods from the shortgrass steppe. + = degree of feeding, ? = potential food.

Taxon	Roots	Foliage
Acarina		
Tuckerellidae		
Tuckerella sp	?	
Linotetranidae		
Linotetranus n. sp.	?	
Tenuipalpidae		
Pseudoleptus nr. *palustria*	?	+++
Tetranychidae		
Petrobia latens		+++
Monoceronychus nr. *mcgregori*		+++
Schizotetranychus eremophilus		+++
Eriophyidae		
Eriophyid sp.	?	+++
Tarsonemidae		
Steneotarsonemus sp.	?	+++
Stigmaeidae		
Eustigmaeus spp.		mosses

The adults of the 3-year white grubs (*Phyllophaga* spp.) are called May or June beetles. *Phyllophaga anxia* (LeConte) has caused extensive damage to hay crops in the Sandhills grasslands of Nebraska (Jarvis 1964, 1966). *Phyllophaga crinita* (Burm.) will complete its life cycle in one year in southern or two years in the northern Texas, and can reduce perennial grass cover by 88% in areas of the Texas shortgrass steppe (Ueckert 1979). In Colorado, *P. fimbripes* (LeConte) may cause severe but sporadic damage to perennial grasses at the CPER and may damage grama grasses in other areas of Colorado and New Mexico (Wiener 1979, Wiener and Capinera 1980).

The annual white grub (*Cyclocephala* spp.), as its name implies, may complete its development in a year. Damage is similar to that of the 3-year white grubs, with grasses succumbing to drought stress after their roots have been consumed by grubs. The adults are known as masked chafers. The black turfgrass ataenius (*Ataenius spretulus*) (Hald) is a small, bivoltine white grub that may achieve high densities. Both annual white grubs and the black turfgrass ataenius may be pests in lawns and golf courses (Roselle 1983). Lloyd and Kumar (1977) found that 34% of the soil Coleoptera larger than 1 mm collected on the CPER belonged to the root feeding white grub *Trichorhyssemus riparius* Horn.

TABLE 19.5
Herbivorous belowground macroarthropods from the shortgrass steppe. + = degree of feeding, ? = potential food.

Taxon	Roots	Foliage
Coleoptera		
Scarabaeidae		
Phyllophaga spp.	+++	
Cyclocephala sp.	+++	
Ataenius sp.	+++	
Trichorhyssemus riparius	+++	
Hemiptera		
Lygaeidae		
Blissus spp.		+++
Homoptera		
Pseudococcidae		
Amonostherium lichtensioides	+++	
Anisococcus sp.	+++	
Antonina sp.	+++	
Cryptoripersia arizonensis	+++	
Distichlicoccus megacirculus	+++	
Phenacococcus sp.	+++	
Radicoccus kelloggi	+++	
Rhizoecus sp.	+++	
Syrmococcus pecoensis	+++	
Eriococcidae		
Apezococcus idiastes	+++	
Eriococcus spp.	+++	
Margarodidae	+++	
Margarodes hiemalis	+++	
Ortheziidae		
Orthezia spp.	+++	
Aphididae		
Geoica utricularia	+++	
Tetraneura ulmi	+++	
Forda spp.	+++	

Scale insects and aphids. Kumar et al. (1975) listed 15 species in four families of scale insects (Coccoidea) which feed on belowground portions of plants on the CPER. Of these, Margarodidae and Ortheziidae become attached to roots and are not collected by normal soil microarthropod extraction techniques. Lloyd and Kumar (1977) using a sieving-flotation procedure found *Margarodes hiemalis* Cockerell to account for most of the mesofaunal (> 1 mm) root-sucking arthropods on the CPER. The mealybugs (Eriococcidae and Pseudococcidae), however, are often collected in extremely high numbers, especially the crawler stages. I have found standard soil cores (ca. 100 g dry soil) to contain over 700 pseudococcid crawlers. Kumar et al. (1975) reported 12 species of mealybugs from the CPER (see Table 19.5). At Sidney, Nebraska, an apterous encyrtid wasp parasitoid of mealybugs is often common.

Other belowground Homoptera may be exerting grazing pressure on native and crop plants. The corn root aphid, *Aphis maidiradicis* Forber is tended by ants in the genus *Lasius*. Kumar et al. (1975) report four species of *Lasius* from the CPER. The subterranean aphid, *Geoica tricularia* (Passerini) is also tended by *Lasius* ants, and has been reported attacking the roots of a number of grasses in Nebraska (Vogel and Kindler 1980). Palmer (1952) has reported other root feeding aphids from grasses (*Tetraneura ulmi* (L.), *Forda formicaria* von Heyden) and grain crops (*Forda marginata* Koch).

CONCLUSIONS

Soil microarthropods are generally assumed to be important in decomposition and nutrient cycling. However, very little is known about the biology and life history of microarthropods of semiarid grasslands. For example, the extent to which vesicular-arbuscular mycorrhizae are consumed by mycophagous microarthropods has not been investigated. Previous sampling methods have resulted in underestimations of diversity, density and biomass. Because of these conditions it is difficult to judge the importance of the soil fauna to the shortgrass steppe system; however, see Hunt et al. 1987.

When belowground arthropods are large and the damage they cause severe, such as during outbreaks of white grubs, the importance of belowground arthropods to the shortgrass steppe is clear. My investigations have indicated that the soil microarthropod fauna is affecting native grasslands in two generally unrecognized ways. In the first case, there is a diverse and abundant guild of root-sap-feeding microarthropods that are potentially limiting plant productivity. Root-feeding nematodes have long been suspected of controlling productivity in the shortgrass steppe (Scott et al. 1979), but root-feeding microarthropods have been essentially ignored. In the second case, many of the predatory mites in the shortgrass system, and at least some of the microbial grazers, are consuming a significant part of the nematode population. Again, the influence of this behavior on nutrient cycling, food web stability, and nematode impact on plant growth has not been investigated.

ACKNOWLEDGEMENTS

My investigations into the belowground arthropod fauna of the shortgrass steppe have been supported in part by NSF Grant No. BSR-8418049. I'd like to thank J. Kethley, E.E. Lindquist, D.E. Johnston, S.J. Loring, B.M. OConnor, M. Kaliszewski, and B.C. Kondratieff for their assistance in identifying specimens, and H.W. Hunt, W.K. Lauenroth, T.R. Seastedt, E. Hudgens, G.W. Krantz, and J.L. Capinera for reviewing drafts of this manuscript. Among the many colleagues whose discussions have contributed to my understanding of soil arthropods in the shortgrass steppe, I'd especially like to thank Diana Freckman, John Kethley, John Capinera, John Moore, Tim Seastedt, and David Coleman.

REFERENCES

Bell, R.T. 1970. Identifying Tenebrionidae (Darkling beetles). Grasslands Biome, U.S. International Biological Program Technical Report No. 58.

Bell, R.T. 1971. Carabidae (Ground beetles). Grasslands Biome, U.S. International Biological Program Technical Report No. 66.

Brickhill, C.D. 1958. Biological studies of two species of tydeid mites from California. Hilgardia 27:601-620.

Crossley Jr., D.A., C.W. Proctor, Jr., and C. Gist. 1975. Summer biomass of soil microarthropods of the Pawnee National Grasslands, Colorado. Am. Midl. Nat. 93:491-495.

Daniels, N.E. 1966. The association of a rangeland grub, *Phyllophaga koeleriana* (Coleoptera: Scarabaeidae), with asilid larvae and with mites. Ann. Entomol. Soc. Am. 59:1021.

Gerson, U. 1972. Mites of the genus *Ledermuelleria* (Prostigmata: Stigmaeidae) associated with mosses in Canada. Acarologia 13:319-343.

Harris, L.D., and L. Paur. 1972. A quantitative food web analysis of a shortgrass community. Grasslands Biome, U.S. International Biological Program Technical Report No. 154.

288

Hunt, H.W., D.C. Coleman, E.R. Ingham, R.E. Ingham, E.T. Elliott, J.C. Moore, S.L. Rose, C.P.P. Reid, and C. Morley. 1987. The detrital food web in a shortgrass prairie. Biol. Fert. Soil 3:57-68.

Jarvis, J.L. 1964. An association between a species of *Caloglyphus* (Acarina: Acaridae) and *Phyllophaga anxia* (Coleoptera: Scarabaeidae). J. Kansas Entomol. Soc. 37:207-210.

Jarvis, J.L. 1966. Studies of *Phyllophaga anxia* (Coleoptera: Scarabaeidae) in the sandhills area of Nebraska. J. Kansas Entomol. Soc. 39:401-409.

Knop, N.F., and M.A. Hoy. 1983. Biology of a tydeid mite, *Homeopronematus anconai* (n. comb.) (Acari: Tydeidae), important in San Joaquin Valley vineyards. Hilgardia 51:1-30.

Krantz, G.W. 1978. A manual of acarology. Oregon State Univ. Bookstore, Corvallis.

Kuhnelt, W. 1976. Soil biology with special reference to the animal kingdom. Michigan State Univ. Press, East Lansing.

Kumar, R., R.J. Lavigne, J.E. Lloyd, and R.E. Pfadt. 1975. Macroinvertebrates of the Pawnee Site. Grasslands Biome, U.S. International Biological Program Technical Report No. 278.

Leetham, J.W. 1977. Population structure, abundance, and vertical distribution patterns of soil microarthropods in a Northeastern Colorado shortgrass prairie. Grasslands Biome, U.S. International Biological Program Preprint No. 219.

Leetham, J.W., and D.G. Milchunas. 1985. The composition and distribution of soil microarthropods in the shortgrass steppe in relation to soil water, root biomass, and grazing by cattle. Pedobiologia 22:172-184.

Lloyd, J.E., and R.R. Grow. 1971. Soil macro-arthropods of the Pawnee Site. Grasslands Biome, U.S. International Biological Program Technical Report No. 104.

Lloyd, J.E., and R. Kumar. 1977. Root feeding insects of a shortgrass prairie and their response to grazing pressure and ecosystem stress, p. 267-272. *In:* J.K. Marshall (ed.), The belowground ecosystem: a synthesis of plant associated processes. Range Science Department series No. 26. Colorado State University, Fort Collins.

McDaniel, B. and E. Bolen. 1979. Two new species of the genus *Cryptognathus* Kramer from South Dakota and Texas (Acari:Cryptognathidae). Inter. J. Acarol. 5:93-102.

McDaniel, B., D.K. Morihara, and J.K. Lewis. 1975. A new species of *Tuckerella* from South Dakota and a key with illustrations of all known described species. Acarologia 17:274-283.

Merchant, V.A., and D.A. Crossley, Jr. 1970. An inexpensive high efficiency Tullgren extractor for soil microarthropods. J. Georgia Entomol. Soc. 5:83-87.

Moore, J.C., T.V. St. John, and D.C. Coleman. 1985. Ingestion of vesicular-arbuscular hyphae and spores by soil microarthropods. Ecology 66:1979-1981.

Muraoka, M., and N. Ishibashi. 1976. Nematode-feeding mites and their feeding behaviour. Appl. Entomol. Zool. 11:1-7.

Overmeer, W.P.J. 1985. Alternative prey and other food resources, p.137-145. In: W. Helle and M.W. Sabelis (eds.) Spider mites: their biology, natural enemies and control. Volume 1B. Elsevier, New York.

Palmer, M.A. 1952. Aphids of the rocky mountain region. Thomas Say foundation 5:1-452.

Pimm, S.L. 1982. Food webs. Chapman and Hall, New York.

Rockett, C.L. 1980. Nematode predation by oribatid mites (Acari:Oribatida). Inter. J. Acarol. 6:219-224.

Roselle, R.E. 1983. White grubs in turf. NebGuide G80-522, Cooperative Extension Service, University of Nebraska, Lincoln.

Santos, P.F., and W.G. Whitford. 1981. The effects of microarthropods on litter decomposition in a Chihuahuan desert ecosystem. Ecology 62:654-663.

Scott, J.A., N.R. French, and J.W. Leetham. 1979. Patterns of consumption in grasslands, p.89-116. In: N.R. French (ed.) Perspectives in grassland ecology. Springer-Verlag, New York.

Tanigoshi, L.K. 1982. Advances in knowledge of the Phytoseiidae, p. 1-22. In: M.A. Hoy (ed.) Recent advances in knowledge of the Phytoseiidae. Univ. of California, Division of Agricultural Sciences Publ. 3284, Berkeley.

Ueckert, D.N. 1979. Impact of a white grub (Phyllophaga crinita) on a shortgrass community and evaluation of selected rehabilitation practices. J. Range Manage. 32:445-448.

Vogel, K.P., and S.D. Kindler. 1980. Effects of the subterranean aphid [Geoica utricularia (Passerini)] on forage yield and quality of sand lovegrass. J. Range Manage. 33:272-274.

Walter, D.E. 1987. Trophic behavior of 'mycophagous' microarthropods. Ecology 68:226-229.

Walter, D.E. Predation and mycophagy by endeostigmatid mites (Acariformes: Prostigmata). Exp. Appl. Acarol. (in press).

Walter, D.E., R.A. Hudgens, and D.W. Freckman. 1986. Consumption of nematodes by fungivorous mites, Tyrophagus spp. (Acarina: Astigmata: Acaridae). Oecologia 70:357-361.

Walter, D.E., J. Kethley, and J.C. Moore. 1987. A heptane flotation method for recovering microarthropods from arid soils, with comparisons to the Merchant-Crossley high-gradient extraction method and estimates of microarthropod biomass. Pedobiologia 30:221-232.

Walter, D.E., and R.A. Norton. 1984. Body size distribution in sympatric oribatid mites (Acari: Sarcoptiformes) from California pine litter. Pedobiologia 27:99-106.

Wiener, L.F. 1979. Biology and economic impact of *Phyllophaga fimbripes* (LeConte). Unpublished M.S. thesis, Colorado State Univ.

Wiener, L.F., and J.L. Capinera 1980. Preliminary study of the biology of the white grub *Phyllophaga fimbripes* (LeConte) (Coleoptera: Scarabaeidae). J. Kansas Entomol. Soc. 53:701-710.

Yount, V.A. 1971. Food habits of selected insects in the Pawnee Grasslands. Unpublished M.S. thesis, Colorado State Univ.

Zaher, M.A., and E.A. Gomaa. 1979. Genus *Neophyllobius* in Egypt with descriptions of three new species (Prostigmata-Neophyllobiidae). Inter. J. Acarol. 5:123-130.

20.   NEMATODES IN RANGELANDS

Nancy L. Stanton
Department of Zoology and Physiology
University of Wyoming, Laramie, Wyoming 82071

Rangeland plants are eaten episodically by vertebrates and nearly continuously by a complex of invertebrates. A few of these species occasionally reach epiphytotic levels but the biological mechanisms which trigger these outbreaks remain elusive. Even less well understood is the impact of chronic herbivory by hundreds of species of belowground herbivores. Phytophagous nematodes are one important subset of this complex and their effect on primary production of shortgrass prairie may be quite high. What proportion of net primary production (NPP) do they consume?   Do they reach epidemic levels?   Is there a long coevolutionary history of plant defense countered by herbivore offense?   How efficacious are predators in controlling densities? These questions remain unanswered as do questions of how does the belowground system respond to various range management schemes.

This chapter will focus primarily on the nematode component of the belowground system. I will discuss the functional roles of nematodes and review some recent studies which provide insights into the effects nematodes and their compatriots have on the quantity and quality of primary production. I will also speculate on the feasibility of actually "managing" this belowground component of grasslands for the specific purpose of increasing long-term primary productivity. Finally I will briefly discuss the potential of nematodes as part of IPM programs for control of other rangeland pests.

THE BELOWGROUND SYSTEM

In semiarid grasslands the soil is a complex heterogeneous environment with resources that are pulsed and unpredictable. Root biomass varies significantly within and between years because root production is driven by temperature and soil moisture which fluctuate widely within and between years. Typically, a flush of new root growth in the spring is often followed by a summer lull when moisture becomes limiting. During times of moisture stress, plants shed their new root growth very much like deciduous trees shed their leaves. Ares (1976) found that mid-season drought resulted in death of 30-60% of the new roots. If precipitation events occur in the fall, root growth is again initiated.

Numerous papers in the last few years have emphasized the importance of belowground processes in grasslands. We know that in shortgrass prairie about 85% of NPP ($Kcal/m^2$) is belowground (Coleman et al. 1976). The rhizosphere (root and associated microflora and fauna) is the center of biological activity and the basic functional unit of grasslands. As they grow (anabolize) and senesce (catabolize), roots are the resources that drive the belowground trophic web. The primary root resources include new roots and root hairs, exudates, plant mucilages, mucigels and lysates (Coleman et al. 1984). Roots are leaky, ephemeral organs and it has been estimated that 25-50 % of belowground production is shed into the soil (Shamoot et al. 1968, Barber and Martin 1976, Singh and Coleman 1977). These estimates may be conservative. Sauerbeck and Johnen (1976) labelled spring wheat, *Triticum aestivum*, and a mustard, *Sinapis alba*, with $^{14}CO_2$ and concluded that input of organic matter by roots can be 3-4 times greater than the root residue at harvest. If their figures are correct, then in shortgrass prairie the amount of organic deposition in the rhizosphere could be as much a 20 metric ton/ha, if one assumes a "harvestable" root biomass of 500 g/m^2. Thus, roots and their organic depositions provide a large resource base for the soil flora and fauna.

However, the belowground biomass varies considerably within and between seasons and across grassland types. Temperature and soil moisture appear to be the driving variables. (Some relationships between root biomass and physical parameters are listed in Table 20.1). For example, root biomass is inversely related to temperature (Sims et al. 1978). Data from Sims and Coupland (1979) plotted in Figure 20.1 for 10 western grasslands in North America illustrate this relationship. In native grasslands, a positive relationship exists between live root biomass and nematode biomass (Yeates and Coleman 1982) which suggests that belowground production may be a variable driving soil faunal populations. Thus, we would predict that belowground consumers are a more pervasive force in temperate grasslands with the highest belowground biomass: short- and midgrass prairies.

The aboveground standing crop is much lower than root biomass. Sims et al. (1978) reported root/shoot ratios from 6 in mixedgrass prairies to a high of 13 in shortgrass prairie. The large organic input to the soil relative to aboveground production may be one reason grasslands are resistant to a wide range of disturbances such a grazing and fire. If belowground biomass provides some degree of buffering from disturbances, then the northern grasslands would be less "perturbable" than those in more southern latitudes.

TABLE 20.1
Relationships between root biomass or turnover as a function of physical characteristics of grasslands.

---

1. Root turnover increases with total annual radiation (Sims and Coupland 1979).
2. Root biomass is inversely related to temperature (Sims et al. 1978, Sims and Coupland 1979).
3. Ratio of underground plant biomass to maximum canopy biomass is lower in tropical grasslands than in temperate grasslands (Coupland 1979).
4. Turnover of root biomass is higher in tropical grasslands (Singh and Joshi 1979).
5. Root turnover increases with actual evapotranspiration during the growing season. Root turnover $= C + aX_1 + bX_2$ where C is a constant, X_1 is annual radiation, and X_2 is growing season evaporation (Sims and Coupland 1979).
6. If latitude is constant, root production varies directly with precipitation (Singh and Joshi 1979).
7. Within a grassland type, root biomass increases with increasing latitude (Sims and Coupland 1979).
8. Proportion of belowground biomass decreases with available precipitation although absolute production increases (Marshall 1977).

Ten years ago Heady (1975) suggested that belowground organisms may actually be the limiting element to grassland productivity. He was referring primarily to the decomposers but belowground there are two major trophic pathways: herbivory and decomposition. Each pathway is linked by the primary resource, roots, and by the hundreds of species of secondary consumers preying on members within each trophic pathway. Thus, it is virtually impossible to extricate these into single pathways within an ecosystem context. However, experimental manipulation in the field and in microcosms has yielded some promising results;

evidence suggests that species within both trophic pathways have significant effects on primary production.

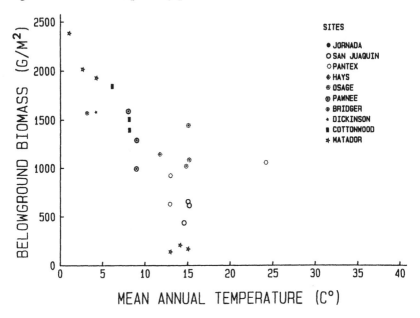

FIG. 20.1 Relationship between mean annual temperature and belowground biomass (ungrazed) for North American grasslands (data from Sims and Coupland 1979, Table 5.5).

SOIL NEMATODES

In grasslands, where they are the most abundant multicellular animals, nematode densities commonly range from one to several million/m^2. Additionally, grasslands appear to support higher densities of nematodes than other types of ecosystems (Sohlenius 1980). Nematodes are taxonomically and functionally diverse with hundreds of species coexisting in rangelands. Orr and Dickerson (1966) found 80 genera and 228 species in 60 samples of Kansas prairie soil. Smolik identified 58 species from only 200 individuals selected at random from shortgrass prairie soil (Stanton et al. 1981). Griffin (1984) lists a number of species of plant parasites which may pose particular problems to native grasses.

The nematode fauna is also trophically diverse. Herbivores constitute from 25 to 31% of the nematode fauna with fungivores, microbivores, omnivores, and predators constituting the remainder (Solhenius 1980). Functional diversity is also high and the trophic structure and function of belowground dwellers is more complex and "interwoven" than the trophic structure aboveground. The question of "managing" this belowground fauna has not been addressed because our knowledge of belowground systems is extremely limited. We would be hard pressed even to identify the pestilent elements because the fauna is not only taxonomically rich but the systematics of most groups remains undeveloped. Thus, labelling all nematodes as pests would be comparable to designating all insects as pests.

Plant Parasitic Nematodes

Only a handful of studies have measured the effect of these herbivores on native grassland production. Smolik (1977) applied a systemic nematicide, S-methyl 1-(dimethylcarbamoyl)-N-[(methylcarbamoyl) oxy] thioformimidate, to potted grassland sod and within 30 days measured a 28-59% increase in clipped weights. However, the possibility was not eliminated that this compound could have stimulated plant growth either directly or indirectly via fertilization. Stanton et al. (1981) applied the nematicide carbofuran to plots of shortgrass prairie in a two-year experiment and found that at the end of each growing season root biomass was significantly higher by 130-150 g/m^2 with treatment, which was a 25% increase in standing crop. Although this nematicide has no fertilizing effect it did change the species composition of the fungi, so changes in decomposition rates may have been confounding and responsible for at least some of the higher root biomass.

van Berkum and Hoestra (1979) summarized a number of studies in which carbofuran was applied to crops and the results are even more dramatic: e.g., soybean yield was increased by 200% (Sasser 1975) and pasture grass yield was 250% higher (Hoveland et al. 1977).

In a laboratory experiment, without the confounding effects of nematicide, Stanton (1983) measured the effect on blue grama, *Bouteloua gracilis*, of belowground grazing by a phytophagous nematode, *Helicotylenchus exallus*. This species accounted for a 13% reduction in NPP over the 28 weeks of the experiment.

More tightly controlled microcosm experiments also provide evidence that native plants may be affected by nematodes. Ingham and Coleman (1983) inoculated gnoto-biotic blue grama with a herbivorous nematode, *Tylenchorhynchus claytoni*, and found that shoot biomass was 17% lower at the end of the experiment. In a growth chamber experiment (Ingham and Detling 1986), sideoats grama, *Bouteloua curtipendula*, was inoculated with *T. claytoni* and root biomass and tiller number were significantly reduced by the nematode. The reduction in NPP was considerably more than the amount directly consumed by the nematode population. Smolik (1977) and Stanton (1983) both reported that growth reduction due to nematodes was 10 times greater that the amount directly assimilated by them. Ingham and Detling (1986) concluded that nematodes may consume less than aboveground grazers but that their impact on total production could be considerably greater. One reason may be that nematodes provide access or transport for other root pathogens such a viruses, bacteria, and fungi (Baker and Cook 1974).

If future experiments corroborate these results, then phytophagous nematodes may be credited as a major herbivore of grasslands and they may exert a stronger selective force on grasslands than do their more conspicuous aboveground counterparts. All of the studies cited above (and many hundreds more dealing with cultivars) have demonstrated that the presence of root-feeding nematodes on susceptible plants significantly reduces plant standing crop. If this loss (12-25%) could be translated into forage availability (and that is a big "if"), then this increased production would be at least 1 metric ton/ha. At the very least, range managers should recognize that belowground herbivory may be an important source of variability for productivity estimates.

Microbivores

This scenario of belowground herbivory is complicated by the presence of a large suite of microflora: bacteria and fungi quickly colonize the surface of growing roots (Baker and Cook 1974). Microbe grazers (primarily protozoa, nematodes, and microarthropods) are ubiquitous and abundant. The massive organic deposition in the rhizosphere forms a "hot bed" of activity for microflora (the primary decomposers) and their grazers. Consumption rates of bacteria by protozoa and nematodes are impressive. Bryant et al. (1982) calculated that 8300 bacterial cells must be consumed to produce one new amoeba. Bacterial feeding nematodes may consume 5000 cells/minute or 6.5 times their body weight per day (Duncan et al. 1974). These high rates of consumption by the millions of animals/m^2 may have dramatic impacts on the ecosystem as illustrated by the results of a number of experiments.

Microcosm experiments have demonstrated that grazing of the microflora by soil fauna accelerates mineralization and even the quantity and quality of the primary producers. For example, blue grama was grown with varying amounts of inorganic N and with and without soil amoebae (Elliott et al. 1979). After 45 days, the seedlings in the presence of amoebae had significantly higher N in their shoot tissue.

In another experiment, Ingham et al. (1985) grew blue grama seedlings in the presence or absence of microbivorous nematodes with a slow-release organic N fertilizer. Within 20 to 50 days soil P_i, soil NH^+_4 -N, and soil NH^-_3 -N were enhanced by the faunal grazers, and these differences in soil nutrient status were reflected in increased total shoot P and N.

This type of trophic manipulation has been extended into field studies of decomposer communities in which biocides have been used to eliminate particular taxonomic groups and then functional changes in the system were measured. For example, when a mite predaceous on nematodes was removed from a simple desert system, the bacterial feeding nematodes increased to such a level that the microflora and decomposition rates were depressed (Elkins and Whitford 1982). Conversely, in another experiment the presence of phytophagous nematodes increased the microbial populations (Ingham and Coleman 1983), probably because the nematode increased root exudation.

Thus, there appears to be measurable effect of trophic complexity on decomposition and mineralization; the more complex the trophic structure, the faster the catabolic processes. There is good evidence that the consumption of microflora by nematodes, protozoa, and microarthropods increases the availability of nutrients for root uptake. Most of the belowground fauna are involved in some link of decomposition/mineralization and either directly or indirectly act as facilitators of belowground catabolism. Unfortunately, it is empirically difficult to separate catabolic and anabolic processes and the food webs that control them. Therefore, it is impossible to measure realistically the relative effects of decomposers versus herbivores on grassland productivity.

MANAGEMENT

The results of these experiments suggest there may be several desirable objectives relative to the belowground system when planning a management program that is designed to increase long-term productivity of grasslands. Some of these considerations should include ways of reducing root consumption and synchronizing mineralization with maximum root growth. Gosz and Fisher (1984) noted that in forests, peak mineralization of roots is not synchronized with peak rates of plant nutrient assimilation; mineralization lags behind root anabolism. The soil fauna may well be able to induce a closer synchrony of mineralization with root growth, primarily for perennials. During the short period of root elongation grazers may keep some microbial populations in exponential growth phase and also serve as their prime dispersal agents. Microbivores appear to be more tolerant of dry conditions than do plant parasites, and with available resources they can also remain active over a longer period of time during the growing season (Stanton et al. 1984). Thus, periods of mineralization may be accelerated and extended over the growing season by the soil fauna. Plant growth in shortgrass prairie should rarely be nutrient limited.

Nematicides

All nematode control schemes have been devised for cultivated crops. Most involve the use of nematicides and none, as yet, are specific for plant parasites and leave the catabolic processors intact. For example, nematicides (Smolik 1977, Stanton et al. 1981) inhibit all trophic groups of nematodes including the microflora grazers, and can significantly change the species composition of the microflora. The application of carbofuran to shortgrass prairie reduced fungal species from 102 to 68 and resulted in a 70% decline in colony sources (Stanton et al. 1981). This type of change in decomposers may ultimately result in a long term change in the rates of decomposition and mineralization.

The National Academy of Science (1968) proposed that the "ideal" nematicide would be systemic and translocated to the roots. No harmful residues would remain either inside or outside the plant. Morton (1986), in a recent review is "optimistic that such an ideal nematicide is feasible." However, even with a safe nematicide and a significant increase in rangeland production, chemical control may not be cost effective. A variety of other control methods, which I will briefly review, have been used in cultivated systems, but most would be impractical for extensive use on rangelands.

Application of Organic Matter

The application of organic matter in a variety of forms is known to reduce phytophagous nematodes (e.g., Alam et al. 1977), but it is not clear exactly what the mechanisms are in all cases. Norton (1979) reported on significant negative correlations between soil organic matter and stylet nematodes. Increased organic matter may favor growth of organisms antagonistic to some nematode taxa. For example, fungi predaceous on nematodes may increase. Sayre (1986) reports that over 140 species of fungi have been found that are antagonistic to either egg, larval, or adult nematodes. Mites can be voracious predators on nematodes (see Walter, Chapter 19) but there are a variety of other taxa which take their toll including predaceous nematodes, tubellarians, tardigrades, collembolans, amoebae (Sayre 1980), and bacteria (Sayre and Starr 1985).

The primary decomposition of this organic matter involves a complex array of microflora, some of which produce nematotoxic chemicals. Organic acids reduce mobility of larva and egg production of root knot nematodes (Sitaramaiah and Singh 1978) and these organic acids increase with organic amendments to the soil (Sitaramaiah and Singh 1978). Tannins also degrade into nematotoxic polyphenols. Extracts from decomposing rye (Patrick et al. 1965) were found to be toxic to plant parasites at 380-440 ppm, but saprophagous nematodes were far more resistant to the extract. Norton (1979) reviewed several studies which reported that organic acids were toxic to nematodes and a bacterium, *Clostridium butyricum*, present in decaying matter produces proprionic and butyric acids.

Other chemicals can also achieve nematode control. Eno et al. (1955) applied anhydrous ammonia to soil and phytophagous nematodes seemed particularly susceptible. Population density was reduced by all levels of ammoniacal nitrogen from 136 to 741 ppm. In alkaline soil, *Tylenchulus semipenetrans* was inhibited by additions of nitrogenous material. Larvae of this plant parasite were adversely affected by high NH^+_4 applications. Microbivorous and fungivorous nematodes were not affected (Mankau 1963).

Urea at high concentrations (0.2 to 1.0 g/kg soil) was found to be toxic to several genera of plant parasites (Miller 1976). However, even applications as low as 70 ppm of nitrite, organic nitrogen, and ammonium resulted in decreases in *Pratylenchus penetrans* (Walker 1971). In a recent review, Rodriguez-Kabana (1986) suggested that "It should be possible through careful choice of appropriate organic carriers to enhance nematicidal properties of inorganic N fertilizers, avoid phytotoxic effects, and provide excellent plant nutrition." We are just not exactly sure how this can be achieved.

The application of organic debris to soils may have a number of cascading effects. Some organic material may be nematotoxic, and/or during decomposition nematicides specific for herbivores are produced by microflora, and/or an environment more conducive to nematode predators and competitors is created. However, the amount of organic matter necessary to achieve measurable results is impractical for rangelands. Even in cultivated systems, Norton (1979) states that "the quantity of amendments necessary to bring about effective control often exceeds practical supply."

Resistant Plants

Many cultivars have been bred for nematode resistance but little work has been done with native grassland species. In fact, in 1979 a report on IPM programs by the Intersociety Consortium for Plant Protection suggested that management research in plant resistance should have priority over research in cultural, chemical, or biological control (Bird 1980). Conventional wisdom suggests that ectoparasites, the type most common in grasslands, are broad-niched and selection for resistance in host plants has not occurred (Roberts 1982). However, in cultivated forage grasses a number of species and varieties have been found to be resistant to a migratory ectoparasite, *Belonolaimus longicaudatus* (Boyd and Perry 1969). *Species* of *Digitaria* that were resistant to this sting nematode yielded twice as much harvested biomass as did a susceptible species of the same genus (Boyd and Perry 1969).

Griffin et al. (1984) reported a high degree of variability among grasses in susceptibility to the endoparasite *Meloidogyne chitwoodi*. For example, blue bunch wheatgrass was resistant, while "Nugaines" winter wheat yielded almost 70,000 eggs/g of roots. But no attempt was made to determine the actual incidence on native species. Screening of native species for nematode resistance may pay off if productivity of pasturelands can be improved by reseeding or interseeding with resistant varieties.

The mechanisms conferring resistance are still unknown but probably multifaceted. They may include secondary chemicals, root nutrient content, hormone levels, and root morphology. The importance of each one at this point is purely speculative.

Secondary chemicals. Many plant species are known to produce natural nematicides. Extracts and leaf homogenates from the leaves and roots of a variety of cultivars have been reported to reduce nematode numbers. Some of the species tested include raspberry (Taylor and Murrant 1966); ginger, garlic, and chili peppers (Sukul et al. 1974); white pine, dogwood, tobacco, sunflower, tomato, geranium, and bluegrass (Miller 1978). However, this topic of natural resistance has not been pursued very extensively within natural grassland ecosystems. In shortgrass prairie, fringed sage-wort, *Artemisia frigida*, usually hosts about half the number of plant parasites/volume of root as does blue grama, *Bouteloua gracilis* (Stanton et al. 1984). Stanton (unpublished) found that nematodes exposed to roots extracts from blue grama, fringed sagewort, and *Allium cernum* had significantly higher LD50's than did those in water. In *A. cernum* extracts (1 g ww/50 ml H_2O), 80% were dead within 6 hours and all were dead within 12 hours. After 24 hours of exposure, 50% were alive in fringed sagewort roots extracts and 65% were alive under exposure to blue grama root extracts. Interspecific root interdigitation of grasses with shrubs and forbs containing secondary compounds may to a degree control the build-up of plant parasites--a natural polyculture.

Few studies have actually examined secondary chemicals within the roots. Rodman and Louda (1984) measured concentrations of isothiocyanate-yielding glucosinolates in various plant parts of bittercress, *Cardamine cordifolia*; roots had the highest level of any of the plant parts examined. Nicotine is synthesized in roots (Baker and Cook 1974), and roots may prove to be the origin for other nitrogen-based secondary chemicals.

It is quite possible that grass roots are not as palatable as one would suppose. After millions of years of exposure to chronic herbivory why not evolve nematode repellents? Redak (Chapter 4) reports on a large number of secondary chemicals found in grass leaves: alkaloids, cyanogenic glycosides, phenolics, tannins, polyphenolics, and proteinase inhibitors. Counter to conventional wisdom, grasses may very well have a rich battery of chemical defenses in both shoots and roots.

An extensive body of research on cultivars relates nematode resistant varieties to root chemistry. For example, in tomatoes resistance is directly related to phenolic content (Wallace 1961, Singh and Choudhury 1973). Nematode infection can also induce increases in phenolic content (Epstein 1972). Attack of citrus by *Radopholus similis* increased the content of bound phenolics up to 300% in tolerant varieties, but decreased the bound phenolics in susceptible varieties by 19 to 34% (Feldman and Hanks 1968).

Giebel (1974) found a correlation between phenols and plant resistance. Within the plant most phenols occur as glycosides and are activated by the nematode secretion of beta-glycosidase into the host tissue. Hydrolysis of the glucosides results in phenolic-induced host necrosis and, thus, resistance because nematodes cannot feed on necrotic tissues (Endo 1971). Wilski and Giebel (1966) found that potatoes with low levels of phenolic glucosides are not resistant to nematodes because necrotic tissue is not formed in response to herbivory. Endo (1971) speculated that nematodes with low levels of beta-glucosidase induce lower levels of toxic phenols within the host and nematode biotypes can develop on previously resistant hosts. For some systems there is good evidence of close coevolutionary ties between herbivores and hosts (Stone 1980).

However, all of these studies involved endoparasitic nematodes which are reputed to be far more restricted in host preferences than the ectoparasites (Stone 1980); the latter are the most common form in grasslands. Roberts (1982) suggested that resistance to ecoparasites has not developed in cultivars or wild plants. These root browsers are eurytrophic and he purports that these grazers exert such a low selective pressure on plants that resistance offers little advantage to the host unless the resistance genes are linked to other loci that confer some increase in fitness. However, a reduction in NPP by 13 to 26% as reported above does seem to be a significant force. To date complete nematode resistance has not been found in any native grassland species (Griffin, per. comm.), but it is likely that species and genotypes vary in their tolerance and palatability. The question is certainly amenable to testing.

One must not slip into the trap of viewing phenols and other secondary compounds as single-action compounds. Plant phenols, for example, prevent smudge infections; a toxic dicumarol is an anticoagulant in vertebrates; and preocenes cause premature metamorphosis in insects (Salisbury and Ross 1985). Thus, phenolics probably provide a broad-based defense against a variety of herbivores and diseases as well as having a number of effects of absolutely no evolutionary significance whatsoever.

Nutrient content. Although secondary chemicals are now commonly accepted as herbivore deterrents, a number of other factors could be equally deterring. The nutrient content of roots may impair reproduction, e.g., reproduction of endoparasites has been found to be limited by deficiencies in N, P, K , and Fe in watercress (Bird 1960). Applications of P and N to sour cherry effected differential changes by nematode species (Kirkpatrick et al. 1964). The study concluded that population levels were dependent upon the nutritional status of the host and that each nematode species had different nutritional requirements. However, the nutrient contents of the roots were not analyzed so the results are suggestive but not conclusive. Changes in mineral composition of foliage can also occur with nematode feeding. MacDonald (1979) examined several studies which reported changes in Ca, N, P, K, Mg, Fe, Na and Zn but there was no consistent pattern of change attributable to parasitism (MacDonald 1979). In the roots, simple sugars have been reported to increase (Epstein 1972), while carbohydrates and minerals decrease (Nasr et al. 1980).

Seastedt (1985) has argued that the nitrogen content of live roots in eastern prairies is very low; thus, they are almost marginal as resources for tissue-feeding herbivores. In fact, dead roots have a higher N content than live roots because of the colonizing microflora (Seastedt 1985). However, it is not known if the low N content is a function of the high C/N ratio of the roots or the absolute amount of N within the cells. Only the latter would affect nematodes. However, MacDonald (1979) reported that for seven different studies the amino acid content of nematode-parasitized plants was from 100 to 400% higher than controls and this increase was found throughout the plant. The ecological significance of these changes is not known.

Hormones. Nematodes are known to secrete saliva into their hosts, and the saliva has been found to contain amylase, invertase, cellulase, pectinase, and beta-glucosidase among others. Enzyme composition and amounts vary among species of plant parasites and life stages (Decker 1981). Plants respond to these secretions in a number of ways: cells hypertrophy, root growth ceases, cell walls dissolve, more lateral roots form, and giant cells form. It is not known if nematodes also produce hormones, but auxins within plants are known to alter nematode-plant interactions (Giebel 1974). However, the nature of the interactions between nematodes and plants mediated by hormones is not well understood. Small doses of gibberellic acid have been found to increase nematode populations and plant growth, whereas supra-doses suppressed both (Badra et al. 1980). The induction of giant cell formation is suggestive of hormone disorders (Pegg 1985). Cytokinin is reported to increase susceptibility to nematode attack because resistant plants increase in susceptibility after cytokinin application (Pegg 1985).

Tomato roots inoculated with root knot nematodes produced 3-6 times as much ethylene as controls. Both ethylene production and gall growth were stimulated by ACC, IAA, and ethrel (Glazer et al. 1985a). Ethylene is known to inhibit lignification, and Glazer et al. (1985a) speculated that the nematode in some way stimulated ethylene production which allowed expansion of giant cells and parenchymatous material without lignification. Thus, a protective and edible gall is provided for the nematode. Species of root-knot which produce higher levels of ethylene also produce larger galls (Glazer et al. 1985b). Cultivars resistant to nematode attack release only small amounts of ethylene. In galled roots there is also an increase in free amino acids, RNA, DNA, lipids, and minerals. Endo (1971) suggested that not only do nematode secretions cause gall formation, but they also induce accumulation of nutrients within the gall.

The physiological state of the plant affects nematode reproductive rate in a number of ways. The nutrient content of the root, the amount of secondary chemicals, and the amount and type of plant hormones may influence the belowground herbivores. (See Giebel 1974 for a hypothetical scheme of resistance in solanaceous plants to an endoparasitic nematode). Unfortunately, none of this work has been done with native grasses *in situ*.

Cultural Practices

The following physical parameters are also known to have differential effects on species of nematodes (see review by Norton 1979): soil texture, temperature, moisture, aeration, osmotic pressure, and pH. Any manipulation of rangelands such as changing frequency, intensity or timing of grazing; shrub control; changing fire frequency or occurrence; and flooding may change the nematode community. Few experiments have been done to determine how this results. Research in this direction with natural forage species may prove fruitful.

A useful set of studies awaiting attention would be to measure nematode response to general management plans. For example, how does the belowground component respond to different grazing regimes, shrub control, insecticide application, and fertilization? Clearly a critical area for research is to determine how various aboveground manipulations, including grazing pressure, affect the belowground component; and, visa versa, how do belowground processes affect quantity and quality of primary production?

In a review article, Seastedt (1985) stated that belowground herbivore species respond differentially to aboveground grazing. Some species may increase, while some may decrease, as long as the grazing pressure is moderate. The data are fairly conclusive that if aboveground herbivory is intense and regular, root biomass and belowground herbivore numbers will decline. However, with moderate aboveground herbivory several researchers have reported higher populations of phytophagous nematodes (Chapman 1963, Norton 1965, Stanton 1983). The mechanism is not known. Possibilities include proliferation of lateral roots which serve as good feeding sites, increased nutrient content of roots, reduction of constitutive or inducible secondary compounds, or changes in plant hormones. However, if aboveground grazing can be timed to occur one to two weeks after the period of fastest root growth, and if it is intense as opposed to light, then I predict a sharp decline in belowground herbivore densities. Root growth should stop and feeding sites decline, whereas predators such as fungi, mites, and microbivores should continue to increase. Thus, intense short-term grazing sufficient to stop root growth may increase overall productivity of grasses by reducing belowground herbivores and increasing rates of mineralization. With intense grazing there is also a concentrated input of urea and feces which should provide additional stimulation to decomposers and inhibition of root parasites. If soil moisture is adequate to sustain root regrowth, then shoot regrowth should be almost as much as the initial growth. In contrast, continual grazing at moderate intensities may stimulate the belowground herbivores.

At present there is insufficient information to provide recommendations for increasing grasslands production through better management of decomposers. However, enough basic information is available to focus questions for innovative field experiments on soil fauna as regulators of primary production.

NEMATODES AS BIOCONTROL AGENTS

Nematodes could be used in IPM programs, but their full potential of regulatory agents has yet to be explored. There are many nematodes parasitic on insects (Poinar 1979, Nickle 1984), and if problems of mass culturing, storage, and dispersal can be solved, then these parasites provide some hope for specific and environmentally sensitive control agents for such important pests as grasshoppers and white grubs. For example, mermithid infections of some grasshopper populations can be quite high, although infections appear highly localized and quite variable between years (Webster 1972,Webster and Thong 1984, see also Chapter 11). Likewise, foliar nematodes have been found effective for control of some plant species. Northam and Orr (1982) broadcast fourth stage larvae of *Nothanguina phyllobia* on silver leaf nightshade, *Solanum elaeagnifolium*, and concluded that on disturbed sites this nematode was an effective control.

This topic deserves more attention than space permits here because the potential of nematodes as pest control agents of both plants and insects is quite high. But as always, our paucity of knowledge of basic biology imposes impediments to utilization of these organisms as biocontrol agents.

SUMMARY

Phytophagous nematodes significantly reduce primary production in rangelands, but saprophagic nematodes and other invertebrates probably increase rates of decomposition and mineralization and, ultimately, nutrient availability for the primary producers. Any IPM program which incorporates control of root grazers must also take into account the long-term effect of the program on rates of decomposition and mineralization. Control methods such as application of nematicides or organic matter are not cost effective for native grasslands. Interseeding or reseeding with nematode-resistant plants may prove cost effective. An area for research may be the design of grazing management schemes which minimize belowground herbivory and synchronize maximum rates of mineralization with root growth. Nematodes that are parasitic on noxious plants and insect pests provide unexploited opportunities for use in IPM Programs.

REFERENCES

Alam, M.M., S.A. Siddiqui, and A.M. Khan. 1977. Mechanism of control of plant parasitic nematodes as a result of the application of organic amendments to the soil. III--role of phenols and amino acids in host roots. Indian J. Nematol. 7:27-31.

306

Ares, J. 1976. Dynamics of the root system of blue grama. J. Range Manage. 29:208-216.

Badra, T., M.M. Khattab, and G. Stino. 1980. Influence of sub- and supra-optimal concentrations of some growth regulators on growth of guava. Phenol status, nitrogen concentration and number of *Meloidogyne incognita.* Nematologica 26:157-162.

Baker, K.F., and R.J. Cook. 1974. Biological control of plant pathogens. W.H. Freeman and Co., San Francisco.

Barber, D.A., and J.K. Martin. 1976. The release of organic substances by cereal roots into soil. New Phytol. 77:69-80.

Bird, A.F. 1960. The effect of some single element deficiencies on the growth of *Meloidogyne javanica.* Nematologica 5: 78-85.

Bird, G.W. 1980. Nematology--status and prospects: the role of nematology in integrated pest management. J. Nematol. 12:170-176.

Boyd, F.T., and V.G. Perry. 1969. The effect of sting nematodes on establishment, yield, and growth of forage grasses on Florida sandy soils. Proc. Soil and Crop Sci. Soc. Florida 29:288-300.

Bryant, R.J., L.E. Woods, D.C. Coleman, B.C. Fairbanks, J.F. McClellan, and C.V. Cole. 1982. Interactions of bacterial and amoebal populations in soil microcosms with fluctuating moisture content. Appl. Environ. Microbiol. 43: 747-752.

Chapman, R. A. 1963. Development of *Meloidogyne hapla* and *M. incognita* in alfalfa. Phytopathology 53:1003-1005.

Coleman, D.C., R. Andrews, J.E. Ellis, and J.S. Singh. 1976. Energy flow and partitioning in selected man-managed and natural ecosystems. Agro-Ecosystems 3:45-54.

Coleman, D.C., R.E. Ingham, J.F. McClellan, and J.A. Trofymow. 1984. Soil nutrient transformations in the rhizosphere via animal-microbial interactions, p.35-58. *In:* J.M. Anderson, A.D.M. Rayner, and D.W.H. Walton (eds.) Invertebrate-microbial interactions. Cambridge Univ. Press, Cambridge.

Coupland, R.T. 1979. Conclusion, p. 335-355. *In:* R.T. Coupland (ed.) Grassland ecosystems of the world. Cambridge University Press, Cambridge.

Decker, H. 1981. Plant nematodes and their control (phytonematology). Kolos Publ., Moscow.

Duncan, A., F. Schiemer, and R.Z. Klekowski. 1974. A preliminary study of feeding rates on bacterial food by adult females of a benthic nematode *Plectus palustris* De Man 1880. Polish Archiv. Hydobiol. 21:249-258.

Elkins, N.Z., and W.G Whitford. 1982. The role of microarthropods and nematodes in decomposition in a semi-arid ecosystem. Oecologia 55:303-310.

Elliot, E.T., D.C. Coleman, and C.V. Cole. 1979. The influence of amoebae on the uptake of nitrogen by plants in gnotobiotic soil, p. 221-229. *In:* J.L. Harley and R.S. Russell (eds.) The soil-root interface. Academic Press, London.

Endo, B. Y. 1971. Nematode-induced syncytia (giant cells). Host-parasite relationships of Heteroderidae, p. 91-117. *In:* B.M. Zuckerman, W.F. Mai and R.A. Rohde (eds.) Plant parasitic nematodes. Academic Press, New York.

Eno, C.F., W.G. Blue, and J.M. Good, Jr. 1955. The effect of anhydrous ammonia on nematodes, fungi, bacteria, and nitrification in some Florida soils. Proc. Soil Sci. Soc. Am. 19:55-58.

Epstein, E. 1972. Biochemical changes in terminal root galls caused by an ectoparasitic nematode, *Longidorus africanus*: phenols, carbohydrates and cytokinins. J. Nematol. 4:246-250.

Feldman, A. W., and R. W. Hanks. 1968. Phenolic content in the roots and leaves of tolerant and susceptible citrus cultivars attacked by *Radopholus similis*. Phytochemistry 7:5-12.

Giebel, J. 1974. Biochemical mechanism of plant resistance to nematodes: a review. J. Nematol. 6:175-184.

Glazer, I., A. Apelbaum, and D. Orion. 1985a. Effect of inhibitors and stimulators of ethylene production on gall development in *Meloidogyne javanica*-infected tomato roots. J. Nematol. 17:145-149.

Glazer, I., D. Orion, and A. Apelbaum. 1985b. Ethylene production by *Meloidogyne* spp-infected plants. J. Nematol. 17:61-63.

Gosz, J.R., and F.M. Fisher. 1984. Influence of clear-cutting on selected microbial processes in forest soils. p. 523-530. *In:* M.J. Klug and C.A. Reddy (eds.) Current perspectives in microbial ecology. Am. Soc. for Microbiol., Washington, D.C.

Griffin, G.D., R.N. Inserra, and N. Vovlas. 1984. Rangeland grasses as host of *Meloidogyne chitwoodi*. J. Nematol. 16:399-402.

Griffin, G.D. 1984. Nematode parasites of alfalfa, cereals, and grasses, p. 243-321. *In:*W.R. Nickle (ed.) Plant and insect nematodes. Marcel Dekker, Inc., New York.

Heady, H.F. 1975. Rangeland management. McGraw-Hill Book Co, New York.

Hoveland, C.S., R.L. Haaland, and R. Rodriguez-Kabana. 1977. Soil nematodes and forage production of some cold season grasses., p. 123-128. 13th Int. Grassl. Congr., Leipzig.

Ingham R.E., and D.C. Coleman. 1983. Effects of an ectoparasitic nematode on bacterial growth in gnotobiotic soil. Oikos 41:227-232.

Ingham, R.E., and J.K. Detling. 1986. Effects of defoliation and nematode consumption of growth and leaf gas exchange in *Bouteloua curtipendula*. Oikos 46:23-28.

Ingham R.E., and J.A. Trofymow, E.R. Ingham, and D.C. Coleman. 1985. Interactions of bacteria, fungi, and their nematode grazers: effects on nutrient cycling and plant growth. Ecol. Mono. 55:119-140.

Kirkpatrick, J.D. and W.F. Mai, K.G. Parker, and E.G. Fisher. 1964. Effect of phosphorus and potassium nutrition of sour cherry on the soil population levels of five plant-parasitic nematodes. Phytopathology 54:706-712.

MacDonald, D. 1979. Some interactions of plant parasitic nematodes and higher plants. p. 157-178. *In:* S.V. Krupa and Y.R. Dommergues (eds.) Ecology of root pathogens. Elsevier Scientific Publishing Co., Amsterdam, Oxford.

Mankau, R. 1963. Effect of organic soil amendments on nematode populations. Phytopathology 53:881-882.

Marshall, J. K. 1977. Biomass and production partitioning in response to environment in some North American grasslands, p. 73-84. *In:* J.K. Marshall (ed.) The belowground ecosystem: a synthesis of plant-associated processes. Range Science Department Science series No.26. Colorado State Univ., Fort Collins.

Miller, P.M. 1976. Effects of some nitrogenous materials and wetting agents on survival in soil of lesion, stylet, and lance nematodes. Phytopathology 66:798-800.

Miller, P. M. 1978. Toxicity of homogenized leaves of woody and herbaceous plants to root lesion nematodes in water and soil. J. Amer. Soc. Hort. Sci. 103:78-81.

Morton, H.V. 1986. Modification of proprietary chemicals for increasing efficacy. J. Nematology. 18:123-128.

National Academy of Science. 1968. Control of plant-parasitic nematodes. *In:* Principles of plant and animal pest control. Vol 4. Publication 1696. NAS, Washington, DC.

Nasr, T.A., K.A. Ibrahim, E.M. El-Azab, and M.W.A. Hassan. 1980. Effect of root-knot nematodes on the mineral, amino acid and carbohydrate concentrations of almonds and peach rootstocks. Nematologica 26:133-138.

Nickle, W.R. 1984. Plant and insect nematodes. Marcel Dekker, Inc., New York.

Northam, F.E., and C.C. Orr. 1982. Effects of a nematode on biomass and density of silverleaf nightshade. J. Range Manage. 35:536-537.

Norton, D.C. 1965. *Xiphinema americanum* populations and alfalfa yields as affected by soil treatment, spraying and cutting. Phytopathology 55:615-619.

Norton, D.C. 1979. Relationship of physical and chemical factors to populations of plant-parasitic nematodes. Annu. Rev. Phytopathol. 17:279-299.

Orr, C.C., and O.J. Dickerson. 1966. Nematodes in true prairie soils of Kansas. Trans. Kans. Acad. Sci. 69:317-334.

Patrick, Z.A. and R.M. Sayre, and H.J. Thorpe. 1965. Nematocidal substances selective for plant-parasitic nematodes in extracts of decomposing rye. Phytopathology 55:702-704.

Pegg, G.F. 1985. Pathogenic and non-pathogenic microorganisms and insects, p. 599-624. In: R.P. Pharis and D.M. Reid (eds.) Hormonal regulation of development. III. Role of environmental factors. Encyclopedia of plant physiology, new series, Vol. 11, Springer-Verlag, New York.

Poinar, Jr., G.O. 1979. Nematodes for biological control of insects. CRC Press, Boca Raton, FL.

Roberts, P.A. 1982. Plant resistance in nematode pest management. J. Nematol. 14:24-32.

Rodman, J.E., and S.M. Louda. 1984. Phenology of glucosinolate concentrations in root, stems and leaves of Cardamine cordifolia. Biochem. Syst. Ecol. 12:37-46.

Rodriguez-Kabana, R. 1986. Organic and inorganic amendments to soil as nematode suppressants. J. Nematol. 18:129-135.

Salisbury, F.B., and C.R. Ross. 1985. Plant physiology. Wadsworth Publ. Co., Belmont, CA.

Sasser, J.N., K.R. Baker, and L.A. Nelson. 1975. Chemical soil treatments for nematode control on peanut and soybeans. Plant Dis. Rep. 59:154-158.

Sauerbeck, D.R., and B.G. Johnen. 1976. Root formation and decomposition during plant growth. International symposium on soil organic matter studies, p. 1-11. Intern. Atom. Ener. Agen. and Food and Ag. Org. of the U.S., Brunswick, Federal Republic of Germany.

Sayre, R.M. 1980. Promising organisms for biocontrol of nematodes. Plant Dis. 64:527-532.

Sayre, R.M. 1986. Pathogens for biological control of nematodes. Crop Prot. 5:268-276.

Sayre, R.M., and M.P. Starr. 1985. Pasteuria penetrans (ex Thorne, 1940) nom. rev., comb. n., sp. n., a mycelial and endospore-forming bacterium parasitic in plant-parasitic nematodes. Proc. Helminthol. Soc. Wash. 52:149-165.

Seastedt, T.R. 1985. Maximization of primary and secondary productivity by grazers. Am. Nat. 126:559-564.

Shamoot, S., L. McDonald, and W.V. Bartholomew. 1968.Rhizo-deposition of organic debris in soil. Soil Sci. Soc. Am. Proc. 32:817-820.

310

Sims, P.L., and R.T. Coupland. 1979. Natural temperate grasslands: producers, p. 49-72. *In:* R.T. Coupland (ed.) Grassland ecosystems of the world: analysis of grasslands and their uses. Cambridge Univ. Press, Cambridge.

Sims, P.L., J.S. Singh, and W.K. Lauenroth. 1978. The structure and function of ten western North American grasslands. I. Abiotic and vegetational characteristics. J. Ecol. 66:251-285.

Singh, B., and B. Choudhury. 1973. The chemical characteristics of tomato cultivars resistant to root-knot nematodes (*Meloidogyne* spp.). Nematologica 19:443-448.

Singh, J.S., and D.C. Coleman. 1977. Evaluation of functional root biomass and translocation of photoassimilated ^{14}C in shortgrass prairie, p. 123-131. *In:* J.K. Marshall (ed.) The belowground ecosystem. Range Science Department Series No. 26. Colorado State Univ., Fort Collins.

Singh, J.S., and M.C. Joshi. 1979. Tropical grasslands: primary production, p. 197-218. *In:* R.T. Coupland (ed.) Grasslands ecosystems of the world: analysis of grasslands and their uses. Cambridge Univ. Press, Cambridge.

Sitaramaiah, K., and R.S. Singh. 1978. Effect of organic amendment on phenolic content of soil and plant and response of *Meloidogyne javanica* and its host to related compounds. Plant and Soil 50:671-679.

Smolik, J.D. 1977. Effect of nematicide treatment on growth of range grasses in field and glasshouse studies, p. 257-260. *In:* J.K. Marshall (ed.) The belowground ecosystem. Range Science Department Science Series, No. 26. Colorado State Univ., Fort Collins.

Sohlenius, B. 1980. Abundance, biomass and contribution to energy flow by soil nematodes in terrestrial ecosystems. Oikos 34:186-194.

Stanton, N. L. 1983. The effect of clipping and phytophagous nematodes on net primary production of blue grama, *Bouteloua gracilis*. Oikos 40:249-257.

Stanton, N.L., M. Allen, and M. Campion. 1981. The effect of the pesticide carbofuran on soil organisms and root and shoot production in shortgrass prairie. J. Appl. Ecol. 18:417-431.

Stanton, N.L., D. Morrison, and W.A. Laycock. 1984. The effect of phytophagous nematode grazing on blue grama die-off. J. Range Manage. 37:447-450.

Stone, A.R. 1980. Co-evolution of nematodes and plants. Symb. Bot. Upsal. 22:4:45-61.

Sukul, N.C. and P.K. Das, and G.C. De. 1974. Nematicidal action of some edible crops. Nematologica 20:187-191.

Taylor, C.E., and A.F. Murant. 1966. Nematicidal activity of aqueous extracts from raspberry cane and roots. Nematologica 12: 488-494.

Walker, J.T. 1971. Populations of *Pratylenchus penetrans* relative to decomposing nitrogenous soil amendments. J. Nematol. 3:43-49.

VanBerkum, J.A., and H. Hoestra. 1979. Practical aspects of the chemical control of nematodes in soil, p. 53-134. *In:* D. Mulder (ed.) Soil disinfestation. Elsevier Scientific Pub., New York.

Wallace, H.R. 1961. The nature of resistance in Chrysanthemum varieties to *Aphelenchoides ritzemabosi*. Nematologica 6:49-58.

Wilski, A., and J. Giebel. 1966. Beta-glucosidase in *Heterodera rostochiensis* and its significance in resistance of potato to this nematode. Nematologica 12:219-224.

Webster, J.M. 1972. Nematodes and biological control, p. 469-496. *In:* J.M. Webster (ed.) Economic nematology. Academic Press, New York.

Webster, J.M., and C.H.S. Thong. 1984. Nematode parasites of orthopterans, p. 697-726. *In:* W.R. Nickle (ed.) Plant and insect nematodes. Marcel Dekker, Inc., New York.

Yeates, G.W., and D.C. Coleman. 1982. Role of nematodes in decomposition, p. 55-80. *In:* D.W. Freckman (ed.) Nematodes in soil ecosystems. Univ. Texas Press, Austin.

21. RELATION OF BLACK-TAILED PRAIRIE DOGS AND CONTROL PROGRAMS TO VEGETATION, LIVESTOCK, AND WILDLIFE

Daniel W. Uresk
USDA Forest Service
Rocky Mountain Forest and Range Experiment Station
Rapid City, South Dakota 57701

Black-tailed prairie dogs, *Cynomys ludovicianus* (Ord), have been considered to be pests on shortgrass and mixedgrass prairies of central North America since the late 1800's (Merriam 1902). Because prairie dogs compete for forage with livestock, massive prairie dog control programs were initiated and are still a common practice (Schenbeck 1982). Prairie dog control also has been justified to reduce reservoirs of contagious diseases. Throughout these control programs, little attention has been given to economic benefits from increased forage, comparison of effectiveness of rodenticides used for prairie dog control, secondary impacts on nontarget animals, livestock-prairie dog forage relationships, and wildlife utilizing prairie dog colonies. This chapter presents an overview of these aspects as related to black-tailed prairie dogs on western rangelands.

PLANT PRODUCTION

Control of black-tailed prairie dogs has been considered necessary to increase forage production on rangelands. O'Meilia et al. (1982) found no differences in forb production on pastures with steers only and with steers plus prairie dogs. However, availability of blue grama, *Bouteloua gracilis* (H.B.K.) Griffiths, and sand dropseed, *Sporobolus cryptandrus* (Torr.) A. Gray, and other grass species was significantly reduced on pastures with prairie dogs. Other studies of vegetative standing crop and canopy cover were conducted on prairie dog towns being grazed exclusively by prairie dogs, or by prairie dogs with another large herbivore, or on abandoned prairie dog towns (Taylor and Loftfield 1924, Osborn and Allan 1949, Koford 1958, Bonham and Lerwick 1976, Potter 1980, Coppock et al. 1983).

Uresk and Bjugstad (1983) examined plant production in relation to four treatments by harvesting the vegetation under cages. The treatments were: 1) no grazing, 2) grazing by prairie dogs, 3) grazing by cattle, and 4) grazing by cattle and prairie dogs. Peak plant production values on the prairie dog treatment differed by +24% when compared to the cattle only grazing treatment, while it differed by +13% compared to the cattle plus prairie dog treatment. Plant production was similar between the cattle and no grazing treatments. Grasses showed no differences in production among treatments (Figure 21.la). Forb production differed significantly on the prairie dog treatment (+165%), followed by no grazing (+91%), and the cattle plus prairie dog treatment (+76%) when compared to the cattle treatment (Figure 21.lb). The greatest difference in peak plant production occurred with forbs when rangelands were grazed by prairie dogs.

Uresk (1985) evaluated the effects of black-tailed prairie dog control on production of 43 plant species. Pre-control treatments included: 1) ungrazed, 2) prairie dogs only, and 3) cattle plus prairie dogs grazing together. Generally, there was no increase in western wheatgrass, *Agropyron smithii* Rydb., production among treatments 4 years after the prairie dogs were eliminated, indicating that control of prairie dogs on pastures with and without cattle did not affect production of this species. Buffalograss, *Buchloe dactyloides* (Nutt.) Engelm., showed a significant decrease in production when cattle were allowed to graze the area after prairie dogs were removed, which indicates that prairie dog clipping stimulates the growth of buffalograss, a stoloniferous grass. Production estimates of needleleaf sedge, *Carex eleocharis* Bailey, were significantly lower on the cattle-prairie dog treatment than prairie dog only treatment. All other plants, including grasses and forbs as groups, did not show a significant treatment response in production over the 4-year period. Total exclusion of herbivores for 9 or more years may be required to increase forage production for all treatments when the range is in a low condition class (Uresk and Bjugstad 1983, Uresk 1985). Forage improvement is very slow on western rangelands that have been heavily grazed for many years.

The economics of improving forage production by controlling black-tailed prairie dogs with zinc phosphide was evaluated by Collins et al. (1984). Their economic analyses were based on the projections of the USDA Forest Service and private ranchers. Control programs were examined relative to annual maintenance with complete retreatment of initially treated areas to prevent repopulation of prairie dogs. Program costs could only be recovered if control was possible at a low annual maintenance

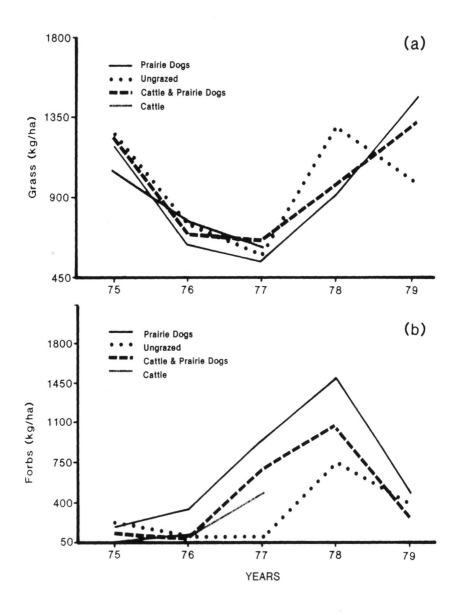

FIG. 21.1 Peak plant production under four treatments, over a 5-year period. Peak plant production is equal to the sum of the peaks for (a) grasses and (b) forbs (from Uresk and Bjugstad 1983).

level. Recovery of initial prairie dog control costs was estimated at 40 and 22 years at a maintenance control level of 5% for both perspectives. Costs could not be recoverable at 3-, 6-, 9-, and 15-year cycles for retreatment of prairie dog colonies. To gain an annual increase of 1 AUM (Animal Unit Month), 6.9 hectares of treated prairie dog colonies were required. Prairie dog control was not economically feasible except at a very low maintenance level. Once prairie dog control programs have been completed, range management practices that keep populations in check need to be followed.

CONTROL OF PRAIRIE DOGS BY RODENTICIDES AND OTHER METHODS

The efficacy of three rodenticide treatments for prairie dog control was evaluated immediately after treatment (Uresk et al. 1986). The treatments included zinc phosphide-treated grain (prebaited with nonpoison grain) and strychnine-treated grain (both with and without prebaiting). Active prairie dog burrows were reduced 95% with zinc phosphide treated grain, 83% with strychnine-treated grain (prebaited), and 45% with strychnine without prebaiting. Zinc phosphide was most effective in reducing active burrows of prairie dogs. Tietjen and Matschke (1982) reported a reduction in prairie dog activity of 96% with the use of zinc phosphide, and Knowles (1986) reported an average 85% reduction (range 65% to 95%) when prairie dogs were poisoned with zinc phosphide. Tietjen (1976) reported that the number of active black-tailed prairie dog burrows was reduced more in prairie dog colonies prebaited than without prebaiting when zinc phosphide-treated grain was used as the control agent.

Other prairie dog control techniques have included shooting, visual barriers, and diethylstilbestrol to inhibit prairie dog reproduction. Garrett and Franklin (1982) tested a burlap visual barrier for limiting expansion of colonies and dispersal of prairie dogs. This visual barrier was effective, but difficult to apply. The use of diethylstilbestrol to inhibit reproduction was effective, and it is easy to apply. However, use of diethylstilbestrol can interfere with reproduction of other animal species and may need to be restricted to specific areas.

RODENTICIDE IMPACTS ON NONTARGET ANIMALS

Although rodenticides have been used for prairie dog control for approximately 100 years (Merriam 1902), only in recent years have nontarget species been of concern when applying rodenticides. Timm (1983b) presents an excellent review of methods of control for 18 rodent species, including additional information for each animal on identification, range, habitat, food habits, general biology, damage, legal status, and economics of damage and control.

Zinc phosphide has been widely used as a rodenticide since 1942-43 when strychnine supplies became limited during World War II (Timm 1983a). More recently, zinc phosphide formulation and application rates have been developed for use in prairie dog control (Tietjen 1976). There has been some concern about the effects of zinc phosphide on nontarget animals. However, Bell and Dimmick (1975) reported that secondary poisoning by zinc phosphide was not a threat to red fox, *Vulpes fulva* (Desmarest), gray fox, *Urocyon cinereoargenteus* (Schrober), or great horned owls, *Bubo virginianus* Gmelin. Kit fox, *Vulpes macrotis* Merriam, survived after feeding on kangaroo rats, *Dipodomys* sp. Gray, killed with zinc phosphide (Schitoskey 1975). Matschke et al. (1983) reported no mortality among nontarget animals when zinc phosphide grain bait was broadcast to control Richardson's ground squirrels, *Spermophilus richardsonii* (Sabine).

Strychnine has been used for prairie dog control since the late 1800's (Merriam 1902). It has caused secondary hazards to nontarget animals (Schitoskey 1975, Hegdal et al. 1981). Wood (1965) reported that densities of five rodent species fluctuated independently of strychnine treatment over a 2-year period after an area was poisoned, but mortality of nonrodent animals did occur. Birds were killed on areas where steam-rolled oats with strychnine were applied on the surface for control of Richardson's ground squirrels (Hegdal and Gatz 1977). No detrimental effects were observed on other rodents or mammalian predators.

Strychnine can harm some bird species. Rudd and Genelly (1956) stated that the hazards of strychnine application in the field were much higher for waterfowl than for gray partridge, *Perdix perdix* (Linnaeus), ringnecked pheasant, *Phasianus colchicus* Linnaeus, quail, sharp-tailed grouse, *Tympanuchus phasianellus* (Linnaeus), and prairie chicken, *Tympanuchus* sp. Gloger. Strychnine is a significant hazard to some seed-eating birds, including horned larks, mourning doves, *Zenaida macroura* (Linnaeus), and blackbirds (Hegdal and Gatz 1977). Hegdal and Gatz (1977) also stated that vesper sparrows, *Pooecetes gramineus* (Gmelin), and western meadowlarks, *Sturnella neglecta* Audubon, were affected, but to a lesser extent. Tietjen (1976) and Matschke et al. (1983) reported no significant mortality for nontarget seedeating birds with application of zinc phosphide, but additional tests were recommended. Apa (1985) observed immediate losses of horned larks on prairie dog colonies treated with strychnine-treated grain (66% reduction), and also on areas prebaited (non-poison grain) followed with strychnine-treated grain (45% reduction). No significant reduction in horned lark numbers was found on comparison areas treated with zinc phosphide. Among the three rodenticide treatments, no impacts were observed one year later. Indirect effects did occur with habitat changes resulting from prairie dog control.

Deisch (1986) evaluated seven invertebrate taxa for immediate and first-year treatment effects from zinc phosphide and strychnine-treated grain (with and without non-poison grain for prebaiting) used for prairie dog control. Spider mites and crickets were not affected by the three rodenticide treatments. Significant immediate and post-1-year impacts by zinc phosphide occurred with ants. Strychnine without prebaiting immediately reduced wolf spiders. Darkling beetles were significantly reduced with zinc phosphide immediately following treatment. However, ground beetles and dung beetles showed no changes in density. Generally, the few long-term impacts were attributed to habitat changes resulting from prairie dog control.

Uresk et al. (in press) reported that zinc phosphide was responsible for reduced densities (79%) of deer mice, *Peromyscus maniculatus* (Wagner), a nontarget species, immediately after treatment for prairie dog control. However, the effect was highly variable and not statistically significant. Strychnine with or without prebait was not associated with significant reductions in deer mice densities. This finding is contrary to the 86% reduction in rodent populations reported by Wood (1965) one month after treatment with strychnine.

Information does not indicate rodenticides affect Leporids. Eastern cottontail, *Sylvilagus floridanus* (J.A. Allen), fecal pellet counts after rodenticide treatment were apparently greater on strychnine-prebaited sites than on strychnine without prebait and zinc phosphide-treated sites (Uresk et al. in press). However, eight months after rodenticide treatment eastern cottontails showed no population differences among rodenticide treatments. White-tailed jackrabbit, *Lepus townsendii* Bachman, abundance was apparently higher on areas treated with zinc phosphide. Eight months later, white-tailed jackrabbit abundance was not different among rodenticide treatments.

CATTLE-PRAIRIE DOG RELATIONSHIPS

Dietary information is essential to assess the role of prairie dogs in prairie ecosystems. Since Kelso (1939) first presented general data on black-tailed prairie dog diets, research has been conducted in Colorado (Tileston and Lechleitner 1966, Bonham and Lerwick 1976, Hansen and Gold 1977), Montana (Fagerstone et al.

1981), and South Dakota (Summers and Linder 1978, Fagerstone et al. 1981, Uresk 1984). Most of these studies were conducted in National Parks or in areas with limited, regulated grazing by cattle. Dietary information by plant species will vary from region to region even though major categories (grasses, forbs, shrubs) consumed by prairie dogs are similar. Intensive management of rangelands must be based on plant species consumed by herbivores in the area of interest.

Black-tailed prairie dog food habits and diets are variable (Fagerstone 1982). Generally, prairie dogs eat the same forage species as cattle, thus causing a decrease for some beneficial plant species preferred by cattle. Yet, this dual use by herbivores did not significantly reduce cattle weights compared to cattle weights in areas without prairie dogs during the summer and winter months (Hansen and Gold 1977, O'Meilia et al 1982). Greater plant species diversity and increased crude protein occur on prairie dog colonies because the vegetation is maintained in an early stage of phenological development.

Prairie dog expansion can be reduced by management for cool season grasses with increased height and density (Cincotta et al. 1986). Prairie dogs did not significantly expand over a 4-year period on areas where cattle were excluded (Uresk et al. 1982). Light grazing or periodic exclusion of livestock would increase height and density of the vegetation and reduce prairie dog expansion. Snell and Hlavachick (1980) and Snell (1985) reported reduced expansion and elimination of prairie dog colonies by deferred grazing. It has been suggested by some researchers that black-tailed prairie dogs were most abundant in areas intensively grazed by livestock (Koford 1958, Uresk and Bjugstad 1983). Black-tailed prairie dogs inhabited areas where vegetation height was reduced by clipping unpalatable plants to ground level (Koford 1958). Increased stocking rates of cattle and grasses with low height and density allows for prairie dog expansion (Uresk et al. 1982, Cincotta et al. in press).

Range condition classification has been widely accepted as a measure of range status (Dyksterhuis 1949, 1985, Smith 1978). Range classes also provide a useful means of evaluating effects of management. Recently, climax or near-climax vegetation and departures from climax have been used to measure range condition. However, management of rangelands for climax condition may not meet all management objectives and goals. Uresk (1986b) reported that range condition class below climax condition was optimum to obtain highest forage yields for cattle management; but for management of nongame animals, a lower range condition class may be more beneficial (Agnew et al. 1986). Most sites in lower condition were associated with heavy herbivore use (i.e., prairie dog colonies). No prairie dogs were associated with range classes

TABLE 21.1
Average abundance (number/5 hectares ± SE) of common birds on
prairie dog towns and on adjacent mixedgrass prairie sites without
prairie dogs in western South Dakota during 1981 and 1982 (Agnew
et al. 1986).

| Species | Sites | |
	Prairie dog colonies	Mixedgrass prairie
Horned lark	97 ± 23	2 ± 1[*]
Western meadowlark	34 ± 7	43 ± 6[*]
Mourning dove	13 ± 4	6 ± 1[*]
Killdeer	7 ± 2	1 ± 1
Barn swallow	7 ± 4	1 ± 1
Burrowing owl	3 ± 1	0[*]
Common grackle	3 ± 2	1 ± 1[*]
Red-winged blackbird	2 ± 1	6 ± 1
Rock dove	1 ± 1	0
Upland sandpiper	1 ± 1	3 ± 1[*]
Lark bunting	1 ± 1	4 ± 1
Grasshopper sparrow	1 ± 1	2 ± 1
Common nighthawk	0	2 ± 1
Unidentified	4 ± 1	5 ± 1
Total	171	73[*]

[*]Significantly different from prairie dog towns at $\alpha = 0.05$.

at or near climax. Thus, the relationship of range classification
of vegetation as related to livestock and wildlife provides
managers a tool to meet management objectives; clearly no one
range condition class can meet all objectives.

WILDLIFE HABITAT ENHANCEMENT AND OTHER USES ASSOCIATED WITH PRAIRIE DOGS

Prairie dogs act as habitat regulators by maintaining
shortgrass and "patchy" plant associations with less mulch cover
and lower vegetation height than ungrazed or lightly grazed areas
(Agnew et al. 1986). These are vegetative features that also
provide quality habitat for other species of small rodents such as

deer mice and grasshopper mice, *Onychomys leucogaster* (Wied-Neuwied). The abundance of these small rodents was significantly higher on active prairie dog colonies with "patchy" habitat. Greater avian densities and species richness were reported on prairie dog colonies by Agnew et al. (1986). Thirty-six avian species were observed on prairie dog colonies compared to 29 on areas without prairie dogs. Densities of avian species were 171 and 73 individuals per 5 hectares on prairie dog colonies and areas without prairie dogs, respectively (Table 21.1). Although the role of prairie dogs as habitat regulators has not been fully addressed, it is evident that prairie dogs influence the abundance and species composition of birds, small mammals, and vegetation.

Black-tailed prairie dogs in western South Dakota have become an important species for sport hunting. In 1985, they provided 46,000 hunter-days of recreation (South Dakota Dept. Game, Fish and Parks, Pierre SD). Prairie dog sport hunting provides a source of state revenue that is expected to increase in the future. Other jobs and revenue related to prairie dog sport hunting include guide service and fee hunting. Currently, a feasibility study is being conducted by the Bureau of Indian Affairs on the use of prairie dog meat as mink food, and as pelts for manufacturing gloves.

REFERENCES

Agnew, W., D.W. Uresk, and R.M. Hansen. 1986. Flora and fauna associated with prairie dog colonies and adjacent ungrazed mixed-grass prairie in western South Dakota. J. Range Manage. 39:135-139.

Apa, A.D. 1985. Efficiency of two black-tailed prairie dog rodenticides and their impacts on non-target bird species. Unpublished M.S. Thesis, South Dakota State Univ.

Bell, H.B., and R.W. Dimmick. 1975. Hazards to predators feeding on prairie voles killed with zinc phosphide. J. Wildl. Manage. 9:816-819.

Bonham, C.D., and A. Lerwick. 1976. Vegetation changes induced by prairie dogs on shortgrass range. J. Range Manage. 29:221-225.

Cincotta, R.P., D.W. Uresk, R.M. Hansen. A statistical model of expansion in a colony of black-tailed prairie dogs. *In*: Eighth Great Plains wildlife damage control workshop proc., April 28-30, 1987, Rapid City, SD. Rocky Mountain Forest and Range Exp. Sta., Ft. Collins, CO. (in press).

Collins, A.R., J.P. Workman, and D.W. Uresk. 1984. An economic analysis of black-tailed prairie dog (*Cynomys ludovicianus*) control. J. Range Manage. 37:358-361.

Coppock, D.L., J.K. Detling, J.E. Ellis, and M.I.Dyer. 1983. Plant-herbivore interactions in a North American mixed-grass prairie. I. Effects of blacktailed prairie dogs on intraseasonal above ground plant biomass and nutrient dynamics and plant species diversity. Oecologia 56:1-9.

Deisch, M.S. 1986. The effects of three rodenticides on nontarget small mammals and invertebrates. Unpublished M.S. Thesis, South Dakota State Univ.

Dyksterhuis, E.J. 1949. Condition and management of rangeland based on quantitative ecology. J. Range Manage. 2:104-115.

Dyksterhuis, E.J. 1985. Follow-up on range sites and condition classes as based on quantitative ecology. Rangelands 7:172-173.

Fagerstone, K.A. 1982. A review of prairie dog diet and its variability among animals and colonies, p. 178-184. In: R.M. Timm and R.J. Johnson (eds.) Fifth Great Plains wildl. damage control workshop proc., Oct. 13-15, 1981. Univ. Nebraska, Lincoln.

Fagerstone, K.A., H.P. Tietjen, and 0. Williams. 1981. Seasonal variation in the diet of black-tailed prairie dogs. J. Mammal. 62:820-824.

Garrett, M.G., and W.L. Franklin. 1982. Prairie dog dispersal in Wind Cave National Park: possibilities for control, p. 185-198. In: R.M. Timm and R.J. Johnson (eds.) Fifth Great Plains wildl. damage control workshop proc., Oct. 13-15, 1981. Univ. Nebraska, Lincoln.

Hansen, R.M., and I.K. Gold. 1977. Black-tailed prairie dogs, desert cottontails and cattle trophic relations on shortgrass range. J. Range Manage. 30:210-213.

Hegdal, P.L., and T.A. Gatz. 1977. Hazards to seed-eating birds and other wildlife associated with surface strychnine baiting for Richardson's ground squirrels. EPA report under Interagency Agreement EPA-IAG-D4-0449.

Hegdal, P.L., T.A. Gatz, and E.C. Fite. 1981. Secondary effects of rodenticides on mammalian predators, p. 1781-1793. In: J.A. Chapman and D. Pursley (eds.) The world furbearer conference proceedings.

Kelso, H. 1939. Food habits of prairie dogs. USDA Circ. 529.

Knowles, C.J. 1986. Population recovery of black-tailed prairie dogs following control with zinc phosphide. J. Range Manage. 39:249-251.

Koford, C.B. 1958. Prairie dogs, whitefaces, and blue grama. Wildl. Monogr. 3.

Matschke, G.H., M.P. Marsh, and D.L. Otis. 1983. Efficacy of zinc phosphide broadcast baiting for controlling Richardson's ground squirrels on rangeland. J. Range Manage. 36:504-506.

Merriam, C.H. 1902. The prairie dog of the Great Plains, p. 257-270. *In:* Yearbook of the USDA. U.S. Gov. Print. Office, Washington, DC.

O'Meilia, M.E., F.L. Knopf, and J.C. Lewis. 1982. Some consequences of competition between prairie dogs and beef cattle. J. Range Manage. 35:580-585.

Osborn, D., and P.F. Allan. 1949. Vegetation of an abandoned prairie dog town in tall grass prairie. Ecology 30:322-332.

Potter, R.L. 1980. Secondary successional patterns following prairie dog removal on short grass range. Unpublished M.S. Thesis, Colorado State Univ.

Rudd, R.L., and R.E. Genelly. 1956. Pesticides: their use and toxicity in relation to wildlife. California Fish and Game Bull. 7.

Schenbeck, G.L. 1982. Management of black-tailed prairie dogs on the National Grasslands, p. 207-217. *In:* R.M. Timm and R.J. Johnson (eds.) Fifth Great Plains wildl. damage control workshop proc., Oct. 13-15, 1981. Univ. Nebraska, Lincoln.

Schitoskey, F. 1975. Primary and secondary hazards of three rodenticides to kit fox. J. Wildl. Manage. 39:416-418.

Smith, E.L. 1978. A critical evaluation of the range condition concept, p. 266-267. *In:* D.N. Hyder (ed.) Proceedings of the first international rangeland congress. Soc. Range Management. Denver, CO.

Snell, G.P. 1985. Results of control of prairie dogs. Rangelands 7:30.

Snell, G.P.. and B.D. Hlavachick. 1980. Control of prairie dogs--the easy way. Rangelands 2:239-240.

Summers, C.A., and R.L. Linder. 1978. Food habits of the black-tailed prairie dog in western South Dakota. J. Range Manage. 31:134-136.

Taylor, W.P., and J.V.G. Loftfield. 1924. Damage to range grasses by the Zuni prairie dog. USDA Bull. 1227.

Tietjen, H.P. 1976. Zinc phosphide--its development as a control agent for black-tailed prairie dogs. USDI, Fish and Wildl. Serv. Spec. Sci. Rep.--Wildl. 195.

Tietjen, H.P., and G.H. Matschke. 1982. Aerial pre-baiting for management of prairie dogs with zinc phosphide. J. Wildl. Manage. 46:1108-1112.

Tileston, J.V., and R.R. Lechleitner. 1966. Some comparisons of the black-tailed and white-tailed prairie dogs in north central Colorado. Amer. Midl. Nat. 75:292-316.

Timm, R.M. 1983a. Description of active ingredients, p. G-31-G-77. *In:* R.M. Timm (ed.) Prevention and control of wildlife damage. Great Plains Agr. Council, Wildl. Res. Comm. Nebraska Coop. Ext., Lincoln.

Timm, R.M. (ed.) 1983b. Prevention and control of wildlife damage. Great Plains Agr. Council, Wildl. Res. Comm. Nebraska Coop. Ext., Lincoln.

Uresk, D.W. 1984. Black-tailed prairie dog food habits and forage relationships in western South Dakota. J. Range Manage. 37:325-329.

Uresk, D.W. 1985. Effects of controlling black-tailed prairie dogs on plant production. J. Range Manage. 38:466-468.

Uresk, D.W. 1986a. Food habits of cattle on mixed-grass prairie on the northern Great Plains. Prairie Nat. 18:211-218.

Uresk, D.W. 1986b. A method for quantitatively estimating range condition classes with multivariate techniques in a mixed grass prairie. Final Report. USDA Forest Service, Rocky Mountain Forest & Range Experiment Station. South Dakota School of Mines, Rapid City.

Uresk, D.W., and A.J. Bjugstad. 1983. Prairie dogs as ecosystem regulators on the northern high plains, p. 91-94. *In:* Seventh North American prairie conference, proc. Aug. 4-6, 1980. Southwest Missouri State Univ., Springfield.

Uresk, D.W., R.M. King, A.D. Apa, M.S. Deisch, and R.L. Linder. Rodenticidal effects of zinc phosphide and strychnine on nontarget species. *In:* Eighth Great Plains wildlife damage control workshop proc., April 28-30, 1987, Rapid City, SD. Rocky Mountain Forest & Range Exp. Sta., Ft. Collins, CO.

Uresk, D.W., R.M. King, A.D. Apa, and R.L. Linder. 1986. Efficacy of zinc phosphide and strychnine for black-tailed prairie dog control. J. Range Manage. 39:298-299.

Uresk, D.W., J.G. MacCracken, and A.J. Bjugstad. 1982. Prairie dog density and cattle grazing relationships, p. 199-201. *In:* R.M. Timm and R.J. Johnson (eds.) Fifth Great Plains wildl. damage control workshop, proc. Oct. 13-15, 1981. Univ. Nebraska, Lincoln.

Wood, J.E. 1965. Response of rodent populations to controls. J. Wildl. Mange. 29:425-427.

22. FUNCTION OF INSECTIVOROUS BIRDS IN A SHORTGRASS IPM SYSTEM

Lowell C. McEwen
Department of Fishery and Wildlife Biology
Colorado State University, Fort Collins, Colorado 80523

The potential of insectivorous birds and other wild vertebrates as regulators of pest insect populations has long been recognized. Laws to protect beneficial birds were proposed as early as 1877 by the United States Entomological Commission (1878). Study of the relationships of birds to man was called economic ornithology and included beneficial and harmful aspects. Before the development of synthetic chemical pesticides, birds were considered to be significant predators of agricultural insect pests by many farmers, and were encouraged for pest control. There was great interest in determining the kinds and amounts of insects eaten by birds. Much of the research involved examination of stomach contents (food habits analysis). It is noteworthy that S.A. Forbes, considered the founder of economic ornithology (McAtee 1933), was also a leading entomologist who defined many of the principles of integrated pest management (IPM) more than 100 years ago (Metcalf 1980).

RECOGNITION AND PROTECTION OF BENEFICIAL BIRDS

In the act establishing a Federal Department of Agriculture in 1862, reference was made to "the introduction and protection of insectivorous birds" (McAtee 1933). A Section of Economic Ornithology and Mammalogy was formed under the Division of Entomology in 1885. The purpose was to investigate and manage both beneficial and harmful activities of wildlife. By 1906, the Section evolved into the Bureau of Biological Survey (Trippensee 1948) with broader responsibilities, but retained emphasis on food habits studies. In 1916, the food habits work became a separate section called Economic Investigations in Ornithology and in 1921 was elevated to a Division of Food Habits Research. The findings of thousands of bird stomach analyses were summarized in American Wildlife and Plants by Martin et al. (1951).

In 1918, the Migratory Bird Treaty Act of 1916 was enabled. This legislation placed "all migratory insectivorous birds and songbirds, as well as threatened species, under federal protection" (Clepper 1966). The Bureau of Biological Survey was transferred to the U.S. Department of Interior in 1940 and combined with the Bureau of Fisheries to establish the U.S. Fish and Wildlife Service (USFWS). The USFWS is currently responsible for the protection and management of migratory birds in the United States.

EARLY STUDIES OF BIRD PREDATION ON INSECTS

Professor S.A. Forbes examined the stomachs of more than 5,000 birds of 32 species and published some of the earliest papers on food habits beginning in 1880. He estimated that birds destroyed as much as 70% of the annual insect crop in Illinois (Henderson 1927). Another pioneer worker was Samuel Aughey of Nebraska, who studied bird predation on grasshoppers in the 1870's. He found that 202 species of birds fed on grasshoppers and the majority of them had 25 or more grasshoppers in their stomachs (Henderson 1927, McAtee 1933). Bird predation on grasshoppers was also investigated by H.C. Bryant (1914). He estimated that 13 species of birds consumed approximately 120,445 grasshoppers per day per square mile. Consumption ranged from an estimated 100 grasshoppers/day by house sparrows to 78,500/day by western meadowlarks. Extensive studies were also done by Knowlton and Harmston (1943) in Utah; they examined 1,601 stomachs of 42 bird species for presence of orthopterans. All 42 species contained at least one and, overall, 654 (40.8%) had grasshoppers/crickets in numbers ranging from one to 68 per stomach. They estimated a statewide average bird density of 2.5 birds/acre taking "billions of injurious insects" each year.

The most prolific of bird food habits researchers was F.E.L. Beal who examined and reported on 37,825 bird stomachs (McAtee 1933). Professor Beal observed the high intake of grasshoppers by many species and reported, for example, that one Franklin's gull contained 70 entire grasshoppers and the jaws of 56 more (Forbush 1907). In the review of Biological Survey and other food habits studies, grasshoppers were mentioned as an important animal food for at least 120 species of American birds (Martin et al. 1951). Perhaps the most well-known bird/grasshopper predation event was the saving of the Mormon settlers' crops from hordes of Mormon crickets (long-horned grasshoppers) by flocks of gulls. The crops were saved by "thousands of gulls" that "entirely freed" the fields of the crickets several times (Forbush 1907; see Chapter 9).

Forbush attempted to estimate insect control by birds on a national scale. He concluded that birds reduced insect damage by 28% for a savings value of $444 million (Henderson, 1927). Several cases of destruction of grasshopper egg beds by various species of birds were reported by Wakeland (1958); predation on nymphs also may be heavy.

NUTRITIONAL VALUE OF GRASSHOPPERS FOR BIRDS

The nearly universal dependence of rangeland birds on grasshoppers as an animal food source is understandable because of several attributes. Grasshoppers are exceptionally high in nutritional value, containing 50-75% crude protein (Ueckert et al. 1972, DeFoliart 1975), they are distributed throughout most habitats, and are available in several sizes. Modern birds evolved considerably later than grasshoppers. Thus, grasshoppers were an available food source throughout the evolution of birds and may share an interdependence with the co-evolved rangeland vegetation (McEwen and DeWeese 1987). Insects are especially significant for successful avian reproduction since "the young of nearly all smaller land birds are fed largely on an insect diet and do not shift to plant food until they are fairly well-grown" (Martin et al. 1951). The growth and survival of young gallinaceous birds is very closely tied to the abundance and availability of ground-dwelling insects (Potts 1980). Any review of food eaten by wild birds in agricultural habitats will show the great importance of grasshoppers. This is particularly true of rangeland birds (McEwen et al. 1972).

ROLE OF BIRDS IN RANGELAND IPM SYSTEMS

Although speculation on the value of birds for pest control generates interest in the subject and support for protection of beneficial species, field experiments are needed to help develop specific IPM practices. More work has been done in forest ecosystems than in rangeland or cropland systems. Takekawa (1982) reviewed the use of avian predators in preventing and controlling forest insect pest outbreaks. He cited several successful experiments including placement of nest boxes to increase populations of beneficial birds. McFarlane (1976) reviewed the potential value of birds in agricultural IPM systems.

He concluded that the effectiveness of birds in controlling insect pests was underestimated, that birds were especially important in regulating endemic insect populations, and that avian predation could be increased and made more effective with modification of agricultural practices. Use of insect control methods that did not kill or repel beneficial birds or cause their reproductive failure would be an evident benefit in an IMP system. Rickleffs (1973), McFarlane (1976), Luff (1981), Takekawa (1982), and others have discussed predator-prey ecological theory as it applied to birds and insect populations. There is general agreement that bird predation characteristics lend to stability of insect prey populations and reduce or prevent outbreaks.

Wiens and Dyer (1975) discussed the possible function of the avifauna in rangeland ecosystems. They showed that birds are clearly minor components in terms of total ecosystem biomass and energy flow, but postulated that birds may exert an unknown degree of regulation on insect populations. They also stressed the need for intensive research on relationships between avian predators and their prey. McEwen and DeWeese (1987) found an inverse relationship between grasshopper densities and breeding bird densities in mixed sagebrush-grassland in Montana. Joern (1986) conducted one of the very few quantitative experiments on effects of birds on grasshopper populations. Bird predation reduced adult grasshopper numbers in open plots by a mean of 27.4% in comparison to plots where birds were excluded.

Hudleston (1958) experimented by releasing known numbers of African locust nymphs, then observing numbers taken by birds. He found that birds destroyed nearly all the released "hoppers" (400 to 500 per release) in one to three days. He also observed complete elimination of bands of 100,000 and 200,000 "hoppers" in 10 days to two weeks due to bird predation. Greathead (1966) tried to quantify bird predation on locusts in Africa. He called the effects erratic and dependent on factors such as habitat and season, but believed birds were effective in controlling small gregarious populations. Birds did not have the capacity for significant reduction of very large infestations.

Grasshopper control methods that are highly specific for the target insects would be the most desirable from the standpoint of maintaining alternate prey for generalist predators, especially for birds and other small wild vertebrates. Target-specific pest suppression tactics would not have as severe an impact on the total prey base needed by beneficial predators as the broad-spectrum insecticides in current use. The common species of birds nesting in shortgrass habitat are known to be opportunistic facultative feeders with the ability to switch to alternate prey. This is a desirable characteristic for maintaining predator populations when the target prey density is low (Luff 1983).

POTENTIAL OF BIRDS FOR CONTROLLING GRASSHOPPERS IN SHORTGRASS PRAIRIE

The importance of birds as mortality factors in pest insect population dynamics remains largely unquantified. If synthetic organic pesticides had not been developed, and if economic entomology had remained on its pre-chemical course of primarily utilizing biological and cultural pest controls, no doubt we would have more quantitative information on the significance of bird predation. Wiens (1977) is correct in stating that the levels of data on prey standing crops and bird predation rates are "too coarse" to permit detailed analysis. Also, the recruitment and turnover rates of both predator and prey need to be known if valid models are to be constructed. Thus, there are voluminous food habits data documenting the universality of grasshoppers as animal food for birds. On the other hand, there are very few data on the regulating effect of bird predation on grasshopper populations.

Viewed from the evolutionary perspective, it would seem likely that bird predation provides a significant damping force on the increase of grasshoppers. Considering the long period of coevolution of rangeland vegetation, grasshoppers, and rangeland birds, some mutual co-regulation and interdependence would be expected. It may be useful to consider an untestable hypothesis: permanent removal of all land birds from North America would result in higher grasshopper numbers followed by a significant increase in economic infestations. Since we cannot and would not want to conduct such an experiment, we must study the question by obtaining quantitative data on the various mortality factors for grasshopper eggs, nymphs, and adults to determine the relative contribution of bird predation. Peterson (1980) discussed the effects (economic and ecological) of the hypothetical removal of all insectivorous birds from the continent. He suggested that use of pest-resistant plants and pesticides would adequately replace birds for insect control. In the absence of quantitative data, he proposed that preservation of birds in western ecosystems cannot be justified on a functional basis, only on aesthetic and moral grounds. Field experiments, such as the work by Joern (1986), are much needed to clarify the function of birds in rangeland ecosystems and the economic value of birds.

Examination of available information on bird/grasshopper relations, however, does support a general hypothesis that bird predation is a limiting factor on grasshopper populations. As stated earlier, authors have documented predation by finding grasshoppers in the stomachs and gastrointestinal tracts of many bird species (Bryant 1914, Henderson 1927, Martin et al. 1951, Hudleston 1958, Wakeland 1958, Boyd 1976, McEwen et al. 1986).

Since we have data on many of the predator/prey parameters such as bird and grasshopper population densities, reproductive rates, and energetic needs, we can calculate some potential interactions and effects on grasshopper numbers.

Shortgrass prairie has fewer species of breeding birds and lower densities than more mesic habitats. Wiens and Dyer (1975) reported a mean of 4.3 breeding species and 282.3 individual birds/km^2 (SD = 102.2; CV = 35.5%) from 19 locations. The mean number of individuals is the equivalent of 2.8 birds/ha. Others have reported similar or higher densities in shortgrass habitat: Ryder (1980) found a 5-year mean of 2.7 birds/ha; Johnson (1980) estimated a range of 1.25 to 5.0 birds/ha; Finzel (1964) reported a mean of 5.2 breeding birds/ha over a 3 year period. These reported bird densities are minimum values and do not take into account larger, wider ranging species such as raptors, or associated species such as swallows, killdeer, plovers, and nighthawks, that are all known to feed on grasshoppers. Grasshopper densities are more variable than bird numbers and can range from a low of less than 1/m^2 to greater than 60/m^2 over extensive areas, and can exceed 1,000/m^2 in the nymphal stage at local sites (Parker et al. 1955). Typical grasshopper densities present in shortgrass (not in the outbreak stage) average 1.94 to 3.28/m^2 (Bhatnager and Pfadt 1973).

Predation rates by birds vary with grasshopper availability. Bryant (1914) found that western meadowlarks had a mean of 16 grasshoppers per stomach or 96.2% of their total diet when grasshopper densities were abnormally high (outbreak stage). In a year when grasshopper densities were "normal," western meadowlarks still had a mean of 7 per stomach, or 83.1% of their diet. Baldwin (1972) examined utilization of grasshoppers by lark buntings and found that while grasshoppers comprised 10.1% of available insects of four families, they comprised 48.3% of the lark buntings' diet, indicating preference and selectivity by the birds for grasshoppers.

Taking some reported values (Bryant 1914, Henderson 1927, Baldwin 1972) for common breeding bird species in shortgrass prairie, we find that horned larks may have from 8 to 42 grasshoppers per bird stomach, meadowlarks from 23 to 48, lark buntings from 1 to 14, and McCown's longspurs from 1 to greater than 5. Birds have a high metabolic rate (Sturkie 1965) and consume a high proportion of their body weight daily. Thus, the number of insects in a bird's stomach represent one point in time and not the total number consumed in a day. Extensive study of lark buntings in shortgrass habitat led to the conclusion that one bird would eat 65 grasshoppers in a typical day from May to July along with several hundred other insects (Baldwin 1972). At the same location, Boyd (1976) calculated that a pair of horned larks consumed 156 grasshoppers per day in addition to 705 other insects. These calculations were based on analysis of a large number of stomach contents, size and weight of prey, metabolic trials with caged birds, and energy requirements of the birds.

From the foregoing, it is clear that the bird predation potential on grasshoppers in shortgrass prairie is high. To estimate this potential, we can assume a minimum of 3 birds per hectare (of the common breeding species) that capture at least 70 grasshoppers/bird/day or 210 grasshoppers/ha/day. This is a conservative estimate based on the preceding review and does not include predation by other common avian species that are larger and more wide ranging nor does it consider the large number of grasshoppers captured and fed to nestlings and fledglings. A grasshopper density of $5/m^2$ (approaching the economic damage level) would equal 50,000/ha. Over the 4-month period, May through August, bird predation could readily account for 25,200 grasshoppers or 50% of the population. At the lower grasshopper densities that may be found on many shortgrass ranges, predation rates might decline, but birds conceivably could take an even higher percentage of the grasshoppers present. It is difficult not to consider birds as important contributors to biological control of range grasshoppers. Bird predation supplements other predators on grasshoppers such as small mammals of several species, and robber flies. Considering that there are also several pathogens and weather factors that can cause widespread grasshopper mortality, we should not disregard an omnipresent force such as bird predation that works in concert with other mortality factors to assist in regulation of grasshopper populations.

REFERENCES

Baldwin, P.H. 1972. The feeding regime of granivorous birds in shortgrass prairie in Colorado, USA, p.237-47. *In*: S.C. Kendeigh and J. Pinowski (eds.) Productivity, population dynamics and systematics of granivorous birds. Polish Science Public, Warszawa.

Bhatnager, K.N., and R.E. Pfadt. 1973. Growth, density, and biomass of grasshoppers in the shortgrass and mixed-grass associations. U.S. IBP Grassland Biome Rep. 225, Colorado State Univ., Fort Collins.

Boyd, R.L. 1976. Behavioral biology and energy expenditure in a horned lark population. Unpublished Ph.D. Dissertation, Colorado State Univ., Fort Collins.

Bryant, H.C. 1914. Birds as destroyers of grasshoppers in California. Auk 31:168-177.

Clepper, H. 1966. Origins of American conservation. Ronald Press Co., New York.

DeFoliart, G.R. 1975. Insects as a source of protein. Bull. Entomol. Soc. Am. 21:161-163.

DeWeese, L.R., L.C. McEwen, L.A. Settimi, and R.D.Deblinger. 1983. Effects on birds of fenthion aerial application for mosquito control. J. Econ. Entomol. 76:906-911.

Finzel, J.E. 1964. Avian populations of four herbaceous communities in southeastern Wyoming. Condor 66:496-510.

Forbush, E.H. 1907. Useful birds and their protection. Massachusetts State Bd. of Agr., Wright and Potter Print Co., Boston, MA.

Greathead, D.J. 1966. A brief survey of the effects of biotic factors on populations of the desert locust. J. Appl. Ecol. 3:239-250.

Henderson, J. 1927. The practical value of birds. MacMillan Co., New York.

Hudleston, J.A. 1958. Some notes on the effects of bird predators on hopper bands of the desert locust (*Schistocerca gregaria* Forsakal). Entomol. Mon. Mag. 94:210-214.

Joern, A. 1986. Experimental study of avian predation on co-existing grasshopper populations (Orthoptera: Acrididae) in a sandhills grassland. Oikos 46:243-249.

Johnson, R.R., L.T. Haight, M.M. Riffey, and J.M. Simpson. 1980. Brushland/steppe bird populations, p. 98-112. *In*: R.M. DeGraff and N.G. Tilghman (compilers) Management of western forests and grasslands for nongame birds. USDA For. Serv. Gen. Tech. Rep. INT-86.

Knowlton, G.F. and F.C. Harmston. 1943. Grasshoppers and crickets eaten by Utah birds. Auk 60:589-591.

Luff, M.L. 1983. The potential of predators for pest control. Agr. Ecosystems Environ. 10:159-181.

Martin, A.C., H.S. Zim, and A.L. Nelson. 1951. American wildlife and plants. McGraw-Hill, New York.

McAtee, W.L. 1933. Economic ornithology, p. 111-129. *In*: Fifty years' progress of American ornithology. Am. Ornith. Union, Lancaster, PA.

McEwen, F.L., and G.R. Stephenson. 1979. The use and significance of pesticides in the environment. J. Wiley and Sons, New York.

McEwen, L.C. 1982. Review of grasshopper pesticides vs. rangeland wildlife and habitat, p. 362-382. *In*: J.M.Peek and P.D. Dalke (eds.) Proc. wildlife-livestock relationships symp. Univ. Idaho, Moscow.

McEwen, L.C., C.E. Knittle, and M.L. Richmond. 1972. Wildlife effects from grasshopper insecticides sprayed on shortgrass range. J. Range Manage. 25:188-194.

332

McEwen, L.C., and J.O. Ells. 1975. Field ecology investigations of the effects of selected pesticides on wildlife populations. U.S. IBP Tech. Rep. 289, Colorado State Univ., Fort Collins.

McEwen, L.C., and L.R. DeWeese. 1987. Wildlife and pest control in the sagebrush ecosystem: ecology and management considerations. p. 76-85. In: J.A. Onsager (ed.) Integrated pest management on rangeland. State of the art in the sagebrush ecosystem. USDA Agric. Res. Serv., Washington, D.C. ARS-50.

McEwen, L.C., L.R. DeWeese, and P. Schladweiler. 1986. Bird predation on cutworms (Lepidoptera: Noctuidae) in wheat fields and chlorpyrifos effects on brain cholinesterase activity. Environ. Entomol. 15:147-151.

McFarlane, R.W. 1976. Birds as agents of biological control. The Biol. 58:123-140.

Metcalf, R.L. 1980. Changing role of insecticides in crop protection. Annu. Rev. Entomol. 25:219-256.

Mitchell, J.E., and R.E. Pfadt. 1974. A role of grasshoppers in a shortgrass prairie ecosystem. Environ. Entomol. 3:358-360.

Parker, J.R., R.C. Newton, and R.L. Shotwell. 1955. Observations on mass flights and other activities of the migratory grasshopper. USDA Tech. Bull. 1109.

Peterson, S.R. 1980. The role of birds in western communities, p. 6-12. In: R.M. DeGraff and N.G.Tilghman (compilers) Management of western forests and grasslands for nongame birds. USDA For. Serv. Gen. Tech. Rep. INT-86.

Potts, G.R. 1980. The effects of modern agriculture, nest predation and game management on the population ecology of partridges (Perdix perdix and Alectoris rufa), p. 2-79. In: A. Macfayden (ed.) Adv. Ecol. Res. Vol. 11.

Rickleffs, R.E. 1973. Ecology. Chiron Press Inc., Portland, OR.

Rodell, C.F. 1977. A grasshopper model for a grassland ecosystem. Ecology 58:227-245.

Ryder, R.A. 1980. Effects of grazing on bird habitats, p. 51-66. In: R.M. DeGraff and N.G. Tilghman (compilers) Management of western forests and grasslands for nongame birds. USDA For. Serv. Gen. Tech. Rep. INT-86.

Stromborg, K.L., L.C. McEwen, and T. Lamont. 1984. Organophosphate residues in grasshoppers from sprayed rangelands. Chem. Ecol. 2:39-45.

Sturkie, P.D. 1965. Avian physiology, 2nd Ed., Cornell Univ. Press, Ithaca, NY.

Takekawa, J.Y., E.O. Garton, and L.A. Langelier. 1982. Biological control of forest insect outbreaks: the use of avian predators. Trans. N. Am. wildl. and nat. res. conf. 47:393-402.

Trippensee, R.E. 1948. Wildlife management. Vol. 1. McGraw-Hill, New York.

Ueckert, D.N., S.P. Yang, and R.C. Albin. 1972. Biological value of rangeland grasshoppers as a protein concentrate. J. Econ. Entomol. 65:1286-1288.

U.S. Entomological Commission. 1878. First annual report of the United States Entomological Commission for the year 1877. U.S. Govt. Print. Off., Washington, D.C.

Wakeland, C. 1958. The high plains grasshopper. USDA Tech. Bull. 1167, Washington, D.C.

Wiens, J.A. 1977. Model estimation of energy flow in North American grassland bird communities. Oecologia 31:135-151.

Wiens, J.A., and M.I. Dyer. 1975. Rangeland avifaunas: their composition energetics and role in the ecosystem. p. 146-181. *In*: D.R. Smith (coordinator) Symposium on management of forest and range habitats for nongame birds. USDA For. Serv. Gen. Tech. Rep. WO-1.

BIOLOGICAL AND ECONOMIC MODELS FOR RANGELAND PEST MANAGEMENT

23. MODELING THE PEST COMPONENT OF RANGELAND ECOSYSTEMS

Jesse A. Logan
Natural Resource Ecology Laboratory and
Department of Entomology
Colorado State University, Fort Collins, Colorado 80523

The perspective I will adopt in this chapter is that of a synoptic evaluation of insect pest models, with the view of focusing more narrowly on the specific attributes of rangeland systems that offer both challenge and promise for successful modeling applications. In order to develop this perspective, I will first provide an evaluation of modeling contributions in Integrated Pest Management (IPM) research. Then I compare the attributes of more intensively managed agricultural systems to those of rangeland systems. Next, the implication of these attributes for modeling rangeland systems will be discussed. The final section provides a brief summary of the previous sections.

MODELING APPLICATIONS IN PEST MANAGEMENT

Clearly, the major applications of modeling in pest management has occurred in intensively managed crop systems (Ruesink 1976, Getz and Gutierrez 1982). This is not to say that important contributions in rangeland IPM research are lacking (e.g. Bellows et al. 1983, Capinera et al. 1983a, 1983b, Hilbert and Logan 1983, Kemp and Onsager 1986). Nonetheless, the overwhelming majority of modeling resources, both monetary and intellectual, have been expended in analysis of intensively managed systems. Since this is the case, IPM modeling applications in agroecosystems provide a valuable template for evaluating the potential role of modeling in rangeland pest management.

Reasons for the disproportionate emphasis on modeling applications in intensively managed agricultural systems are neither obtuse nor esoteric. They are simply the dollar value of the resource and the potential offered for management intervention. Based on available statistics (USDA 1984, USDA-ARS 1983) the

334

average return on land in crop production is approximately $100./ha ($250./ac) while that for rangeland is $1.20/ha ($3./ac); therefore, intensively managed systems provide the opportunity for anthropogenic intervention that is not possible in natural resource systems. However, the total dollar return from rangeland production is also high ($23.5 billion/yr) and the true value of rangeland systems to society is more difficult to measure than for conventional cropping systems. One important measure of value is total acreage in production. For the United States, this amounts to over 324 million ha (800 million ac) in commercial rangeland production (USDA-ARS 1983). Internationally this figure is much larger with rangeland comprising almost half of the earth's terrestrial surface. Since grazingland represents one of the world's major ecosystems, it has high intrinsic value. Most resource managers would agree that we have an obligation to maintain rangeland systems for future generations, and that the value of such systems is greater than that of an adjacent wheat or corn field because of watershed, wildlife, and other values. Therefore, the currency that measures the value of rangeland systems must be stated in both intrinsic and extrinsic terms, whereas that for conventional cropping systems is largely extrinsic. Another reason for the success of modeling approaches in agroecosystems is that the profit margin is large enough to support expensive monitoring systems often required to drive high technology models (Lemmon 1986). A counter, and equally defensible, argument is that the comparatively narrow profit margin for rangeland systems necessitates enlightened decisions with respect to pest management strategies and tactics. This argument can well be used to justify the application of sophisticated pest management models; the real challenge is to provide the efficient means for development of models that have long-term life expectancy and wide geographic applicability.

In summary, the value of rangeland systems warrants the investment required for development of insightful management tactics, particularly if both intrinsic and extrinsic attributes are included in the evaluation. The economic constraints of rangeland production provides both a challenge and an opportunity for development of enlightened IPM strategies. Finally, the narrow profit margin demands rigorous evaluation of tactics utilized in implementation of specified strategies. An important question is: does modeling provide a viable tool to assist in development of insightful rangeland IPM programs? In my opinion, the answer is yes, and this opinion is based upon the demonstrable contributions that modeling has made to IPM programs in agroecosystems.

A large body of entomological literature deals with quantification of the impact that management tactics have on insect pests. A case in point is pesticide evaluation, and the

Entomological Society of America devotes one entire periodical to publication of these results (Insecticide and Acaricide Tests). The preponderance of this literature, however, deals only with the proximate, direct effect of pesticides on one or a few species. It is relatively easy to demonstrate that a chemical has an effect on a specified pest population; typically it is much more difficult to evaluate this effect within the ecology of complex field associations. A good example is evaluation of the long-term (years post spray) effect of pesticide application for grasshopper control. Although grasshoppers are probably the most studied of all North American rangeland insect pests, a general consensus on the long-term impact of chemical control is noticeably lacking (e.g., see Pfadt, chapter 12, or Blickenstaff et al. 1974). Applications of modeling techniques are providing valuable insights to more global questions regarding the impact of pesticide applications, such as development of resistance (Gould 1986, May 1985) and the direct and indirect interactions of complex arthropod food webs (Gutierrez et al. 1984). In general, computer models have provided an invaluable tool for placing a specific question within the overall context of a complex system.

Validated simulation models have provided a highly cost-effective means to estimate ecological parameters that would be difficult if not impossible to obtain otherwise. A good example of simulation models providing data for parameter estimation is provided in Logan et al. (1985). In this work, a computer simulation model for sitona weevil (*Sitona discoideus* Gyllenhal) was constructed from laboratory data. This model was then used in conjunction with field observations to estimate life table parameters that had previously been impossible to measure. The subsequent life systems model had important ramifications for sitona weevil pest management (Logan et al. 1985). Generalization of this approach has wide-ranging applications for many pest insects that, like sitona weevil, have one or several cryptic life stages.

Perhaps the most important (to date) contribution that simulation modeling has made to IPM research is that it provides a focus for research organization. In applications of computer modeling as a research tool (e.g., developed in close conjunction and interaction with field scientists), it has been found that techniques evolved primarily for model development are of themselves valuable, apart from the computer model (Holling 1978a). In particular, procedures necessary for model development have (1) facilitated communications between specialists in different disciplines, (2) provided a means to synthesize current knowledge, and (3) provided a framework for organizing extant data. This has helped to point out gaps in the data base, and the completed model provides a means to evaluate the importance of specific processes or variables. The iterative and synergistic relationship between modeling and field research is well recognized and has been capitalized upon by many IPM research projects.

The last modeling application I will discuss is that of prediction. Prediction, particularly with respect to computer modeling, has several connotations. The most obvious of these is empirical prediction. This is the case in which, given a set of measurable factors in the real world at some point in time, a prediction is made about the state of the target system at some future point in time. Computer models have played an important predictive role in pest management applications. One widespread application of empirical prediction is the use of computer models to predict pest phenology. For an example of this approach, and the estimated economic benefits gained, see Croft and Knight (1983).

Prediction is also used in a hypothetical sense. The question here is different from empirical predictions, and can be phrased as: *If* the real world were put together in a specified fashion, *then* what would the consequences be? In the hypothetical sense of predictions, computer modeling is playing an increasingly important and beneficial role in IPM research. A computer model based on the best information available provides an objective measure of the current level of understanding. It also provides a vehicle to test conjecture and hypothesis by cost efficient means. Such applications have provided real insights into the structure and function of complex ecological associations through qualitative and quantitative comparison of model results to real world responses.

CHARACTERISTICS OF RANGELAND IPM SYSTEMS

As a continuation of the theme of the previous section, the discussion in this section will utilize a comparative prospective. What, then, are the differences between agroecosystems and rangeland systems? One could argue that the differences are more apparent than real. After all, both systems are comprised of plant producers, animal consumers, and the management objective of maximizing economic return. Undeniably, the two systems differ. Water, for example, is the major limiting factor in the shortgrass steppe system. It may be much less so in an adjacent agroecosystem, owing either to irrigation or other water management practices. Differences such as the availability of water, although real, pose no particular problem with respect to modeling. However, there is another class of differences that are real and do pose significant modeling problems.

The most obvious substantive difference between rangeland systems and agroecosystems is the relative degree of complexity. Rangeland systems are typically much more complex in both the number of species represented and the trophic associations between them. Complexity of rangeland systems is further exacerbated by spatial considerations. Spatial complexity is introduced through both a patchy environment and the absolute scale of contiguous habitat to be considered. The extent of complexity in rangeland systems introduces significant conceptual and procedural problems to model development.

The time frame of interest is typically much longer in rangeland systems than agroecosystems. Agroecosystems are usually ephemeral in nature. For example, the time frame of most crop systems is on the order of one growing season (4 to 7 months). Under a rotation system, the current biological community would be entirely replaced during the next growing season. This situation is contrasted to rangeland systems where community integrity is maintained over 10's, 100's, or even 1000's of years. This long time frame allows the possibility of important IPM questions of an ecological time scale (e.g., succession) or even an evolutionary time scale (e.g., co-evolution of plant defenses and insect herbivory). Experience in modeling such long-term phenomena in IPM applications is noticeably lacking. In fact, such questions have only recently begun to be modelled for rangeland systems (Ojima et al. 1986).

As mentioned in the previous section, the obligation to maintain the integrity of rangeland systems precludes IPM tactics that increase the probability of systems degradation. Concern over the natural state of the system implies understanding both the direct and the indirect effects of a control action. Most insecticides, for example, are in fact biocides with impacts not only on the target insect species, but also beneficial insects and other organisms that perform important ecosystem functions such as nutrient cycling (e.g., Brown 1978). Analysis of the multitude of impacts that application of management tactics may have on ecosystem structure and function is a difficult task indeed, and IPM models have seldom been used to address such questions.

MODELING APPROACHES FOR RANGELAND IPM

Successful extension of IPM modeling applications from agroecosystem to rangeland will necessitate methodologies for dealing with the increased complexity of rangeland systems, incorporating multiple time scales and long-term frames of reference, and accurately predicting the consequences of management actions within the constraints imposed by these two factors.

System Complexity

The most obvious attribute of the comparative complexity of rangeland systems is the number of species represented. Pest models applied in agroecosystems are typically designed with one particular pest species in mind. As a consequence of focusing on a single pest species, pest models usually have been developed in an *ad-hoc* fashion. This *ad-hoc* approach has worked relatively well for the simple communities typical of agroecosystems. However, in the much more complex rangeland system, such an approach may be untenable. *Ad-hoc* model development has often resulted in a static, inflexible approach to what is really a shifting mosaic of management practices and biological associations. This inflexibility has often resulted in models that are difficult to interface with other models (either other pests in the same crop or the same pest in other crops). Such models have been characteristically immutable and dynamically fragile (i.e., changes in one portion of the model result in unanticipated changes in another section of the model). Pest models in the past also have frequently lacked transportability, both conceptually as in reparameterization for a geographic location different from that where the model was developed, and physically in transporting the model from one computer to the next.

As an alternative to the focus on a particular pest species, models can also be developed by focusing on the basic model structure that is common to many (most) pest models. For example, accurate representation of insect phenology is central to most pest models (Logan and Hilbert 1983, see also chapter 24). Generalized modeling techniques have been developed (e.g., Sharpe et al. 1977, Welch et al. 1978) that are capable of effectively modeling phenology for a wide range of insect species. By utilizing such generalized procedures, it is possible to build a model by assembling algorithms describing these key processes in basically a building block fashion (see Gutierrez et al. 1984 for a good example of this technique). Such a modeling strategy would be capable of describing a wide range of insect life systems, the individual models differing only in various parameter values. Since the basic model structure is constant, linking different models together for description of community trophic structure would be a simple procedure. Structural consistency would also facilitate development of standard techniques for interfacing these pest models with other components of the system (e.g., plant models).

By capitalizing on a generalized model structure, it should be possible to build a computer program capable of designing pest models. In concept, this system would "customize" a generalized model structure for description of a specific pest life system. The customization procedure would be based on extensive dialogue with the resource manager and a broad knowledge base in ecological theory, IPM concepts, and techniques of data analysis and model building. Current and emerging technologies that would be useful in development of such a model design system are concepts from hierarchical theory (Allen and Starr 1982), mega-structured approaches to computer programming utilizing the UNIX operating system (Crecine 1986), and concepts from artificial intelligence (Winston 1984) and expert systems (Waterman 1986, Klahr and Waterman 1986).

Design of an expert system for pest model building does not represent a radical departure from the way models are currently developed. In fact, most pest management modelers have a set of techniques that they routinely use for model construction, and the concept of building a customized model from generic components is at least twenty-five years old (Watt 1961). What would be new, however, is the implementation of a formal modeling protocol that attempts to capitalize on the best of various techniques that have been developed. The widespread availability of an effective expert model design system would subject the underlying modeling paradigms to the scrutiny of the IPM community at large. Performance of various approaches would therefore be evaluated in a much more objective fashion than is typically the case at present, where individual modelers often adopt an advocacy position for a particular modeling approach or philosophy. Through objective evaluation, the expert system could be improved over time, and in fact expert systems are currently being used to evaluate the performance, and facilitate the improvement, of other expert systems.

A second major component of community complexity is trophic structure. Once an efficient means exists for generation of life systems models and linking these models together to form food web descriptions, the task of meaningful analysis remains. A developing body of ecological theory is centered about analysis of food web structure (May 1983, Pimm 1982). The models used in this type of analysis are typically compact mathematical descriptions stated in the esoteric language of differential equations. Therefore, these models are abstract to the point of being unintelligible to most field ecologists. Even for those with the interest and patience to gain the facility for understanding such models, they are often so simplistic that they offend biological sensibilities (Wangersky 1978). Conversely, simulation models are not particularly well suited for direct analysis. A simulation model is typically a process oriented mechanistic description of an ecological association. Such models are formulated in a manner that allows inclusion of a great deal of descriptive complexity. Therefore, they provide a realistic, understandable representation of the system, and have been appealing to field biologists. Unfortunately the strength of the simulation approach is also its weakness, in that, due to the ease of including descriptive complexity, simulation models are often unwieldy and dynamically fragile. We therefore appear to be caught on the horns of a dilemma: models that are capable of capturing biological realism are intractable, and those capable of providing general, tractable descriptions are biologically unreasonable (Levins 1966).

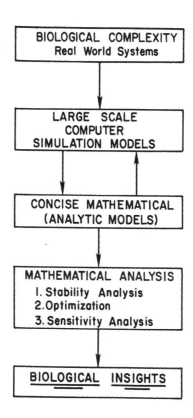

FIG. 23.1 Relationship between simulation models, analytic models, and the subject biological system. This relationship capitalizes on the descriptive power of simulation techniques for initial representation and the inferential power of analytic models for mathematical analysis.

Resolution of Levins' (1966) dilemma requires a modeling approach that capitalizes on the unique strengths of both simulation models and the concise mathematical description of analytic models. One possible relationship between these two modeling approaches is shown in Figure 23.1. The natural context for describing real-world problems is through simulation. Analytic models, in turn, provide the tractability required for mathematical analysis of model structure. The process is iterative in nature, as indicated by the double arrow connecting the two types of models. The insights gained through mathematical analysis are used to formulate empirically testable hypotheses. Experimental results then lead to simulation model refinement and modification. While Figure 23.1 may be inadequate by failing to indicate all potential information feedback, it does serve to illustrate that specific types of models should be utilized in a manner that exploits their potential strengths and avoids ill-suited applications.

The key to implementation of the modeling paradigm presented in Figure 23.1 is rigorous derivation of analytic models from their parent simulation models. One approach is to establish the form of an analytic model through a series of successive simplification procedures. Once a mathematical analysis has been made of the analytic models' properties, results could be used to guide numerical analysis of the progressively more complex models from which they were derived. This approach to model analysis is diagrammatically shown in Figure 23.2. An important feature of this Figure is the representation of a pathway leading from empiricism to an analytic model representation (counterclockwise), and one from the analytic model representation back to the empiricism that originally motivated the model (clockwise). A direct inferential pathway is therefore established from empiricism to mathematical analysis and back again. Through such an approach, it may be possible to incorporate the meaningful aspects of ecological theory into analysis of the complex trophic structure typical of rangeland pest associations.

Food-web analysis is not the only complex aspect of rangeland IPM systems that could benefit from the modeling approach represented in Figures 23.1 and 23.2. Spatial complexity is another good example. Pest management in rangelands involves several aspects of spatial complexity. The first is dispersion or movement of pest species (for example, the well known movement of grasshoppers from "hot-spots" to adjacent rangeland). Recent theoretical work that incorporates diffusive processes in prey/predator models indicates that diffusion alone can introduce instabilities in an otherwise stable interaction (per. comm., David Wollkind, Department of Mathematics, Washington State University, Pullman). Such diffusive instability could lead to outbreaks of pest species in even totally homogeneous environments. Of

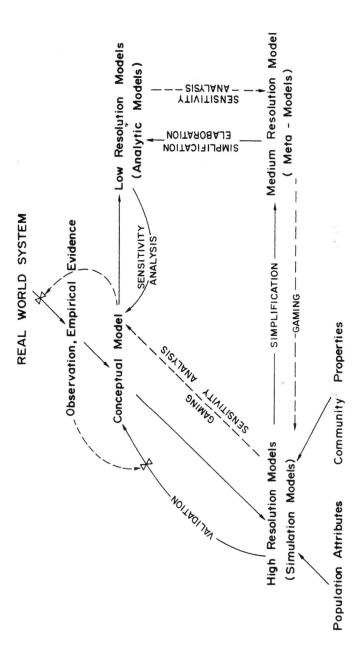

FIG. 23.2   Modeling paradigm that attempts to trace the procedural steps from empiricism to analytic model representation, and back again.

course, rangeland systems are by no means homogeneous. Patchiness and temporal variation in resource distribution are obvious aspects of rangeland systems. Dispersal of organisms through a heterogeneous, patchy environment poses significant modeling problems, and yet analysis of spatial interactions may provide an important key to understanding pest outbreaks. Historically, the difficulty has been that any model that realistically represented the biology of the system was likely too cumbersome to allow significant representation of spatial dynamics.

Through attempts at analysis of phenomena such as food-web structure (Hunt et al. 1986) and spatial complexity (Ludwig et al. 1978) in real-world problems, it is becoming recognized that simplification of simulation models is, in general, a desirable goal (Innis and Rexstad 1983). The real challenge is in development of modeling techniques that allow generation of simple models that maintain a high degree of biological fidelity. Figures 23.1 and 23.2 represent a general scheme for representing ecological attributes that are critical for description of pest associations in rangeland systems, and that is also highly consumptive of computer resources.

Time Frame of Interest

Difficulty in addressing the long-term implications of IPM strategies was noted in the previous section of this paper. The National Science Foundation has recently recognized the importance of such questions by establishment of long-term ecological research sites that are representative of important biomes (Callahan 1984). Three of these sites represent rangeland systems (the Shortgrass Steppe site at the Central Plains Experimental Range in Colorado; the Tallgrass Prairie site in Kansas; and the Jornada site in New Mexico). Application of computer models is central to research at both of these sites (Ojima et al. 1986), and important insights to the procedural approaches necessary for development of long-term IPM models will likely result from this work. Currently the best example of long-term model representation of pest management practices that I am aware of is provided by Holling (1978b) and co-workers with spruce budworm. It is important to note that successful long-term analysis required significant simplification of the original simulation model (Ludwig et al. 1978), and that the spatial component played a critical role in temporal pest dynamics.

A further implication of a long term frame of interest is that any comprehensive model of rangeland IPM systems must incorporate a multitude of vastly different time scales. For example, an IPM model for the shortgrass steppe may require time scales ranging from the annual (or less) reproductive cycle of insect pests, to the climatically controlled seedling reproduction of blue grama (10's of years, Hyder et al. 1975), or to climate itself (100's of years). Approaches to addressing the problem of multiple scales have been well worked out in the physical sciences. One of the first steps taken in model analysis in the physical sciences is to state the problem in terms of dimensionless variables. This relatively straight-forward procedure (Lin and Segel 1974) is seldom taken in IPM models. The result is that problems usually include terms of vastly different time scales and dimensions. Nondimensionalization goes a long way towards resolving these problems and can result in both conceptual clarification and computational simplification. A good example of the potential for model simplification through introduction of time scaled variables is once again provided by modeling insect phenology. In IPM applications it is often necessary to represent explicitly each instar in the life cycle. It is further necessary to preserve the unique developmental rates for each instar since timing of critical ecological events (e.g., diapause, coincident adult emergence, etc.) often depend on the relative relationship between temperature and developmental rates (Logan et al. 1979, Hilbert et al. 1985, Logan and Amman 1986). Therefore, simulation is a complex, multiple scale problem. However, when developmental rates are normalized by using the median rate, all time scales are reduced to a dimensionless variable that represents the proportion of the instar completed. Restating the problem in terms of dimensionless variables reduced an intractable problem to one that was almost trivial (Wagner et al. 1984). Similar scaling procedures applied to the components of a systems model would substantially mitigate difficulties of different intrinsic time scales.

High Intrinsic Value

The high intrinsic value of rangeland systems, combined with the need to address long-term questions, necessitate a broader, more encompassing view of the system than is typical in IPM models. What is really needed is an ecosystem perspective (Woodmansee 1984). Such a perspective has typically been lacking, not only in IPM models, but in animal population models in general (Logan et al. 1983). The reasons for lack of an ecosystem perspective are partially due to the evolution of ecology as a science. When faced with the bewildering array of species present in most natural ecosystem (which literally number in the 10's to 100's of thousands), ecosystem ecologists have tended to aggregate species into functional groups, and then represent the complex system by flows of energy and nutrients between these aggregate compartments (e.g., Odum 1983). Pest management, on the other hand, is concerned with the numbers and dynamics of one or a few specific species.   Apparently, what is needed is a synthesis of the population dynamics and the ecosystem approach.   Such a synthesis has remained an elusive goal!

As an alternative to synthesis or combining the population and ecosystem viewpoints, it may be more productive to recognize the legitimacy of each approach.   Adequate representation may require an acknowledgement of a basic duality in the problem, similar to that in physics where it has become necessary to model explicitly the dual nature of phenomena such as light.   A dual representation would recognize that (1) biochemical transformations performed by a functional group are themselves functions of the individual species attributes which comprise that group, and (2) conversely, the population dynamics of important pest species are in large part a function of ecosystem properties (e.g., plant nutrient status).   At any rate, it is clear that predicting long-term consequences of pest management tactics in rangeland systems will require a broader, more total systems perspective than is typical in most existing IPM models.

SUMMARY

Effective application of modeling approaches in rangeland IPM will require an approach to modeling that is different from the models that have worked well in more traditional IPM applications.   New modeling approaches are required due to the substantive differences between these two systems.   Most notably, rangeland systems are comparatively more complex than agroecosystems in the number of, and the interactions between, species.   Also, the temporal frame of interest is typically much longer, and rangelands have high intrinsic value as one of the earth's major ecosystems.   This new class of pest models should be characterized by a powerful, generalized structure.   By concentrating on the structure of pest models, it would be possible to develop models that are easily interfaced with each other, and with models of other components of the rangeland ecosystem.   The potential importance of ecosystem processes in long-term patterns of pest abundance requires adequate representation of these processes.   This will require either a synthesis or a dual representation of ecosystem and population approaches.   Synthesis is also required between various modeling approaches, in particular between simulation and analytic models.   This latter synthesis is necessary because modeling phenomena such as spatial dynamics are highly consumptive of computer resources.   Analysis of model characteristics, such as sensitivity and stability analysis, also requires tractable model formulation.

What, then, is needed to implement this new class of models? In general, the necessary technology is already in place. Computing power continues to increase at a seemingly unrelenting pace. With the advent of user-friendly networking systems (Jennings et al. 1986), it is now possible to share computer code, essentially instantaneously, and on a world-wide basis. The widespread acceptance of a hardware-independent operating system (UNIX) has reduced difficulties in transferring models between different computers (Crecine 1986). New philosophies of programming have vastly facilitated the ease with which scientists can access sophisticated computer models, without themselves becoming computer sophisticates. Most importantly, ecological modeling is a science that is beginning to mature. There is a general awareness that a variety of philosophies and approaches to modeling have withstood the test of time, and that adopting a rigid ideological stance is self-defeating. Apparently, then, a recently expanded arsenal of computer technologies is ready for application to the problems of pest modeling. It also is becoming recognized that the current generation of IPM models have serious limitations, and there exists a well-qualified cadre of scientists anxious to produce the next generation of IPM models. We appear to on the brink of a new generation of pest models which will no longer be constrained by the generally awkward, inflexible approaches that characterize many current IPM modeling applications.

ACKNOWLEDGMENTS

My current modeling activity in shortgrass steppe ecosystems is supported by the Long Term Ecological Research Program: Shortgrass Steppe (BSR-8114822). I thank W. K. Lauenroth and A. R. Grable for reviewing an earlier version of this manuscript.

REFERENCES

Allen, T.F.H., and T.B. Starr. 1982. Hierarchy: perspectives for ecological complexity. Univ. Chicago Press, Chicago.

Bellows, T.S., J.C. Owens, and E.W. Huddleston. 1983. Model for simulating consumption and economic injury level for the range caterpillar (Lepidoptera: Saturniidae). J. Econ. Entomol. 76:1231-1238.

Blickenstaff, C.C., F.E. Skoog, and R.J. Daum. 1974. Long-term control of grasshoppers. J. Econ. Entomol. 67:268-274.

Brown, A.W.A. 1978. Ecology of pesticides. J. Wiley and Sons, New York.

348

Callahan, J.T. 1984. Long-term ecological research. Bioscience 34:363-67.

Capinera, J.L., W.J. Parton, and J.K. Detling. 1983a. Assessment of range caterpillar (Lepidoptera: Saturniidae) effects with a grassland simulation model. J. Econ. Entomol. 76:1088-1094.

Capinera, J.L., W.J. Parton, and J.K. Detling. 1983b. Application of a grassland simulation model to grasshopper pest management on the North American shortgrass prairie, p. 35-44. *In:* W.K. Lauenroth, G.V. Skogerboe, and M. Flug (eds.) Analysis of ecological systems. State-of-the-art in ecological modeling. Elsevier Sci. Publ. Co., Amsterdam.

Crecine, J.P. 1986. The next generation of personal computers. Science 231:935-943.

Croft, B.A., and A.L. Knight. 1983. Evaluation of PETE phenology modeling system for integrated pest management. Bull. Entomol. Soc. Am. 29:37-43.

Getz, W.M., and A.P. Gutierrez. 1982. A perspective on systems analysis in crop production and insect pest management. Annu. Rev. Entomol. 27:447-466.

Gould, F. 1986. Simulation models for predicting durability of insect-resistant germ plasm: A deterministic diploid, two-locus model. Environ. Entomol. 15:1-10.

Gutierrez, A.P., J.U. Baumgaertner, and C.G. Summers. 1984. Multitrophic models of predator-prey energetics. Can. Entomol. 116:923-963.

Hilbert, D.W., and J.A. Logan. 1983. A population system model of the migratory grasshopper (*Melanoplus sanguinipes*), p. 323-334. *In:* W.K. Lauenroth, G.V. Skogerboe, and M. Flug (eds.) Analysis of ecological systems. State-of-the-art in ecological modeling. Elsevier Sci. Publ. Co., Amsterdam.

Hilbert, D.W., J.A. Logan, and D.M. Swift. 1985. A unifying hypothesis of temperature effects on egg development and diapause of the migratory grasshopper *Melanoplus sanguinipes* (Orthoptera: Acrididae). J. Theor. Biol. 112:827-838.

Holling, C.S. 1978a. Adaptive environmental assessment and management. J. Wiley & Sons, Chichester.

Holling, C.S. 1978b. The spruce-budworm/forest-management problem, p. 143-182. *In:* C.S. Holling (ed.) Adaptive environmental assessment and management. J. Wiley & Sons, Chichester.

Hunt, H.W., D.C. Coleman, E.R. Ingham, R.E. Ingham, E.T. Elliott, J.C. Moore, S.L. Rose, C.P.P. Reid, and C.R. Morley. 1986. The detrital food web in a shortgrass prairie. Biol. Fert. Soils (in press).

Hyder, D.N., R.E. Bement, E.R. Remmenga, and D.F. Hervey. 1975. Ecology of native plants and guidelines for management of shortgrass range. USDA-ARS Tech. Bull. 1503.

Innis, G.S., and E. Rexstad. 1983. Simulation modeling simplification techniques. Simulation (July), p. 7-15.

Jennings, D.M., L.H. Landweber, I.H. Fuchs, D.J. Farber, and W.R. Adrian. 1986. Computer networking for scientists. Science 231:943-950.

Kemp, W.P., and J.A. Onsager. 1986. Rangeland grasshoppers (Orthoptera: Acrididae): modeling phenology of natural populations of six species. Environ. Entomol. 15:924-30.

Klahr, P., and D.A. Waterman. 1986. Expert systems: techniques, tools, and applications. Addison-Wesley, Reading, MA.

Lemmon, H. 1986. Comax: an expert system for cotton crop management. Science 233:29-33.

Levins, R. 1966. Strategy of model building in population biology. Am. Sci. 54:421-431.

Lin, C.C., and L.A. Segel. 1974. Mathematics applied to deterministic problems in the natural sciences. Macmillian, New York.

Logan, J.A., and G.D. Amman. 1986. A distribution model for egg development in mountain pine beetle. Can. Entomol. 118:361-372.

Logan, J.A., and D.W. Hilbert. 1983. Modeling the effects of temperature on arthropod population systems, p. 113-122. *In:* W.K. Lauenroth, G.V. Skogerboe, and M. Flug (eds.) Analysis of ecological systems. State-of-the-art in ecological modeling. Elsevier Sci. Publ. Co., Amsterdam.

Logan, J.A., J.E. Heasley, and C.B. Stalnaker. 1983. State-of-the-art and recommendations for future efforts in modeling the functional role of animals in ecological systems, p. 983-986. *In:* W.K. Lauenroth, G.V. Skogerboe, and M. Flug (eds.) Analysis of ecological systems. State-of-the-art in ecological modeling. Elsevier Sci. Publ. Co., Amsterdam.

Logan, J.A., E.R. Frampton, and S.L. Goldson. 1985. A phenological model for *Sitona discoideus* Gyllenhal (Coleoptera: Curculionidae) and its application in Canterbury. Proc. Australasian Conf. Grassland Insect Ecol. 4:173-180.

Logan, J.A., R.E. Stinner, R.L. Rabb, and J.S. Bacheler. 1979. A descriptive model for predicting spring emergence of *Heliothis zea* (Boddie) populations in North Carolina. Environ. Entomol. 8:141-146.

Ludwig, D., D.D. Jones, and C.S. Holling. 1978. Qualitative analysis of insect outbreak systems: the spruce budworm and the forest. J. Anim. Ecol. 47:315-332.

May, R.M. 1981. Theoretical ecology: principles and applications. Blackwell Sci. Pub., Oxford.

May, R.M. 1983. The structure of food webs. Nature 301:566-568.

May, R.M.   1985.   Evolution of pesticide resistance.   Nature 315:12-13.

Odum, H.T.   1983.   Systems ecology.   John Wiley & Sons, New York.

Ojima, D.S., W.J. Parton, D.S. Schimel, and C.E. Owensby.   1986. Simulating the long-term impact of burning on C, N, and P cycling on a tallgrass prairie.   7th International Symposium on Environmental Biogeochemistry. (in press).

Pimm, S.L.   1982.   Food webs.   Chapman and Hall, London.

Ruesink, W.G.   1976.   Status of the systems approach to pest management. Annu. Rev. Entomol. 21:27-45.

Sharpe, P.J.H., G.L. Curry, D.W. DeMichele, and C.L. Cole.   1977. Distribution model of organisms development time.   J. Theor. Biol. 66:21-8.

USDA.   1985.   Agricultural statistics 1984.

USDA-ARS.   1983.   Range research program of the Agricultural Research Service.

Wagner, T.L.H., H. Wu, P.J.H. Sharpe, and R.N. Coulson.   1984. Modeling distribution of insect development time: a literature review and application of the Weibull function.   Ann. Entomol. Soc. Am. 77:475-483.

Wangersky, P.J.   1978.   Lotka-Volterra population models.   Annu. Rev. Ecol. System. 9:189-218.

Waterman, D.A.   1986.   A guide to expert systems. Addison-Wesley, Reading, MA.

Watt, K.E.F.   1961.   Mathematical models for use in insect pest control. Can. Entomol. 93 (suppl. 19).

Welch, S.M., B.A. Croft, J.F. Brunner, and M.F. Micheles.   1978. PETE: An extension phenology modeling system for management of multi-species pest complexes.   Environ. Entomol. 7:487-494.

Winston, P.H.   1984.   Artificial intelligence.   Addison-Wesley, Reading, MA.

Woodmansee, R. G.   1984.   Comparative nutrient cycles of natural and agricultural ecosystems: a step toward principles, p. 145-56.   In: R. Lowrance, B.R. Stinner, and G.J. House (eds.) Agricultural ecosystems: unifying concepts.   Wiley Interscience, New York.

24. PREDICTIVE PHENOLOGY MODELING IN RANGELAND PEST MANAGEMENT

William P. Kemp
Rangeland Insect Laboratory, Agricultural Research Service
U.S. Department of Agriculture, Bozeman, Montana 59717

Management intensity on rangelands in the western United States has increased significantly in recent years. Concurrent with this is a growing awareness among managers and ranchers concerned with rangeland pests that control activities should be scheduled to maximize efficacy and/or save forage. Of the many insects known to reduce rangeland production, grasshoppers have been studied most extensively. However, few quantitative tools exist for determining whether control tactics are warranted, and if so, when they should be scheduled (see APHIS 1986 as an example of the need for such tools).

The use of such tools, however, assumes that economic injury levels (EIL's) are available for different insect pest life stages and associated control methods. Onsager (1984) presented a method for estimating the EIL's for different life stages of rangeland grasshoppers and various control options. The method is useful in determining whether control is warranted, but it may require repeated sampling of the grasshopper population (made up of several species) to ascertain the overall density and average stage of development.

A logical improvement to the work by Onsager (1984) would be to construct of a general grasshopper development model which would predict the presence of various grasshopper stages. This would allow sampling for a specific grasshopper life stage(s) and evaluation of density to determine if an EIL was exceeded, resulting in optimal timing of sampling as well as elimination of the need for instar determination. Such a model would have direct application to the current population assessments conducted by APHIS (Animal and Plant Health Inspection Service) and would improve the utility of such surveys for long-term trend analysis through the standardization of sampling times.

The objectives of this paper are to: 1) review the state-of-the-art as it relates to predictive insect phenology modeling on

351

rangelands, 2) present a general predictive phenology model for use with rangeland grasshoppers, and 3) review the areas where future research is needed. Though this paper emphasizes grasshopper phenology examples, the major points and model development pertain to many rangeland pest insects and plants.

STATE-OF-THE-ART IN INSECT PHENOLOGY MODELING

Due to the importance of predicting the seasonal occurrence of insect pests, there have been a large number of studies devoted to this topic. It is no surprise that high value systems such as row crops and orchards were among the first to have comprehensive phenology models as central components to integrated pest management activities (Ruesink 1976, Welch et al. 1978). Most of the existing constructs can be grouped into one of two major classes of insect phenology models, rate models and stochastic models.

There are a number of predictive and descriptive models that are based on the nonlinear relationship between temperature and insect development rates derived from laboratory experiments performed at several constant temperatures (Drummond et al. 1985, Logan et al. 1979, Regniere et al. 1981, Regniere 1982). A recent and extremely thorough review of the extant literature by Wagner et al. (1984) on rate models suggested that the biophysical model of Sharpe and DeMichele (1977) is best suited for predicting development rates. Further, Wagner et al. (1985) present a rate summation approach for modeling development times of insect cohorts under variable temperatures. Advantages of rate models, in addition to strong theoretical foundations, are reasonable flexibility and goodness-of-fit capabilities.

Despite the advantages of the rate model approach, often there are problems associated with predicting field populations. Frequently it is difficult to extrapolate experimentally derived insect development rates to field situations. Other variables such as temperature variability, humidity, parasitism, and inherent genetic variability influence insect development.

In response to problems associated with rate models, Read and Ashford (1968) and Kempton (1979) developed stochastic models of insect development and statistical parameter estimation. In their models, the probability that an insect in a specific sample will be in a given stage changes through time. In these and related studies, development was considered dependent only on Julian date. However, later studies under the multinomial assumption and using maximum likelihood (ML) parameter estimation procedures, Osawa et al. (1983) found that a degree day

(DD) based phenology model for balsam fir buds performed better with an independent data set than with a model based on Julian date. Work on the western spruce budworm, *Choristoneura occidentalis* Freeman, by Dennis et al. (1986) and Kemp et al. (1986) applied improved stochastic methods to the insect phenology case and provided parameter comparisons and goodness of fit tests. Stedinger et al. (1985) also expanded upon the work by Osawa et al. (1983) and developed a phenology model for the eastern spruce budworm, *Choristoneura fumiferana* (Clemens), that incorporated spatial variation in larval development from different sampling points within a site which was superimposed upon the multinomial distribution of larval development of larvae in each sample. Advantages of the stochastic modeling approach also include strong theoretical foundations, computational flexibility and ease, and goodness of fit capabilities.

STATE-OF-THE-ART IN GRASSHOPPER DEVELOPMENT MODELING

In spite of phenology modeling done in other pest systems, there is a paucity of predictive development models for rangeland grasshoppers. Further, though there are several descriptive models of grasshopper development, they are usually part of large simulation models (Gyllenberg 1974, Hardman and Mukerji 1982, Hilbert and Logan 1982, Mann et al. 1986, Randell 1972, Rodell 1977, and others) and have not been tested or used for prediction.

Most of the work on predictive phenology modeling has been conducted in Canada. Mukerji and Randell (1975) developed a regression model to predict the embryonic development of *Melanoplus sanguinipes* (F.) eggs in the fall. They found that adult maturity date and current temperatures explained state of fall embryonic development. Results also indicated that soil moisture in the late summer was likely an important factor in determining fall embryonic development. In a related study, Randell and Mukerji (1974) developed a regression model to predict hatching of each decile of *M. sanguinipes* eggs as a function of daily maximum air temperatures (over a 5-day interval) from March 2 through June 29 and embryonic development in the preceding fall. Lastly, Gage et al. (1976) developed a predictive model for the seasonal occurrence of a group of three grasshopper species (*Camnula pellucida* (Scudder), *M. bivittatus* (Say), and *M. sanguinipes*). In this study (Gage et al. 1976), probit equations (Table 24.1) were used to predict the percentage occurrence of each grasshopper stage as a function of accumulated heat units over an arbitrarily determined threshold (50°F, 10°C) (Figure 24.1). Two problems exist in the application of this predictive tool to rangeland conditions. The iteratively estimated developmental threshold was generated for a specific data set and is likely to fail under other circumstances. Further, since individual probits were developed for each stage, relationships between stages are not handled explicitly.

TABLE 24.1
Probit equations for estimating percentage occurrence of each grasshopper stage from heat units accumulated above 50°F (10°C) from 1 April (from Gage et al. 1976).

Stage	Probit Equation
Hatch	$2.9948 + 0.0084*DD_{50}$
Instar 2	$2.6722 + 0.0071*DD_{50}$
Instar 3	$2.2584 + 0.0066*DD_{50}$
Instar 4	$1.8557 + 0.0059*DD_{50}$
Instar 5	$1.1685 + 0.0058*DD_{50}$
Adult	$0.6867 + 0.0051*DD_{50}$

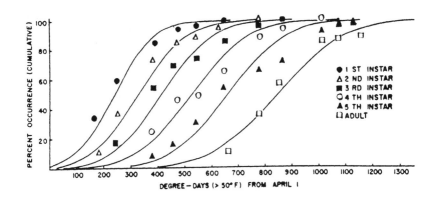

FIG. 24.1 Cumulative (from Gage et al. 1976) percentage occurrence of grasshopper life stages and cumulative heat units based on probit equations in Table 24.1.

In response to the obvious dearth of predictive models for rangeland grasshoppers, Kemp and Onsager (1986) applied the methods of Dennis et al. (1986) and Kemp et al. (1986) and generated phenology models for six individual species of rangeland grasshoppers at Roundup, Montana for each of two years (e.g., Figures 24.2, 24.3). Parameter comparison tests (Dennis et al. 1986) indicated that ML-estimated model parameters were site and species specific. However, the average number of accumulated DD (64°F, 17.8°C base; Putnam 1963) prior to peak instar 3 varied only slightly each year. This fact, together with other similarities in results, suggested that perhaps more general models could be developed for a group of species (similar to the approach used by Gage et al. (1976)). The need for this type of model is further indicated by problems frequently encountered in the field by pest managers. Field-collected data invariably consist of a number of grasshopper species and life stages (Onsager 1986). As noted, a general grasshopper phenology model would be useful to agencies such as APHIS and other pest managers or ranchers who conduct population assessments.

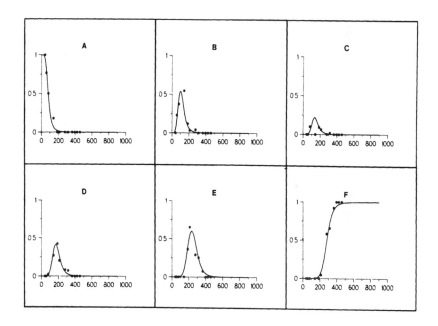

FIG. 24.2 A-F Comparison of raw data (plotted points) and model results (solid line) for the proportion (y-axis) of the *Aulocara elliotti* (Thomas) population in each state (A = instar 1, B = instar 2, C = instar 3, D = instar 4, E = instar 5, F = adult) as a function of accumulated degree days (x-axis); Roundup, MT, 1975.

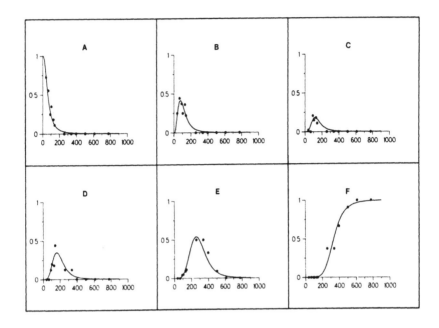

FIG. 24.3 A-F Comparison of raw data (plotted points) and model results (solid line) for the proportion (y-axis) of the *Aulocara elliotti* (Thomas) population in each stage (A = instar 1, B = instar 2, C = instar 3, D = instar 4, E = instar 5, F = adult) as a function of accumulated degree days (x-axis); Roundup, MT, 1976.

METHODS

Field Data

Data used in this analysis were collected from Roundup, Montana. Sampling at Roundup (actual locations of sites were about 30 mi. (48 km) NNW of Roundup at an elevation of 3900 ft (1,186 m)) was done during the rangeland grasshopper population dynamics studies reported by Hewitt (1979) and Onsager and Hewitt (1982). Grasshopper and weather data were collected near Roundup at approximately weekly intervals during the summer months of 1975-1976. The techniques used were published previously (Hewitt 1979, Onsager and Hewitt 1982), but details pertinent to this paper are repeated briefly.

Transect lines for sampling grasshopper densities were established each year prior to hatching, and twenty 0.5 m^2 subsamples were collected weekly along a randomly selected transect. A vacuum quick-trap method of collection was used during the evening or at night when temperatures were sufficiently low to minimize grasshopper activity (Onsager and Hewitt 1982). Insects were extracted in Berlese funnels, and specimens were tabulated according to species and instar.

Temperatures 5 cm above the soil surface (ambient to grasshoppers) were recorded each year starting in May. Methods of Kemp et al. (1983) were utilized to complete weather data sets starting from May 1 of each year of sampling. All temperatures used for DD accumulations were in °F. The accumulation of DD (Allen 1976) with °C data results in reduction (due to scale differences) by a factor of 1.8, when compared with DD accumulations based on °F data.

Model Development

The model assumes that the development of a given grasshopper is a stochastic process that consists of accumulated small increments of development time (see Dennis et al. (1986) for details). The process S(t) is defined as the amount of development time that a grasshopper has accumulated by actual time t. Results of Kemp et al. (1986) and Osawa et al. (1983) show that S(t) and t should be measured in DD rather than in calendar days. The heart of the model is a probability distribution for S(t) that changes as t increases. Dennis et al. (1986) suggest that the logistic probability density function (Patel et al. 1976) is a suitable choice on the basis of accuracy and ease of computations.

The model assumes that for the general grasshopper, the proportion of the population in development stage i at time j is given by the logistic probability density function:

$$p_{ij} = \Pr[a_{i-1} \leq S(t_j) < a_i] =$$

$$1 / \left\{ 1 + \exp\left(-\left[\frac{a_i - t_j}{\sqrt{b^2 t_j}} \right] \right) \right\} -$$

$$1 / \left\{ 1 + \exp\left(-\left[\frac{a_{i-1} - t_j}{\sqrt{b^2 t_j}} \right] \right) \right\}$$

358

where t_j = DD accumulated at collection date j (j = 1, 2,...q), a_i = amount of development needed to complete the ith developmental stage (i = 1, 2,... r-1), b^2 = a positive constant, r = the total number of development stages and $S(t_j)$ = amount of development an insect has accumulated at t_j.

In this study, developmental data for a complex of six grasshopper species (*Ageneotettix deorum* (Scudder), *Amphitornus coloradus* (Thomas), *Aulocara elliotti* (Thomas), *Melanoplus infantilis* (Thomas), *Melanoplus packardii* Scudder, and *M. sanguinipes*) were pooled to develop model parameters (a_i's and b^2) above. The model developed thus contains developmental variation of six spring emerging grasshopper species over two years.

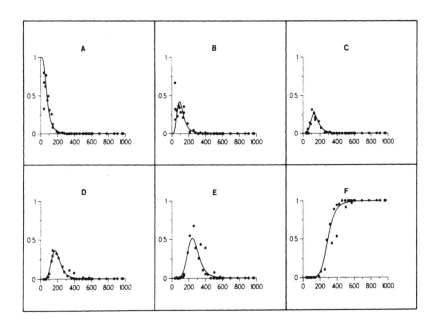

FIG. 24.4 A-F Comparison of raw data (plotted points) and model results (solid line) for the proportion (y-axis) of the general grasshopper population in each stage (A = instar 1, B = instar 2, C = instar 3, D = instar 4, E = instar 5, F = adult) as a function of accumulated degree days (x-axis).

For the general grasshopper, the a_i and b^2 parameters were estimated using maximum likelihood procedures described by Dennis et al. (1986). Further, it was assumed that general grasshopper development was characterized by five nymphal instars and an adult stage ($r = 6$). As DD accumulate, the logistic probability curve moves through the a_i values and this results in changing proportions in each developmental stage (p_{ij}). Therefore, as t_j increases, more of the general grasshopper population is found in the more advanced developmental stages.

RESULTS AND DISCUSSION

Figure 24.4 shows a comparison between model results and field collected data for the general grasshopper population (6 species for 2 years). Despite variation in the field-collected data (due principally to differing yearly survival rates (Kemp and Onsager 1986)), trends in the data are obvious. Figure 24.5 shows model results for the proportion of the general grasshopper population in a given developmental stage, and specific points of interest to managers and the timing of such events. The calculation of the timing of peak instar development (modes) and associated confidence intervals (Table 24.2) are based on additional inferences developed for the Dennis et al. (1986) model by Dennis and Kemp (1987).

Model parameter estimates developed for the general grasshopper population are contained in Table 24.3. The proportion of the population in a development class at any time (t_j) in each year may be determined by using the parameters found in Table 24.1 together with the following equations:

Instar 1;

$$p_{1j} = \left\{ 1/ \; 1 + \exp\left(- \left[\frac{a_1 - t_j}{\sqrt{b^2 t_j}} \right] \right) \right\}$$

Instar i = 2 through i = 5;

$$p_{ij} = 1/\left\{ 1 + \exp\left(- \left[\frac{a_i - t_j}{\sqrt{b^2 t_j}} \right] \right) \right\} - $$

$$1/\left\{ 1 + \exp\left(- \left[\frac{a_{i-1} - t_j}{\sqrt{b^2 t_j}} \right] \right) \right\}$$

Adults;

$$p_{6j} = 1/\left\{1 + \exp\left(\frac{a_5 - t_j}{\sqrt{b^2 t_j}}\right)\right\}$$

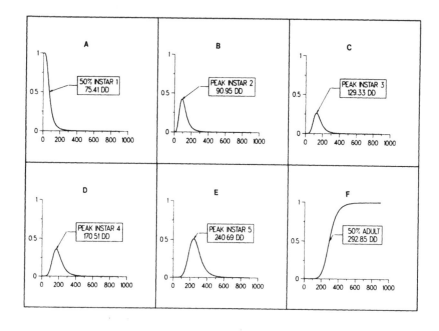

FIG. 24.5 A-F Model results (solid line) for the proportion (y-axis) of the general grasshopper population in each stage (A = instar 1, B = instar 2, C = instar 3, D = instar 4, E = instar 5, F = adult) as a function of accumulated degree days (x-axis) together with timing of specific life stages of interest to managers.

TABLE 24.2
95% confidence intervals for instar peaks (modes) computed for general grasshopper phenology data described in Figure 24.4.

Development Peak	Degree Days (17.8°C Base)
Instar 2	90.95 ± 4.40
Instar 3	129.33 ± 4.98
Instar 4	170.51 ± 5.02
Instar 5	240.69 ± 5.08

A goodness-of-fit test (Dennis et al. 1986) was not conducted on the data and model results presented in this study. In this application of the Dennis et al. (1986) model, the use of the chi-square test was invalid, since more than 20 percent of the cells in the data set had expected frequencies of less than five (Bishop et al. 1975). Early in the year (i.e., <100 DD), the probability of collecting grasshoppers in instars 4 and 5 and adult stages is very low (Figures 24.4-5). Conversely, late in the year (i.e., >500 DD), the probability of collecting grasshoppers in instars 1, 2, 3, or 4 is very small. These small probabilities translate into small expected frequencies for stage of development by date. Though the small expected frequencies prevent the use the chi-square test, they do not detract from the ability of the model to describe the timing of the different developmental stages.

TABLE 24.3
Parameter estimates (± 95% asymptotic confidence limits rounded to three significant places for simplicity) for a general model of rangeland grasshopper phenology.

Parameter	a_1	a_2	a_3
	75.410177 ± 2.504	120.100589 ± 2.775	151.695443 ± 2.951
Parameter	a_4	a_5	b^2
	202.851670 ± 3.098	292.846128 ± 3.480	6.543543 ± 0.376

The model presented is intended to be an initial step toward the development of a general grasshopper phenology model for use on rangeland systems. As data on grasshopper development accumulate, it is expected that model results will be more generally applicable. In this way, managers could schedule sampling for a specific life stage or stages and evaluate only density to determine whether an EIL has been exceeded (see Onsager 1984).

Results of Onsager (1986) indicate that peak instar 3 is an ideal time to sample in order to detect potential economic infestations and provide maximum flexibility among registered insecticide treatment choices. The results of the present analysis (Figure 24.5, Table 24.2) indicate that sampling generally should be conducted near 129 DD. Figure 24.5 also provides general estimates for the application timing of two insecticide treatments. Chemical control with carbaryl or malathion should be scheduled for peak instar 4 and peak instar 5, respectively (Onsager 1986). Figure 24.5 shows that these peak events occurred at approximately 171 and 241 DD, respectively. The methods and results described above could also logically be used by action agencies such as APHIS for standardization of sampling times (to more closely approximate the biological time of the insect) and determine timing of control actions.

FUTURE DIRECTIONS

Clearly, these results are preliminary. However, there are several related and ongoing studies which are intended to improve our ability to predict significant grasshopper life events.

Work continues to improve the phenology model used in this analysis (Dennis et al. 1986, Kemp and Onsager 1986, Kemp et al. 1986). Large annual fluctuations in mortality tend to result in potentially different observed phenological events due to the decrease in recruitment into subsequent stages (Kemp and Onsager 1986). Therefore, analyses are underway to account for the effects of seasonal variations in mortality on both individual and groups of grasshopper species. Further, work is being conducted to account explicitly for spatial variation in grasshopper development from among sampling points within a site. This analysis is similar to that conducted by Stedinger et al. (1985) for the eastern spruce budworm in Maine. Stedinger et al. (1985) first used a general regional development model to predict the mean and variance of larval development in any site within a region. These estimates were then used to estimate the average probability that individuals collected at a given site will be in a specific developmental stage. A second level of variation was then included to account for the differences in numbers of individuals in different life stages that occur from sample point to sample point. Lastly, a third level of variation is employed to describe the sampling distribution of development levels of individuals in a sample at points within sites (Stedinger et al. 1985). Modifications of the methods described by Stedinger et al. (1985) are expected to improve current model based predictions.

One problem associated with this type of model is that it requires fairly extensive geographic data sets in order to determine variation at different levels and to develop the appropriate model parameters. We presently do not have the data to develop an accurate development model for use over a multi-state region; however, data are accumulating rapidly and the initiation of sentinel sites (APHIS 1986) could facilitate data accumulation. The two-week sampling interval described (p. 11, APHIS 1986), however, should be shortened to one-week, even if that results in a reduction in the number of sentinel sites in a given area.

TABLE 24.4
Timing of grasshopper peak instar 3 and end bloom phase of Zabeli honeysuckle near Roundup, MT.

Year	Peak Instar 3 Grasshoppers (Model Estimated)	End Bloom Phase Zabeli Honeysuckle (Observed)
1975	190* (7/9)	182 (7/1)
1976	177 (6/26)	167 (6/16)
Average date	- -	164 (6/13)

*Julian date and calendar (month/day)

In addition to efforts to improve predictive phenology, concurrent analyses are underway to exploit a more extensive data base on plant phenology. Our ultimate goal is to produce a model for land managers and action agencies such as APHIS. Detailed phenological surveys have been conducted throughout Montana and 10 other western states on a number of plants including the common purple lilac, *Syringa vulgaris* L. and Zabeli honeysuckle, *Lonicera krolkowii* Stapf, var. *zabelii* (Rehd.) Rehder, (Caprio 1966, Figure 24.6). Operating on the premise that both plant and insect phenology are influenced by the accumulated DD at a particular site, we are attempting to relate grasshopper development to plant phenological events, since phenology of plants is more easily observed. These analyses are intended to provide an interim tool for use by managers while quantitative models are being developed. Also the analyses will supplement data in the validation of such quantitative models.

In an initial evaluation of the Roundup, Montana grasshopper phenology data (used for the model developed previously), it was found that end bloom phase date of Zabeli honeysuckle (Caprio et al. 1970) preceded by roughly 10 days peak instar 3 for 6 species of grasshoppers (Table 24.4) (Kemp and Onsager 1986). Differences in the phenological events for both the insects and plant between years are obvious (Table 24.4). Climatologists consider 1976 to be an average year (Caprio, per. comm.) and 1975 to be a much later than normal year. This is also indicated by the reduced number of DD accumulated at the Roundup site (Figure 3 in Kemp and Onsager (1986)). However, in spite of the yearly differences in heat input, the timing of peak instar 3 relative to the end bloom phase of Zabeli honeysuckle is remarkably constant. Caprio indicates (unpublished data) that the average date when lilacs begin to bloom (Figure. 24.6) precedes the end bloom phase of Zabeli honeysuckle by about 25 days and the end bloom phase of lilac by 20 days. Therefore, if the relationship indicated by our data from one site for two years is at least marginally correct, we can estimate when to schedule the initial sampling of grasshoppers by adding 35 days to the isophanes shown in Figure 24.6 for a climatologically normal year. In warm years this interval will be shorter and in cooler years will be longer than 35 days, respectively. Obviously, the relationship between lilacs and peak instar 3 is not as simple as this, and data from 1 site cannot be expected to work well for all sites and grasshopper species complexes across a state as variable as Montana. What Figure 24.6 does do, rather, is to provide an example of the type of analyses that can be done with data bases that are already available. Also, it presents information to managers in a form that is easily understood and applied (see Hewitt (1980) as an example). Work continues to validate such implied relationships and develop phenology maps for key life stage events for grasshoppers in Montana.

In conclusion, there is no question that a major challenge to integrated pest management on rangeland is the further development of predictive capabilities for insect phenology. No single method is likely to be sufficient to meet all needs and circumstances under which these predictive tools are to be applied. What I have described above may be considered crude, though I think it is fairly representative of the state-of-the-art. The integration of several technologies will be central to progressive improvements in our predictive capabilities and will put tools in the hands of users in a reasonably short length of time.

ACKNOWLEDGMENTS

I thank George B. Hewitt for use of grasshopper development data from Roundup, MT and Joseph M. Caprio for information on lilac and honeysuckle phenology. I thank Jeffrey A. Holmes for technical assistance and Vera L. Christie for manuscript preparation.

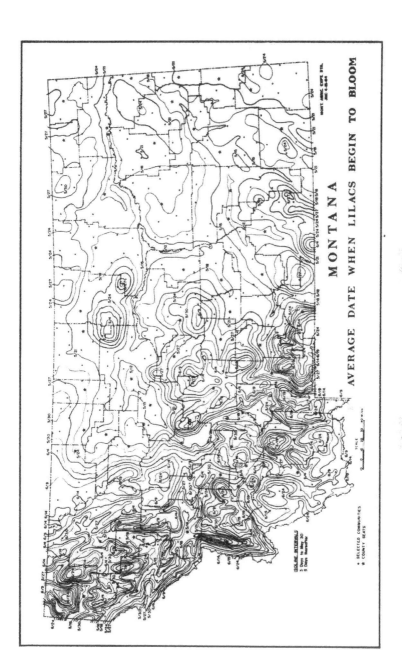

FIG. 24.6 Map of average date when lilacs begin to bloom (reproduced by permission from Caprio (1966)).

366

REFERENCES

Allen, J.C. 1976. A modified sine wave method for calculating degree days. Environ. Entomol. 5:388-396.

APHIS. 1986. Rangeland grasshopper cooperative management program: Final environmental impact statement. April 4, 1986. USDA, APHIS, Hyattsville, MD; 103 pp. plus Appendicies.

Bishop, Y.M.M., S.E. Fienberg, and P.W. Holand. 1975. Discrete multivariate analysis: theory and practice. MIT Press, Cambridge, MA.

Caprio, J.M. 1966. Pattern of plant development in the western United States. Montana Agr. Exp. Sta. Bull. 607.

Caprio, J.M., M.D. Magnuson, and H.N. Metcalf. 1970. Instructions for phenological observations. Montana Agr. Exp. Sta. Circ. 250.

Dennis, B., and W.P. Kemp. 1987. Further inferences in insect phenology modeling. (Submitted)

Dennis, B., W.P. Kemp, and R.C. Beckwith. 1986. A stochastic model of insect phenology: estimation and testing. Environ. Entomol. 15:540-546.

Drummond, F.A.R., G. Van Driesche, and P.A. Logan. 1985. Model for temperature-dependent emergence of overwintering *Phyllonorycter crataegella* (Clemens) (Lepidoptera: Gracillariidae), and its parasitoid, *Sympiesis marylandensis* Girault (Hymenoptera: Eulophidae). Environ. Entomol. 14:305-311.

Gage, S.H., M.K. Mukerji, and R.L. Randell. 1976. A predictive model for seasonal occurrence of three grasshopper species Saskatchewan (Orthoptera: Acrididae). Can. Entomol. 108:245-253.

Gyllenberg, G. 1974. A simulation model for testing the dynamics of a grasshopper population. Ecology 55:645-650.

Hardman, J.M., and M.K. Mukerji. 1982. A model simulating the population dynamics of the grasshoppers (Acrididae) *Melanoplus sanguinipes* (Fabr.), *M. packardii* Scudder, and *Camnula pellucida* (Scudder). Res. Popul. Ecol. 24:276-301.

Hewitt, G.B. 1979. Hatching and development of rangeland grasshoppers in relation to forage growth, temperature, and precipitation. Environ. Entomol. 8:24-29.

Hewitt, G.B. 1980. Plant phenology as a guide in timing grasshopper control efforts on Montana rangeland. J. Range Manage. 33:297-299.

Hilbert, D.W., and J.A. Logan. 1982. A simulation model of the migratory grasshopper (*Melanoplus sanguinipes*), p. 323-333. *In*: W.K. Lauenroth, G.V. Skogerboe, and M. Flug (eds.) Analysis of ecological systems: state-of-the-art in ecological modelling. Elsevier, New York.

Kemp, W.P., D.G. Burnell, D.O. Everson, and A.J. Thomson. 1983. Estimating missing daily maximum and minimum temperatures. J. Climate Appl. Meteorol. 22:1587-1593.

Kemp, W.P., B. Dennis, and R.C. Beckwith. 1986. A stochastic phenology model for the western spruce budworm (Lepidoptera: Tortricidae). Environ. Entomol. 15:547-554.

Kemp, W.P., and J.A. Onsager. 1986. Rangeland grasshoppers (Orthoptera: Acrididae): modeling phenology of natural populations of six species. Environ. Entomol. 15:924-930.

Kempton, R.A. 1979. Statistical analysis of frequency data obtained from sampling an insect population grouped by stages, p. 401-418. *In*: J.K. Ord, G.P. Patil and C. Taillie (eds.) Statistical distributions in ecological work. Int. Cooperative Pub. House, Fairland, MD.

Logan, J.A., R.E. Stinner, R.L. Rabb, and J.J. Bacheler. 1979. A descriptive model for predicting spring emergence of *Heliothis zea* populations in North Carolina. Environ. Entomol. 8:141-146.

Mann, R., R.E. Pfadt, and J.J. Jacobs. 1986. A simulation model of grasshopper population dynamics and results for some alternative control strategies. Wyoming Agr. Exp. Sta. Sci. Mono. 51.

Mukerji, M.K., and R.L. Randell. 1975. Estimation of embryonic development in populations in *Melanoplus sanguinipes* (Orthoptera: Acrididae) in fall. Acrida 4:9-18.

Onsager, J.A. 1984. A method for estimating economic injury levels for control of rangeland grasshoppers with malathion and carbaryl. J. Range Manage. 37:200-203.

Onsager, J.A. 1986. Current tactics for suppression of grasshoppers on range, p. 60-66. *In*: J.A. Onsager (ed.) IPM on rangeland: state of the art in the sagebrush ecosystem. USDA, ARS 50.

Onsager, J.A., and G.B. Hewitt. 1982. Rangeland grasshoppers: average longevity and daily rate among six species in nature. Environ. Entomol. 11:127-133.

Osawa, A., C.A. Shoemaker, and J.R. Stedinger. 1983. A stochastic model of balsam fir bud phenology utilizing maximum likelihood parameter estimation. Forest Sci. 29:478-490.

Patel, J.K., C.H. Kapadia, and D.B. Owen. 1976. Handbook of statistical distributions. Marcel Dekker, New York.

Putnam, L.G. 1963. The progress of nymphal development in pest grasshoppers (Acrididae) of western Canada. Can. Entomol. 95:1210-1216.

Randell, R.L. 1972. Some recent advances in the application of high-speed computing equipment to grasshopper forecasting and the study of grasshopper ecology in Saskatchewan, p. 391-396. *In*: C.F. Flemming and T.H.C. Taylor (eds.) Proc. int. study conf. on the current and future problems of acridology. Center for Overseas Pest Research, Overseas Development Admin., and Foreign and Commonwealth Office, London.

Randell, R.L. and M.K. Mukerji. 1974. A technique for estimating hatching of natural egg populations of *Melanoplus sanguinipes* (Orthoptera: Acrididae). Can. Entomol. 106:801-812.

Read, K.L.Q. and J.R. Ashford. 1968. A system of models for the life cycle of a biological organism. Biometrika 55:211-221.

Regniere, J. 1982. A process-oriented model of spruce budworm phenology (Lepidoptera: Tortricidae). Can. Entomol. 114: 811-825.

Regniere, J., R.L. Rabb, and R.E. Stinner. 1981. *Popillia japonica*: simulation of temperature-dependent development of the immatures, and prediction of adult emergence. Environ. Entomol. 10:290-296.

Rodell, C.F. 1977. A grasshopper model for a grassland ecosystem. Ecology 58:227-245.

Ruesink, W.G. 1976. Modeling of pest populations in the alfalfa ecosystems with special reference to the alfalfa weevil, p. 80-89. *In*: R.L. Tummala, D.L. Haynes, and B.A. Croft (eds.) Modelling for pest management. Michigan State Univ. Press, East Lansing, MI.

Sharpe, P.J.H., and D.W. DeMichele. 1977. Reaction kinetics of poikilotherm development. J. Theor. Biol. 64:649-670.

Stedinger, J.R., C.A. Shoemaker, and R.F. Tenga. 1985. A stochastic model of insect phenology for a population with spatially variable development rates. Biometrics 41:691-701.

Wagner, T.L., H. Wu, P.J.H. Sharpe, R.M. Schoolfield, and R.N. Coulson. 1984. Modeling insect development rates: a literature review and application of a biophysical model. Ann. Entomol. Soc. Am. 77:208-225.

Wagner, T.L., H. Wu, R.M. Feldman, P.J.H. Sharpe, and R.N. Coulson. 1985. Multiple-cohort approach for simulating development of insect populations under variable temperatures. Ann. Entomol. Soc. Am. 78:691-704.

Welch, S.M., B.A. Croft, J.F. Brunner, and M.F. Michels. 1978. PETE: an extension modeling system for management of multispecies pest complex. Environ. Entomol. 7:487-494.

25. APPLICATION OF INVENTORY CONTROL THEORY TO ECONOMIC THRESHOLDS

Roger Mann
Department of Agricultural and Resource Economics
Colorado State University, Fort Collins, Colorado 80523

Edward B. Bradley
Department of Agricultural Economics
University of Wyoming, Laramie, Wyoming 82071

Many conceptual approaches have been used by economists to model mathematically when and how much pesticide application is economically optimal for controlling pest populations. The economic threshold generally is defined as the minimum pest population density at which control would be economically justified. For our purposes, we shall define the economic threshold as the pest population density at which control should be implemented to maximize profit over time.

In most ways the economic problem of rangeland pests is the same as for other agricultural pests. Important differences between rangeland and annual crops include:

1. The pest destroys an input into production rather than the product itself.
2. Rangeland is ecologically complex relative to cultivated lands.
3. Livestock production on rangeland is land-extensive rather than land-intensive.
4. More of the problem originates on public lands.
5. Pest damage can lead to a capital loss.
6. Pest control is often publicly subsidized.

Factors (1), (2), and (5) complicate measurement of pest damage. Measurement is also complicated by public ownership of rangeland, since a market price for grass may not be observable. Factors (3) and (4) are often given as reasons for more public involvement in rangeland pest control. Factor ·(3) also limits the control cost per land unit that can be economically justified and tends to encourage large-scale control programs so that economies of scale can be realized.

THE ECONOMIC THRESHOLD

The concept of the economic threshold as a decision guide to pest control decisions remains a popular component of integrated pest management programs. The literature on the topic is rich, but it appears that analytical complexity has often replaced sound economics as the distinctive characteristic of control decisions. For examples of early works on the economic threshold see Stern (1973), Norton (1976), Tolpaz and Borosh (1974), Headley (1972), and Hall and Norgaard (1973).

The models presented by Norton (1976) and Pedigo et al. (1986) suffer from a problem common in threshold modeling. There is no consideration of time in the model. The proper control decision should maximize profit per unit time, which is equivalent to minimizing the sum of damage and control costs per unit time. In the case of the aforementioned two articles, a pest density below the economic injury level could allow the pest population to cause damage indefinitely even though control cost might be far less than the accumulated damage.

The purpose of this paper is to develop the simplest possible model of a pest control decision and to suggest a general body of theory which is highly adaptable to the pest control problem. This general theory is called inventory control theory. It is the very close analogy between the pest control problem and the inventory control problem which allows for the use of the same methods with little loss of utility.

Inventory control theory is not a single mathematical method, but rather a group of methods used to describe and optimize decisions regarding inventory holding and ordering. Similarly, no single mathematical model can determine economic thresholds for all possible pests in all possible situations. A group of methods is needed to address the wide range of problems.

THE SIMPLEST POSSIBLE FORMULATION

In nearly all pest control situations, the farmer or rancher is faced with two undesirable alternatives: incurring damage from the pest or the cost of control. Nearly any aspect of the economic threshold decision must relate to these two costs. The profit-maximizing plan will minimize the sum of these costs over time.

Consider the following simple example. Suppose a pesticide treatment is perfectly effective, the pest population starts at zero and always grows by 1 member per day, each pest causes 0.01 cent of damage per day, and control costs $1.00. Table 25.1 summarizes the control cost, damage cost, and total cost per year under an increasing number of controls over the year.

TABLE 25.1
Example of pest control problem.

Number of Controls per Year	Total Control Cost	Total Damage Cost	Total Cost per Year
1	$1.00	$6.66	$7.66
2	2.00	3.33	5.33
3	3.00	2.22	5.22
4	4.00	1.66	5.66
5	5.00	1.33	6.33
6	6.00	1.11	7.11

We can see that the cost-minimizing solution is about three controls per year. The actual cost minimizing solution is to control every 141.4 days (2.58 times/year). At the optimum, total cost per year is $5.16. The next section details the inventory control model which can be used to find this solution.

Note from Table 25.1 that damage cost per year declines with each additional control treatment, but at a declining rate. The optimum occurs where the added control cost just equals the reduction in damage cost.

APPLICATIONS OF INVENTORY MODELING TO ECONOMIC THRESHOLDS

The inventory problem normally involves minimizing the sum of two or more cost components per unit of time. A good introduction to inventory models is provided by Taha (1976) and many other operations research texts. In the most simplistic situation, a cost K is incurred whenever an order is placed to stock or restock an item. In addition, a holding cost of h amount per item per unit of time is charged, with items depleted from inventory at a linear rate B per unit of time. More of the item is ordered and immediately delivered whenever inventory reaches zero. This situation is depicted graphically in Figure 25.1. The maximum inventory level, designated by y, occurs at ordering time. With the linear depletion rate B, the average inventory level equals y/2 and the length of time between orders, denoted by T, equals y/B. Thus, the total cost of acquiring and holding inventory per unit of time is:

(1) $TCU = \dfrac{K}{y/B} + \dfrac{hy}{2}$

The order quantity which minimizes total cost per unit time (y^*) is found by setting marginal cost equal to zero.

(2) $\dfrac{dTCU(y)}{dy} = -KB/y^2 + h/2 = 0$

by use of algebra

(3) $y^* = \sqrt{\dfrac{2KB}{h}}$

Furthermore, the optimum time between orders (t^*) is

(4) $t^* = y^*/B$

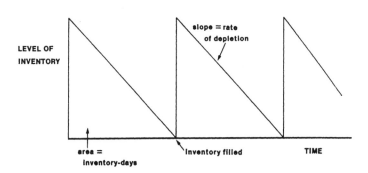

FIG. 25.1 The inventory control problem.

EXTENSION TO THE ECONOMIC THRESHOLD

Let the "economic threshold" be the pest population density at which control practices should be applied so that control plus damage costs per unit of time are minimized. This may also be shown to prescribe the profit-maximizing level of pest control activities.

The problem of optimal timing of pesticides is best visualized by considering the control costs and the crop damage costs of allowing pest populations to exist. The control costs are incurred at points in time and the damage cost is incurred during the periods between pesticide applications. Consider a pest population controlled at a population of y. As y is allowed to increase, control is occurring less frequently and the control cost per unit of time is decreasing.

Allow the growth rate of the pest population to be denoted as B. Inventory models can accommodate a more complex specification of a population growth rate, but knowledge beyond a linear form may often be unavailable. Since B is expressed as population growth per unit time, y/B equals length of time between controls, or

(5) $t = y/B$

The cost of control per unit time expressed in terms of population and growth rate may be given as

(6) $K/t = K/(y/B)$

The other component of cost associated with the pest management problem is the damage incurred by the pest population. Assume that damage cost per pest per unit of time equals h.

It is further assumed that the population is essentially eliminated in the area immediately following control. Although this assumption is unrealistic due to pesticide resistance, it completes the analogy with the simple inventory problem given in equations (1) to (4). The amount of pest damage incurred per unit time over the period may be expressed as $hy/2$. A graphical depiction of the pest problem is given in Figure 25.2. A realistic mortality function will be introduced in the next section.

Given this problem formulation, the optimal pest density at which to initiate control is given by equation (3) and the interval between controls by equation (4). The total cost per unit time incurred by the pest population and control operations can be found by substitution of y^* for y in equation (1). y^* is the economic threshold for control of the pest. Implementing control whenever this population density occurs will yield the smallest total cost of control and damage per unit time.

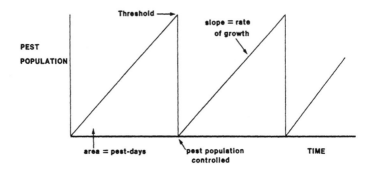

FIG. 25.2 The pest control problem.

AN EXTENSION TO VARYING PESTICIDE QUANTITY

Perhaps the most important modification to the model would allow for variation in the quantity of the pesticide applied such that quantity of pesticide to apply is an additional decision variable.   To allow for variation in the quantity of pesticide applied, the entire system of equations (1) to (8) must be modified.   For greater realism the population after control is assumed to be positive so that

(7) $y = z + Bt$

where z denotes some positive population remaining after control. Let the population z be a positive fraction of the population y such that

(8) $z = y(1 - x)$

where x = proportion mortality caused by the control treatment.
Further, let

(9) $x = F(q)$ and

(10) $K = C + pq$

where C denotes the fixed cost of control, p is the price per unit of pesticide, q is the quantity or amount of pesticide applied, and x is a positive function of q.

The average amount of damage cost incurred over the period between controls equals

(11) $h(y + z)/2$

Substituting from equations (8) and (9), we obtain total cost per unit of time as a function of y and q.

(12) $TCU = [B(C + pq)/F(q)y] + [hy(2 - f(q))]/2$

It is possible to determine a minimum for this total cost function by use of calculus. The derivative of the total cost function with respect to y yields optimal y for a given q. The derivative with respect to q yields another first order condition such that there are two equations in the two unknowns y and q.

Taking the partial derivative of (12) with respect to y and setting the partial equal to zero to find a minimum yields

(13) $y = \sqrt{\dfrac{2B(C + pq)}{hF(q)(2 - F(q))}}$

Thus, optimal (y^*) may be determined for any q.

In many circumstances, the quantity to be applied is known in advance, perhaps due to the label or other conventions. In this case, quantity is not a decision variable and X may be substituted for F(q) in equation (13), and the time between controls may be determined by $t = yX/B$.

The models presented in this paper represent the most simplistic applications of inventory control theory to the pest control problem. Inventory control theory also includes dynamic and probabilistic models and many other variations to the problem of minimizing costs per unit time.

Inventory models not only provide many techniques for economic threshold evaluation, but also constitute a viable theoretical framework for developing pest control models. The theory and methodologies employed have many extensions to the rangeland pest control problem. A general model for determining economic thresholds must allow for the variety of circumstances likely to arise in model specification. As a group of general models and techniques, inventory theory is highly adaptable for determining economic thresholds for pest control.

376

REFERENCES

Hall, D.C. and R.B. Norgaard. 1973. On the timing and application of pesticides. Amer. J. Agr. Econ. 55:198-201.

Headley, J.C. 1972. Defining the economic threshold, p. 100-108. *In*: Pest control strategies for the future (N.R.C.). National Academy of Science, Washington, D.C.

Norton, O.A. 1976. Analysis of decision making in crop protection. Agro-ecosystems 3:72-44.

Pedigo, Larry P., S. H. Hutchins, and L.G. Higley. 1986. Economic injury levels in theory and practice. Annu. Rev. Entomol. 31:341-68.

Stern, V.M. 1973. Economic thresholds. Annu. Rev. Entomol. 18:259-80.

Tolpaz, H., and I. Borosh. 1974. Strategy for pesticide use: frequency and applications. Amer. J. Agr. Econ. 56:769-775.

Taha, H.A. 1976. Operations research: an introduction. MacMillan, New York.

26. FACTORS AFFECTING THE ECONOMIC THRESHOLD FOR CONTROL OF RANGELAND GRASSHOPPERS

L. Allen Torell
Department of Agricultural Economics and Agricultural Business
New Mexico State University, Las Cruces, New Mexico 88003

Ellis W. Huddleston
Department of Entomology, Plant Pathology, and Weed Science
New Mexico State University, Las Cruces, New Mexico 88003

The economic threshold concept has been widely accepted among entomologists as the basis for making rational pest management decisions. Stern et al. (1959) first defined the economic threshold as the amount of injury that would justify the cost of artificial control measures[1]. Headley (1972) provided a precise mathematical definition of the threshold model, which became the basis for additional model refinements by Hall and Norgaard (1973), Talpaz and Borosh (1974), and Hall and Moffitt (1985).

Economists have developed complex mathematical models to consider optimal pest management strategies. These models explicitly consider optimal timing and pesticide application rates. By economists' definition, at the start of the season the farmer or rancher determines the optimal timing of pesticide application and the profit maximizing application rate. The economic threshold is then the smallest level of pest infestation for which the optimal application rate is greater than zero (Plant 1986).

Entomologists have generally had a more pragmatic view of the economic threshold and view the economists rigorous definition as being obtained at the expense of biological and practical reality (Mumford and Norton 1984). They consider the threshold level to be that level of pest infestation at which the cost of applying an artificial pest control treatment equals the cost of damage to the crop if pesticide was not applied. Implicitly, if pesticide is applied, a predetermined rate is used. Plant (1986) refers to this definition of economic threshold as the discrete-choice threshold.

377

For rangeland pests such as grasshoppers, the discrete-choice threshold model is adequate. While the type of material to use in a control treatment (malathion, carbaryl, acephate, *Nosema locustae*) must be chosen, the amount of material to use has not been of major concern. Acceptable application rates have been established by research, and are regulated by the Environmental Protection Agency. These established rates are generally used whenever rangeland pests are artificially controlled.

Further, during a particular year, optimal timing of control is not an economic decision variable. Since forage losses occur during even the earliest nymphal stages, and rangeland pests have not been shown to be more susceptible to pesticide at a later developmental stage, the most economical time to control is as early in the season as an effective control treatment can be implemented after pests have hatched. Realistically, this may be hard to do with weather variations, heterogeneous grasshopper development, and variable species composition (see Onsager, Chapter 13).

In this paper, a discrete-choice economic threshold model is developed for control of rangeland grasshoppers. This model has the flexibility to consider many alternative specifications about how long a treatment might last, pest population dynamics, pest mortality rates, timing of application, control costs, the economic value of forage potentially saved and treatment efficiency. The dynamic nature of the economic threshold is emphasized and economic threshold levels for control of rangeland grasshoppers are presented for alternative model assumptions.

THE THRESHOLD MODEL

Assume the primary loss to grasshoppers is decreased forage yields, and ignore such secondary losses as decreased forage quality, migration to cropland and gardens, and the nuisance of high insect populations. Assuming any forage lost to grasshoppers could have been used in livestock production, a rancher must decide whether the potential value of forage saved by initiating a control program more than offsets control costs. The forage has a value to the rancher as derived from the livestock production process[2]. This derived forage value for livestock production must be weighed against the cost of a control program, with one option being to feed the forage to rangeland insects.

Pest Population Dynamics

Let X be some known (or projected) level of rangeland pest density before any artificial control program, and measured at some specific stage of insect development. A single homogeneous birth-death process without migration is assumed, such that if left untreated, the inter-year population will grow (or decline) at rate r_1. The pest population level satisfies the differential equation

(1) $dX = r_1 X dt$

which gives the untreated pest population level at year t as

(2) $X_u(X,t) = Xe^{r_1 t}$.

During the year, the grasshopper population will mature and destroy additional forage, but as has been shown by Onsager (1984), during any given year the number of grasshoppers will decline between progressive developmental stages because of natural mortality (given no significant recruitment from hatch following insecticide application and with no migration from adjoining rangeland areas).

If control efficacy is α ($0 \leq \alpha \leq 1$), and the number of pests killed by a treatment application is linearly related to the initial pest infestation level, the density of the pests immediately after control would be $Z = (1 - \alpha)X$. Some pests will survive to the adult stage and lay eggs, which will hatch during the following spring and continue the dynamic cycle of the pest population.

If the treated pest density in the subsequent year remains the same, or grows from level Z, control of the pest could result in forage savings for a number of years. A second rate of treated population growth (r_2) is assumed for the treated pest population. It is likely r_2 will be greater than r_1 if some biological control exists in the form of natural predators or parasites that would be eliminated with the pesticide application, or if survivors are released from competition for food and space.

The pest density on the treated area at a point in time following control is given by:

(3) $X_c(X,t) = Ze^{r_2 t} = (1 - \alpha)Xe^{r_2 t}$.

Pest Damage Function

Define D(a) to be the amount of forage consumed during year t when the pest density at the beginning of the treatment period is a. Further, let T denote the length of benefit from a rangeland pest control program. Over the season, the potential amount of forage saved by initiating a rangeland control program is computed as the difference between what X_u pests would have consumed and what X_c pests would harvest or waste. The total amount of forage saved over T years by initiating a control program is given by:

$$(4) \quad \sum_{t=0}^{T} S(X_u(X,t),X_c(X,t)) = \sum_{t=0}^{T} [D(X_u(X,t)) - D(X_c(X,t))],$$

where t=0 is the year of control treatment initiation.

Two potential benefits from initiating a control program could be realized. First, pest densities and corresponding forage consumption would be reduced below untreated levels for some period of time into the future. This saving would correspond to area A in Figure 26.1. Area B would be saved because additional increases in the untreated pest population above X in future years would be prevented.

It is also possible that little if any benefits would be realized from initiating a grasshopper control program. Natural mortality could cause the pest population to crash on its own, minimizing any positive benefit from a control program.

The Economic Threshold Model

The present value of forage potentially saved over the T year treatment life is given by:

$$(5) \quad \sum_{t=0}^{T} vS(X_u,X_c)(1+i)^{-t}$$

where v is the per unit value of the forage saved, and i is the interest rate used in discounting forage savings realized in future years.

Assuming that if a control program is to be initiated a known but constant amount of pesticide will be applied, the cost of pesticide application is some constant amount C, which includes both pesticide and application costs. The net present value of forage saved by controlling a density of X grasshoppers is then given by:

$$(6) \quad \theta = \sum_{t=0}^{T} vS(X_u,X_c)(1+i)^{-t} - C.$$

If θ is positive, the value of forage saved over the T year treatment life exceeds the C dollar treatment cost, indicating a positive economic benefit is realized. If θ is negative, artificial control measures are not economically justified.

The economic threshold is defined to be that level of pest infestation (X^*) where the net present value of forage saved over the T year treatment life just equals the cost of the control program, i.e.,

$$(7) \quad \theta^* = \sum_{t=0}^{T} vS(X_u(X^*,t),X_c(X^*,t))(1+i)^{-t} - C \equiv 0.$$

The threshold level depends upon the specification of the exogenous variables of the model, including C, v, i, T, α, r_1, and r_2. A *ceteris paribus* change in any exogenous model variable will alter the economic threshold. It is not a constant such as the 8 grasshoppers/sq. yd. used as an action threshold by USDA Animal and Plant Health Inspection Service (APHIS) (USDA 1986).

Below, we consider a realistic specification of the model for determining the economic threshold level for rangeland grasshoppers. The threshold levels for alternative model assumptions and exogenous variable specifications are also presented.

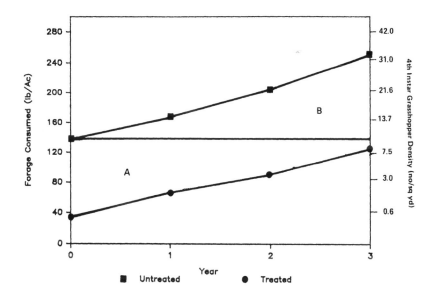

FIG. 26.1 Potential time trend of forage losses and pest population increases from treated and untreated pest densities.

DETERMINANTS OF THE ECONOMIC THRESHOLD

There are many factors that influence treatment success and the economics of grasshopper control programs, including (1) potential forage destruction, (2) pest population dynamics, (3) efficacy of control treatment, (4) cost of treatment, and (5) the potential value of forage saved by a control treatment. Many of these factors are extremely variable and, to a large degree, not well defined.

Forage Destruction

The rate at which a grasshopper infestation destroys forage is a variable function of grasshopper population composition, stage of development, and rate of survival over time (Onsager 1984). It has been estimated that grasshoppers consume about one-half their body weight in green forage per day (Capinera and Sechrist 1982). In addition to that which is actually consumed, grasshoppers waste additional forage because they eat only a portion of each leaf or stem that they cut.

Most destructive grasshopper species have five nymphal instars, with the first three instars responsible for only 15 to 20% of total seasonal forage consumption (Onsager 1983). Onsager (1984) concluded that because the majority of grasshoppers must attain the 3rd instar before one can confidently diagnose an economic infestation on rangeland and bring control measures to bear, forage losses by the early instars must be conceded. He defined the relevant destructive period to begin with the appearance of 4th instar nymphs, and concluded that by this stage control tactics were definitely feasible.

Hewitt and Onsager (1983) estimated the seasonal loss of forage to grasshoppers to be about 12 pounds of air-dry forage per acre (13.4 kg/ha) for each grasshopper per square yard over the acre. In recent studies, Onsager (1983, 1984) has shown forage destruction to be a function of grasshopper survival, and survival to be a function of density. He estimated that an "average" rangeland grasshopper consumes or wastes 9, 22, and 53 mg of forage per day in the 4th instar, 5th instar, and adult stages, respectively. Higher grasshopper densities destroy more forage, but forage losses increase at a decreasing rate because of reduced survival as grasshopper densities increase.

Onsager (1984) estimated the average daily survival rate of grasshoppers (S) as a function of density of 4th instar nymphs/m^2 (X) according to the formula,

$$\ln S = -0.0028909 - 0.0064462(\ln X) - 0.0012987(\ln X)^2.$$

He then used this function to plot the number of survivors against time, and defined the area under the survival curve as grasshopper days (GHD). The GHD for each development period was then multiplied by the appropriate average daily rate of forage destruction, with the summation of these three products giving the potential forage destruction per m^2 from a 4th instar grasshopper infestation of level X.

Extrapolation from the Onsager Model. In our model development, we have used the functions and relationships defined in the aforementioned model of Onsager (1984). Inherent in our threshold model are the same forage consumption and survival rates estimated by Onsager, which were developed for Montana rangeland conditions where cool season grasses predominate. For cool season grasses, late season forage regrowth does not normally occur. However, for warm season grasses, regrowth late in the season may often occur and offset forage losses from grasshoppers. Preliminary data for New Mexico, where above-average rainfall occurred, indicated low grasshopper densities may not be as harmful as would be predicted by the Onsager model. Regional differences in model specification may need to be made.

The potential amount of forage saved by initiating a control treatment is computed as the difference in forage destroyed by the treated and untreated pest densities. Implicit in this equation is the assumption that the forage losses estimated for X grasshoppers is indeed available on the rangeland parcel in question, and if it were not harvested by grasshoppers, it would have been used in livestock production or maintained for beneficial plant reproduction and vigor.

Using data from Onsager (1984), three functions were developed for the three relevant grasshopper development stages:

$$\ln GHD_{4th} = 2.218 + 0.933(\ln X)$$
or,
$$(8) \quad GHD_{4th} = 9.190X^{.933};$$

$$\ln GHD_{5th} = 2.231 + 0.819(\ln X)$$
or,
$$(9) \quad GHD_{5th} = 9.313X^{.819};$$

$$\ln GHD_{Adult} = 4.406 + 0.398(\ln X)$$
or,
$$(10) \quad GHD_{Adult} = 81.937X^{.398}$$

These equations are merely a functional redefinition of Onsager's model.

In equations (8)-(10), X represents the number of 4th instar nymphs per square yard. Forage reductions are estimated at

grasshopper density X by multiplying the estimated GHD for each
development stage by the appropriate average daily rate of forage
destruction. Forage losses are computed using the equations:

$$D(X)= \begin{cases} (11)\ A_1 GHD_{4th}(X) + A_2 GHD_{5th}(X) + A_3 GHD_{Adult}(X),\ X \geq 2 \\ (12)\ (A_1/2)GHD_{4th}(2)X + (A_2/2)GHD_{5th}(2)X \\ \qquad + (A_3/2)GHD_{Adult}(2)X,\ X < 2 \end{cases}$$

where $A_1 = 0.0959$, $A_2 = 0.2345$ and $A_3 = 0.5650$, and represent the
average daily forage destruction rate per GHD measured in
pounds/acre for 4th, 5th, and adult stages, respectively[3].
Redefinition of the Onsager (1984) model resulted in equation
(11), which can be used to estimate potential forage losses from X
grasshoppers when X is greater than $2/yd^2$ ($1.67/m^2$). However, it
is believed that at grasshopper densities below $2/yd^2$ the equation
overestimates forage losses.

Equation (12) reflects a linear equation that computes
average forage losses for densities less than $2/yd^2$ based on the
rate of destruction of $2/yd^2$. This was believed to be a more
realistic estimate of forage loss at lower densities, which is
important when computing potential losses from grasshoppers
remaining after treatment. Figure 26.2 summarizes total forage
loss estimated from the two equations, and shows forage losses at
various 4th instar grasshopper densities.

Pest Population Dynamics

Entomologists have tried, without much success, to predict
grasshopper outbreaks. Several factors, including available food
supply and environmental influences, have been identified as
important determinants of grasshopper populations (see Capinera,
Chapter 11).

Pfadt (1977) found favorable weather conditions to be an
important reason for an outbreak of grasshoppers in the shortgrass
plains of Wyoming during 1974. Favorable weather conditions
during the fall of 1973 resulted in the adults living longer and,
therefore, having the opportunity to produce more eggs.

MacCarthy (1956) also found weather to be an important
consideration. He found that temperature, especially daily minima
during the previous June and August and daily maxima during the
previous August and the current June, seemed to influence
grasshopper populations more than other factors.

Parker (1939) proposed a model with the hypothesis that
prior to an outbreak a gradual increase in the grasshopper
population occurs. He proposed that for several years the

population increases by about 2X, followed by an increase of 3X or 4X during the outbreak year. Work by Pfadt (1977) on the shortgrass plains supported this model.

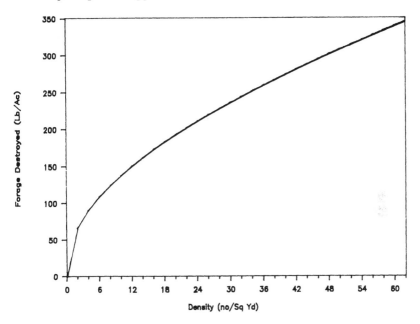

FIG. 26.2 Forage consumed by various 4th instar grasshopper densities.

Efficacy of Control Treatments

The efficacy of insecticides applied to control rangeland grasshoppers will vary, depending on the temperature and amount of precipitation on the day of application, terrain, uniformity of application, and type of vegetation. In general, recommended insecticides applied under acceptable conditions can be expected to provide at least a 90% reduction in grasshopper density; 95 to 97% reductions often occur (Foster et al. 1981, Mann et al. 1986).

The long-term effectiveness of grasshopper control programs is uncertain. Based on many years of field observations, APHIS stated an effective control program should suppress grasshopper populations to below economic levels (defined by APHIS to be a density of less than eight grasshoppers/sq. yd.) for a period of 3 or more years under average conditions (USDA 1980).

Blickenstaff et al. (1974) studied 46 control areas in Wyoming and found 3.8% of the areas were retreated each year during a 5-year period. He found the percentage of retreatment decreased with increasing size of the area initially treated. In certain cases where untreated control checks were available, the apparent lack of demonstrated control beyond the season of application may have been the result of (1) reinvasion of the treated area from adjoining areas, (2) disproportionate natural decline of populations in untreated comparison areas, (3) occurrence of 2-year life cycles at high elevations, (4) extended hatching periods, (5) ability of a small residual population to increase rapidly, or (6) failure to treat infested areas in their entirety (but see Pfadt, Chapter 12).

Nosema locustae is a biological control alternative for grasshoppers. However, the efficacy of *N. locustae* has been variable. Only slight reductions in grasshopper survival have been evident in some field trials, while others have shown substantial mortality. Because *N. locustae* bran bait is likely to be more expensive, to take longer for control, and to perform better in northern climates, Mann et al. (1986) recommended its use only when susceptible species are present and environmental concerns restrict use of chemical control measures.

Cost of Control Treatments

Cost of control will vary depending upon the type of control agent used and the size of the area treated. Generally, the larger the area treated the lower the per-acre cost will be. Recently, the cost of cooperative control programs executed by APHIS have varied from about $2 to $3.75/acre ($4.95 to $9.28/ha) (unpublished data available from APHIS). These programs have traditionally been on a voluntary cost-share basis, with ranchers paying at least some of the cost on private land.

Value of Forage

The value of forage potentially saved by controlling rangeland grasshoppers varies from year to year and from ranch to ranch. Forage has a derived value in the livestock production process, and will vary depending on livestock prices, cost of production, management ability, alternative forage sources, and seasonal forage limitations. As an example, Torell et al. (1985) estimated the marginal value of spring and summer forage on a northern Nevada ranch to be $4.20/AUM[4] when the seasonal forage limitation was winter forage. This limitation arises because of seasonal use limitations on Bureau of Land Management (BLM) and U.S. Forest Service (USFS) allotments, and harsh winter weather that forces removal of livestock from range.

If an AUM of range forage was to substitute directly for hay and help to alleviate the winter feed shortage, the marginal AUM value was estimated to be about $27/AUM, equivalent to the value of purchased hay. If the ranch increased hay production, such that spring forage limited optimal production levels, the estimated marginal value of an AUM of spring/summer forage was estimated to be $7.70/AUM. This indicates that the value of forage potentially saved by a grasshopper control program will vary depending upon the seasonal availability of forage lost to grasshoppers.

Recent evaluations of public land grazing fees have stimulated much interest in the value of public land forage. The BLM and USFS studied private land lease arrangements in the 16 western states, plus two counties in Texas. They concluded that the market value of public land grazing ranged from an average of $5.20/AUM in the desert states of Nevada and Arizona to $9.50/AUM in North and South Dakota and Nebraska (USDA/USDI 1986, p.15).

Fowler et al. (1985) derived similar estimates of forage value for three pricing regions in New Mexico. Average value estimates ranged from $1.56/AUM in the desert southwest area of the state to $4.74/AUM in the grasslands of the northeast.

The value of forage is extremely variable, yet very important in determining the economics of rangeland pest control. Important considerations in assigning forage value in pest control evaluations include seasonal forage limitations and alternative forage sources. If purchasing hay to replace lost forage is the likely alternative the rancher will follow, the value of hay is an appropriate value to use. If the rancher could lease forage from a neighboring ranch, the private lease rate is an appropriate value to use. If forage would likely go unused if not harvested by rangeland insects, zero is an appropriate forage value to use.

ECONOMIC THRESHOLD FOR RANGELAND GRASSHOPPERS

Average Conditions

Based upon a literature review, the following assumptions were made in defining the parameters of the economic threshold model under "average" conditions:

(1) Total cost of the grasshopper control program was estimated to be $3/acre ($7.43/ha) with the rancher's share being $1/acre ($2.47/ha) (C=$3).

(2) Insect control was assumed to be initiated at or before the start of the 4th instar development stage.

(3) Efficacy of treatment (α) was considered to be 90%.

(4) Implementing a control program was assumed to exert a positive reduction on grasshopper density for 3 years (T=3).

(5) Forage saved in years 2 and 3 was discounted to present value at a 7% discount rate (i=.07).
(6) The value of forage potentially saved (v) was set at $8/AUM.
(7) If control was not initiated, the untreated pest population would remain constant at level X over the 3-year planning horizon (r_1=0).
(8) The treated population will double each year (r_2=.693).
(9) All populations have one generation per year.

Given this model definition, the economic threshold was estimated to be 23 4th instars/sq. yd. (Table 26.1). This would be equivalent to 11.5 adult grasshoppers/sq. yd. based on a daily survival rate of 96.22%, as estimated for this density using the Onsager (1984) model. Control of 23 4th instar nymphs would just cover the total $3/acre treatment cost when the forage saved is worth $8/AUM.

Various Treatment Costs

Obviously, as treatment cost goes up, a higher level of forage loss would have to be negated to justify implementing a grasshopper control treatment. When treatment cost is $3.50/acre, the economic threshold was estimated to be 30 4th instar grasshoppers/sq. yd., which is equivalent to 13.9 adult grasshoppers/sq. yd. (Table 26.1). At a treatment cost of $1/acre, which is the rancher's estimated share of cost, the economic threshold level is less than three 4th instar grasshoppers/sq. yd.

Various Forage Values

Grasshopper control programs were not economical when the forage potentially saved was only valued at $4/AUM and when only forage saving benefits were considered. In this case, the economic threshold was estimated to be greater than 60 4th instars/sq. yd.. At the other extreme, when forage was valued at $14/AUM, even five 4th instar nymphs could be economically controlled.

Various Discount Rates

The rate at which future benefits were discounted did not significantly affect the economic threshold. Increasing the discount rate from 4 to 16% had little impact on estimated threshold levels (Table 26.1). This occurred because the largest forage-saving benefit occurred during the year of treatment, which was not discounted, and because only a 3-year treatment life was considered.

TABLE 26.1
Economic threshold levels for grasshopper control programs, various model specifications.

Model Assumption	Economic Threshold (No./ sq. yd.)	
	4th instar	Adult
1. Average conditions[a]	23	11.5
2. Various Treatment Costs($/Acre)		
$1.00	<3	<2.4
$2.00	8	5.2
$2.50	15	8.4
$3.00	23	11.5
$3.50	30	13.9
3. Various Forage Values ($/AUM)		
$4.00	>60	>22.6
$6.00	38	16.5
$8.00	23	11.5
$10.00	14	8.0
$12.00	8	5.2
$14.00	5	3.6
4. Various Discount Rates (%)		
4%	22	11.1
7%	23	11.5
10%	23	11.5
13%	24	11.9
16%	25	12.2
5. Various Growth Curves For Untreated Pests		
Decrease by 1/2 each year ($r_1 = -.695$)	56	21.6
Remain Constant ($r_1 = 0$)	23	11.5
Double each year ($r_1 = .695$)	5	3.6
6. Various Growth Curves For Treated Pests		
Remain Constant ($r_2 = 0$)	11	6.6
Double each year ($r_2 = .695$)	23	11.5
Triple each year ($r_2 = 1.10$)	38	16.5
7. Various Treatment Lengths		
1 year(year of treatment)	>60	22.6
2 years	35	15.6
3 years	23	11.5
4 years	20	10.4

TABLE 26.1 (Cont.)

Model Assumption	Economic Threshold (No./ sq. yd.)	
	4th instar	Adult
8. Various Treatment Efficacy Rates (%)		
99%	7	4.7
95%	12	7.1
90%	23	11.5
85%	33	14.9
80%	48	19.4
75%	>60	>22.6
9. Various Stages of Control		
Before 4th Instar	23	11.5
Before 5th Instar	25	12.2
Before Adult Stage	30	13.9
1/2 of Adult Stage[b]	>60	>22.6

[a]conditions: treatment cost = $3/acre, forage value = $8/AUM, discount rate = 7%, constant untreated pest density, treated pest densities will double each year, treatment efficiency = 90%, and treatment occurs just before the 4th instar nymphal stage. These conditions were sequentially altered to compute economic threshold levels for alternative assumptions about treatment cost, forage value, discount rate, etc.

[b]If treatment did not take place until halfway through the adult stage it is likely that many adults would have already laid eggs. Therefore, it was assumed that in this case the treatment would last only 1 year.

Various Growth Curves for Untreated Pests

Knowing the direction and magnitude of grasshopper populations in the future is an important consideration in determining threshold levels, but probably the least known component of the threshold model. As shown in Table 26.1, if the untreated pest density was going to "crash" on its own (decrease by 1/2 each year), a high density of 56 4th instar grasshoppers would be required to justify economically a control treatment. This is in contrast to the case where populations are on an upswing (doubling each year) where only five 4th instars could justifiably be controlled. In this latter case, the major benefit is not current year forage savings but rather the forage potentially saved in future years from the growing pest population.

Various Growth Curves for Treated Pests

As the growth rate of the treated pest population increased, the benefits of a control treatment were decreased and threshold levels increased (Table 26.1).

Various Treatment Lengths

As there was an increase in the length of time that a control treatment suppressed grasshopper populations, the number of grasshoppers needed to justify a control program decreased. If grasshopper populations are only suppressed during the year of treatment, it was determined that control programs could not be economically justified (based solely on forage saving benefits). The economic threshold in this case was estimated to be in excess of 60 4th instars/sq. yd. (Table 26.1).

Threshold levels were estimated to be 35, 23, and 20 4th instar grasshoppers/sq. yd. for a 2, 3, and 4-year treatment life, respectively. These calculations included a doubling of the pest population each year so that the amount of forage saved each year was decreasing. After 4 years (year 1 is considered to be the year of treatment), the pest population would be back to pre-treatment levels and no additional benefits from the control treatment would be realized.

Various Treatment Efficacy Rates

Treatment efficacy directly affects potential forage availability and pest population growth potential. As treatment efficacy increases, more pests are killed and the amount of forage saved increases correspondingly. A treatment efficacy of 100% saves twice the forage during the first year of treatment as does a 65% rate (Figure 26.3). In addition, forage availability in future years is greatly reduced at lower efficacy rates because more grasshoppers remain to reproduce.

The economic threshold level for grasshopper control programs is greatly influenced by treatment success. If a 99% efficacy rate was expected, the economic threshold level is estimated to be seven 4th instar nymphs/sq. yd. (Table 26.1). If the treatment was only expected to kill 75% of the pests, the control treatment could not be economically justified, with the threshold level estimated to be in excess of 60 4th instar grasshoppers/sq. yd.

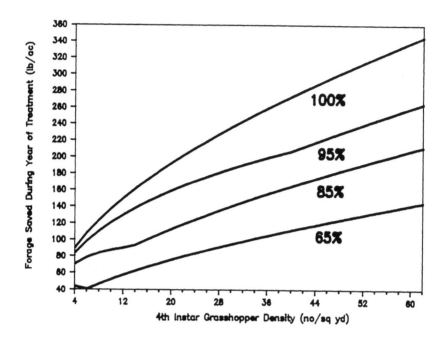

FIG. 26.3 The effect of treatment efficacy on potential forage savings from grasshopper control programs. Discontinuities in the line slopes are caused by the discontinuous function given by equations (11) and (12).

Various Stages of Control

Like treatment efficacy, the stage of development when grasshoppers are controlled affects potential forage saving benefits, and if control is delayed until after egg laying has occurred, future grasshopper populations may also be affected. Because grasshoppers have not been shown to be more susceptible to pesticides at a later maturity, the "optimal" time to treat is as early in the year as possible after an economic infestation has been identified and acceptable levels of hatch have occurred.

Early control of the grasshopper population saves the most forage (Figure 26.4), but as pointed out by Onsager (1984), the 4th instar stage is probably as early as an economic infestation can be identified and control measures implemented. While early treatment is economically best, economic threshold levels are not greatly affected by stage of control if treatment occurs before the adult developmental stage.[5]

Threshold levels were estimated to be 23, 25, and 30 4th instar nymphs/sq. yd. for treatment before the 4th instar, 5th instar, and adult stages, respectively. If control did not take place until halfway through the adult stage when many eggs had already been laid, implementation of a control treatment could not be justified. In this case, forage savings benefits during the year of treatment would be minimal (Figure 26.4) and treatment would not yield positive benefits beyond the year of control.

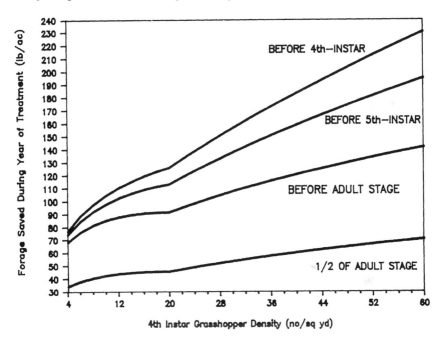

FIG. 26.4 The effect of stage of control on potential forage savings from grasshopper control programs. Discontinuities in the line slopes are caused by the discontinuous function given by equations (11) and (12).

Provided treatment is initiated before the adult stage, threshold levels are about the same. This resulted for two reasons. First, 4th and 5th instar forage consumption rates are relatively low (9 mg and 22 mg/day) when compared to the 53 mg/day eaten by adults. Second, because treatment efficacy was assumed to be the same regardless of treatment timing, second and third year benefits were considered to be equivalent.

CONCLUSIONS

The economic basis of grasshopper control is extremely variable. It varies from where "you cannot afford to control grasshoppers" to where "you cannot afford not to", depending on the potential value of forage saved, treatment cost, treatment life, treatment efficiency, and pest population dynamics. Other factors found to be less important in determining the economics of grasshopper control were discount rate and stage of pest development when treated.

Many factors found to be important in determining the economics of grasshopper control are not well known. Pest population dynamics, for example, were found to be a very important economic consideration, but yet very little can presently be done to predict when grasshopper outbreaks will occur.

Much research is needed before sound economic decisions about grasshopper control strategies can be made. Particularly important is additional information on grasshopper populations (i.e., predicting trends of grasshopper populations), forage losses, forage value, conditions for obtaining high treatment efficacy, and duration of control treatments. Regional variations of these factors, especially between cool and warm season grasses, may be of major importance in determining economic threshold levels.

NOTES

[1]The authors developed their decision rule based on economic damage and the economic injury level (EIL). These two terms, the economic threshold and EIL, have been interchanged repeatedly by various authors. Pedigo et al. (1986) clarifies the semantics and definitions that have been used in historical definition of the threshold concept. We chose to use the term interchangeably in this paper.

[2]Saving the forage may also beneficially affect plant reproduction and vigor, which influences future livestock production.

[3]Onsager (1984) estimated an "average" rangeland grasshopper consumes 9, 22 and 53 mg of forage per GHD in the 4th instar, 5th instar, and adult stages, respectively. Using these estimates, A_1 was computed as follows: first, the 9 mg consumed per 4th instar was divided by 454,000 to convert to pounds consumed by each 4th instar. This number was then multiplied by 4,840, the number of square yards in an acre, which would be the number of grasshoppers over the acre at a density of 1 per square yard. A_2 and A_3 were calculated in a similar manner.

[4]An AUM (Animal Unit Month) is considered to be the amount of forage required by a mature cow or the equivalent for one month.
[5]This is true only if cost of treatment, treatment efficiency and treatment life are held constant. If early treatment meant treatment life would be shortened, for example, then stage of control would obviously affect threshold levels (see Onsager, Chapter 13).

ACKNOWLEDGMENTS

This paper is New Mexico State University Scientific Paper Number SP 272. Funding for this research was provided by the NMSU Range Improvement Task Force and a special grant from the USDA-CSRS integrated pest management program.

REFERENCES

Blickenstaff, C.C., F.E. Skoog, and R.J. Daum. 1974. Long-term control of grasshoppers. J. Econ. Entomol. 67:268-274.

Capinera, J.L. and T.S. Sechrist. 1982. Grasshoppers (Acrididae) of Colorado: identification, biology, and management. Colorado State Univ. Agr. Exp. Sta. Bull. 584S.

Foster, R.N., T.A. Henderson, E.W. Huddleston and R.G. Bullard. 1981. Effect of Triton X-190 and water on malathion and carbaryl related rangeland grasshopper mortalities, 1978. Insect. Acar. Tests 6:136.

Fowler, J.M., L.A. Torell, J.M. Witte, and R.D. Bowe. 1985. Private land grazing transactions in New Mexico 1983-84, implications for state trust land grazing fees. New Mexico State Univ. Range Improvement Task Force (RITF) Report 18.

Hall, D.C., and R.B. Norgaard. 1973. On the timing and application of pesticides. Amer. J. Agr. Econ. 55:198-291.

Hall, D.C., and L.J. Moffitt. 1985. Application of the economic threshold for interseasonal pest control. West. J. Agr. Econ. 10: 223-227.

Headley, J.C. 1972. Defining the economic threshold, p. 100-108. In: Pest control strategies for the future. National Academy of Sciences, Washington, D.C.

Hewitt, G.B., and J.A. Onsager. 1983. Control of grasshoppers on rangelands in the United States--a perspective. J. Range Manage. 36:202-207.

MacCarthy, H.R. 1956. A ten-year study of the climatology of *Melanoplus mexicanus* (Sauss.) (Orthoptera: Acrididae) in Saskatchewan. Can. J. Agr. Sci. 36:445-62.

Mann, R., R.E. Pfadt, and J.J. Jacobs. 1986. A simulation model of grasshopper population dynamics and results for some alternative control strategies. Univ. Wyoming. Sci. Mono. 51.

Mumford, J.D., and G.A. Norton. 1984. Economics of decision-making in pest management. Annu. Rev. Entomol. 29:57-174.

Onsager, J.A. 1983. Relationships between survival rate, density, population trends and forage destruction by instars of grasshoppers (Orthoptera: Acrididae). Environ. Entomol. 12:1099-1102.

Onsager, J. 1984. A method for estimating economic injury levels for control of rangeland grasshoppers with malathion and carbaryl. J. Range Manage. 37:200-203.

Parker, J.R. 1939. Grasshoppers and their control. USDA Farmers' Bull. 1828.

Pedigo, L.P., S.H. Hutchins, and L.G. Higley. 1986. Economic injury levels in theory and practice. Annu. Rev. Entomol. 31: 341-368.

Pfadt, R.E. 1977. Some aspects of the ecology of grasshopper populations inhabiting the shortgrass plains, p. 73-79. In: H. M. Kulman and H. C. Chiang (eds.) Insect ecology-papers presented in the A. C. Hodson lectures. Univ. Minn. Agr. Exp. Sta. Tech. Bull. 310.

Plant, R.E. 1986. Uncertainty and the economic threshold. J. Econ. Entomol. 79:1-6.

Stern, V.M., R.F. Smith, R. Van Den Bosch, and K.S. Hagen. 1959. The integrated control concept. Hilgardia 29:81-101.

Talpaz, H., and I. Borosch. 1974. Strategy for pesticide use: frequency and applications. Amer. J. Agr. Econ. 56:769-775.

Torell, L.A., E.B. Godfrey, and D.B. Nielsen. 1985. Forage utilization cost differentials in a ranch operation: a case study. J. Range Manage. 39:34-39.

United States Department of Agriculture (USDA), Animal and Plant Health inspection Service. 1986. Rangeland Grasshopper Cooperative Management Program, Final Environmental Impact Statement as Supplemented-1986. Washington, D.C.

United States Department of Agriculture (USDA), Animal and Plant Health Inspection Service. 1980. Rangeland grasshopper cooperative management program, final environmental impact statement. Washington, D.C.

United States Department of Agriculture (USDA), Forest Service, and United States Department of the Interior, Bureau of Land Management. 1986. Grazing fee review and evaluation, a report from the secretary of agriculture and the secretary of the interior.

27. ASSESSMENT OF ALTERNATIVE GRASSHOPPER CONTROL STRATEGIES WITH A POPULATION DYNAMICS SIMULATION MODEL

Roger Mann
Department of Agricultural and Resource Economics
Colorado State University, Fort Collins, Colorado 80523

Robert E. Pfadt
Department of Plant, Soil, and Insect Sciences
University of Wyoming, Laramie, Wyoming 82071

James J. Jacobs
Department of Agricultural Economics
University of Wyoming, Laramie, Wyoming 82071

Competition between grasshoppers and cattle for available forage can reduce weight gains of cattle. Producers can respond either by controlling grasshopper population or by changing cattle management activities. This paper presents a grasshopper simulation model which can be used to assess forage damage in relation to grasshopper control. Control options include a protozoan pathogen of grasshoppers, *Nosema locustae*, which may be delivered with or without carbaryl insecticide on wheat bran bait. Malathion insecticide is the conventional control considered. Management strategies include reduced stocking rate or increased feeding of supplemental forage to cattle.

The simulation model presented in this paper uses species-specific relationships of forage preference, destruction, and hatching time. The model is useful for assessing impacts under alternative grasshopper control strategies. It allows instant and/or sustained mortality, and can assess the impacts of control timing on population and forage losses in future years. Population, environment, and grasshopper control strategies can be manipulated to determine their impacts on forage destroyed by grasshoppers. The model is written in FORTRAN IV and a print-out is available in Mann, Pfadt, and Jacobs (1986).

Two relatively independent subcomponents of the population model are isolated and discussed separately: (1) a total population function determines total number of grasshoppers in the population, and (2) population distribution functions determine proportions of the population in the instars and egg life stages.

THE TOTAL POPULATION FUNCTIONS

To start the simulation[1], an initial population of eggs must be specified for the site being analyzed. A hatching distribution function, discussed under "Population Distribution Functions," determines the proportion of eggs which have become grasshoppers. The grasshopper population dies at a constant proportion each day as hypothesized by Onsager and Hewitt (1982)[2]. The population on any day without control is given as:

(1) $P_d = [P_{d-1} + (I \times R_e^d \times HA_d)] \times R$

where: P_d = population on day d
I = initial population of eggs
R_e = daily survival rate of the unhatched eggs, $0 \leq R \leq 1$
d = day of the simulation, an exponent when associated with R_e
HA_d = proportion hatching on day d
R^d = daily survival rate of the total hatched population.

The daily egg survival rate (R_e) must be used in the simulation to account for egg loss due to parasitism, predation, weather and other factors[3].

Equation (1) does not apply on the day of control, because of the mortality caused by the insecticide applied. Population on the day of control is given as:

(2) $P_{d=d_s} = [P_{d_s-1} + (I \times R_e^d \times HA_d)] \times (1-M) \times R$

where: d_s = the day of control
M = the proportion of mortality from the control treatment[4].

Population of grasshoppers on the day after d_s is given by:

(3) $P_{d>d_s} = [P_{d-1} + (I \times R_e^d \times HA_d)] \times R_x$

where: $d>d_s$ = a population of grasshoppers on the day after control and

R_x = the daily survival rate of the total hatched population after the treatment.

R_x can be set at a lower level for a given number of days after control to allow for the residual effect of an insecticide. The mortality induced by *Nosema* is considered by allowing R_x to decrease starting 20 days after the control date[5].

The number of eggs laid during the year is the sum of eggs laid by the population on each day, except that on each day the accumulated number of eggs is multiplied by the daily survival rate of eggs. Eggs laid per day per female adult grasshopper during the oviposition period determines total eggs laid[6,7].

The initial egg population for the next year is determined by multiplying eggs that survived during the year by the overwinter survival rate of the eggs[8].

POPULATION DISTRIBUTION FUNCTIONS

The proportions of the grasshopper population in the egg, nymph, and adult stages are determined by a set of population distribution functions within the model. Progression through the life stages follows an adjusted normal distribution for each instar in each of three species groups.

The model allows for three separate species groups to be considered. This is done to allow for substantially different mean hatching dates for each of the populations[9].

The variance of instar proportions, fecundity rates, percent mortalities, initial populations, and survival rates can be varied by species group. For each population 26 species may be grouped into any of three species groups according to hatching date. Forage loss per instar is available for each species.

The normal distributions allow for overlapping of life stages on any given day. This is important in timing grasshopper control since egg hatching and egg laying can overlap. The population of eggs, which are unaffected by control measures, to a large degree can determine the population remaining after control.

The normal distributions for instars have two parameters (means and variances) which must be specified. The mean date of occurrence of a life stage distribution is determined by the mean hatch date plus the number of days required to progress to the mean of the life stage. Thus, the mean date of occurrence of the first instar is determined by adding the estimated duration of the first instar to the mean hatch date.

The speed of progression through life stages is determined largely from temperatures experienced by the grasshoppers. Regression equations estimated from previously published data

(Putnam 1963) determine the duration of each lifestage. Inspection of Putnam's data strongly suggested that duration of an instar as a function of temperature could be estimated as a log-linear relationship (Table 27.1).

TABLE 27.1
Average duration of grasshopper instars as a function of temperature[a].

	R^{2b}	F
Log(ONE) = 6.04 - .0548 TEM (.35) (.0039)	.98	195
Log(TWO) = 5.98 - .0507 TEM (.53) (.0061)	.95	69
Log(THREE) = 5.34 - .0424 TEM (.57) (.0064)	.92	43
Log(FOUR) = 5.50 - .0427 TEM (.66) (.0075)	.89	33
Log(FIVE) = 5.82 - .0430 TEM (.42) (.0047)	.95	83

N = 6   Observations are means of replicates

Definition of Variables
TEM = Temperature experienced by a group of grasshoppers in °F
ONE = Average duration of first instar in days
(NUMBER) = Average duration of (Number) instar in days

[a] Data from Putnam (1963). Standard errors given beneath each estimated parameter in parentheses.
[b] R^2 is the R^2 value for the equation. F is the F statistic for the equation.

Eggs, five nymphal instars, non-egg-laying adults, and egg-laying adults are considered as life stages. The fifth instar and egg-laying adult stages are separated by the preovipositional period.

Temperature input to the simulation model is provided by a quadratic temperature function for the summer period estimated with average historical data for Sheridan, Wyoming. Daily expected air temperature is adjusted upward to account for the temperature experience of the grasshoppers[10].

The model allows for three groups of grasshopper species. Variance of the distributions is assumed to be the same for each instar in a species group. In the current version of the model, estimated mean hatch dates for each species are drawn from Capinera and Sechrist (1982). A weighted mean hatch date for each species group is calculated. The mean hatch date can also be estimated from published methods (Shotwell 1965) or by some other method applicable to the location under consideration.

Variance of the normal distributions for each group is the root of the product of the sum of squared differences from the weighted mean hatch date multiplied by the species' proportion in the group. Five days are added to the variance of each group to allow for within-group differences.

Estimates of the variance of instar distributions of single species were made. Data collected by R.E. Pfadt in eastern Wyoming were used to approximate variance by the formula[11]:

$$(4) \quad V = \sqrt{\sum_{d=1}^{D} (d-\bar{d})^2 p_d}$$

where: V = estimated variance in days

d = day the population of grasshoppers of a species of instar i was observed

D = last day grasshoppers of instar i were observed

$\bar{d}$ = mean date of occurrence of grasshoppers of instar i, = $\sum_{d=1}^{D} (p_d d)$

p_d = proportion or grasshoppers of instar i observed on day d relative to all days on which grasshoppers of i were observed.

An estimate of the variance of instar distributions for each species group was made and used in the simulation model. The formula for the variance is the same, except that the definition of variables changes as follows:

d = estimated mean hatch date of any species in the group from Capinera and Sechrist (1982)

D = estimated mean hatch date of the latest hatching species in the group

$\bar{d}$ = mean date of occurrence of all grasshoppers in the group

p_d = proportion of grasshoppers of the species in the group.

An addition to this variance of five days was made to account for the variance of the individual species occurring first and last in the species group.

Hatching variance of grasshoppers is affected by seasonal weather factors through the temperature experience of eggs. Physical site characteristics and moisture are also important since they affect the temperatures experienced by eggs and adults. Phenotypic variability in the population could affect the variance. These factors should be considered by any user of the simulation model.

Because the means for instars among the groups are different, the proportions in each of the instars will not necessarily sum to one. Consequently, a loop in the simulation model counts the estimated proportions for each day, sums them, inverts this sum, and multiplies the inverted sum times each instar's "first round" proportions to force the proportions to sum to one:

$$(5) \quad \overline{\overline{PR}}_{di} = (1/ \sum_{i=1}^{I} \overline{PR}_{di}) \times \overline{PR}_{di}$$

where: $\overline{\overline{PR}}_{di}$ = the second round estimate of the proportion in instar i

$\overline{PR}_{di}$ = is the first round estimate.

This correction to the simulation then allows total population Pd to be multiplied by each $\overline{PR}$di to find the population in each instar. Note that second round estimates retain the same relative size to each other as first round estimates since on each day all first round estimates are multiplied by the same constant. The relative proportions of second round estimates are the same as for original first round estimates.

The population of instar i for any day is given by:

$$(6) \quad P_{di} = PR_{di} \times P_d$$

where: Pdi = the population of grasshoppers in instar i on day d.

OTHER POPULATION PARAMETERS

With the population in each instar determined, forage losses are estimated utilizing data supplied by Hewitt and Onsager (1982). Forage lost on a day for a species group equals the sum of losses for all instars, or:

$$(7) \quad F_d = \sum_{i=1}^{I} (L_i \times P_{di})$$

where: F_d = forage lost on day d by the species group
L_i = forage loss per grasshopper in instar i per day.

The forage loss L_i per grasshopper per day is a weighted average. It is determined by multiplying the proportion of a species in a species group by its forage loss per day, and summing over all the species in the species group. Total forage loss is determined by summing over the three species groups.

Diet of the grasshoppers is considered with the formula:

$$(8) \quad PD_j = \frac{PB_j^{x_j}}{PB_j^{x_j} + \sum_{i=1}^{I} PB_i^{x_i}} \quad j \neq i$$

where: PD_j = the proportion of plant type j in the diet, $0 \leq PD_j \leq 1$
PB_j = the proportion of plant type j in the biomass, $0 \leq PB_j \leq 1$
x_j = a power greater than zero. The smaller x_j, the more preferred the plant
i = all other plants, i.e., except j.

The percent of plant types in the biomass must be included in the grasshopper model to determine percentage in the diet[12].

One of the three species groups allows for overwintering grasshoppers. Overwintering grasshoppers appear in the early spring as late instars. Eggs laid by this group hatch in late summer of the next year. For this group, an initial population of overwintered adults, an emergence date, an initial population of eggs carried over from the previous year, and a late summer hatch date must be specified. Each year grasshoppers emerge twice. The mean occurrences of the instars are adjusted so that the early population consists of fifth instars and adults, and the late population consists of first through fourth instars. Eggs laid by the early population, multiplied by the overwinter survival rate, become the source of the late population in the next year. The ending population of first through fourth instar grasshoppers, multiplied by the overwinter survival rate, becomes overwintered fifth instars in the next year.

The model allows for changing the survival rate of grasshoppers several times over the season. The day at which a survival rate is to change must be used in the simulation model along with the survival rate during the period.

A maximum average longevity is specified, causing all grasshoppers to die 100 days after mean hatch date.

RESULTS OF THE SIMULATION

The grasshopper model was used to compare conventional insecticide with wheat bran bait as treatments for controlling grasshoppers. The wheat bran bait was treated with the insecticide carbaryl and/or the grasshopper pathogen *Nosema locustae*. Conventional insecticides and bait treatments were applied aerially.

Variables of major interest determined by the model are eggs laid and forage destroyed. Table 27.2 provides estimates of these variables under grasshopper control scenarios for a population moderately susceptible to wheat bran baits[13]. Susceptibility refers to the extent to which grasshoppers in the population will eat wheat bran, since not all grasshopper species will eat bran.

Scenario 1 gives results for the population with no control. The uncontrolled population has a slight tendency to increase. Egg numbers at the beginning of year two are 6% [(1093 x 0.6)/617.75] more than the initial 617,750 eggs. If contained on a hectare, this population results in a maximum density of about 20 hatched grasshoppers per square meter on June 21 with less than 10 per square meter during August. The population is comparable to the recognized control threshold of eight adults per square yard.

Scenarios 2 to 5 demonstrate the effect of varying the timing of conventional control. With control on July 8 (7/8), some of the eggs are still unhatched and subsequently not affected by control. With control on 8/7, forage losses accumulate through July, and some of the early hatching group have laid eggs by the time control occurs. These eggs cause reinfestation in the next year. Control in the period between 7/8 and 7/23 is found to provide minimum forage loss over three years. With control on 8/15, forage losses are increased three times over losses with control on

7/23.   Figure 27.1 is a graph of a population over time under conventional control.

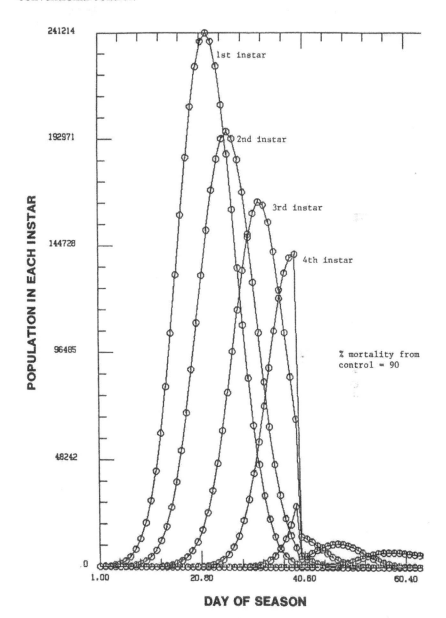

FIG. 27.1 Plot of simulated grasshopper population over time. Mean hatch date is day 20. Day of control is day 40.

TABLE 27.2
Results of grasshopper population dynamics under alternative control scenarios where the population is assumed to be moderately susceptible to wheat bran[a].

Scenario No.	Control Type	Control Date	Eggs/Laying Female /Day	Survival Rates After Control		Percent[c] Mortality from Control	Year	Accumulated Forage Destroyed (kg) to the End of Month					Total Eggs Laid (1000s)	% Forage Loss Saved
				Year 1	Year 2			June	July	Aug.	Sept.	Oct.		
1	no control	-				-	1	28	133	216	241	246	1093	
							2	29	142	231	257	264	1162	
							3	31	150	247	278	283	1236	
2	conventional	7/8	5	.955	.955	95	1	28	59	77	83	84	164	66
							2	4	21	41	48	50	178	74
							3	4	23	45	53	55	194	75
3	conventional	7/23	5	.955	.955	95	1	28	104	111	113	113	155	54
							2	5	20	32	36	37	165	71
							3	5	21	35	39	39	175	76
4	conventional	8/7	5	.955	.955	95	1	28	133	161	162	163	330	34
							2	10	44	67	73	74	356	53
							3	12	46	70	77	78	376	60
5	conventional	8/15	5	.955	.955	95	1	28	133	184	185	185	480	25
							2	14	62	95	104	106	506	43
							3	14	66	101	110	112	534	49
6	conventional	7/23[e]	5	.955	.955	95	1	28	104	105	105	105	101	57
							2	3	13	20	22	22	106	75
							3	3	14	21	23	23	112	81
7	conventional	7/23[f]	5	.955	.955	95	1	55	135	150	155	157	386	36
							2	21	60	87	97	100	421	50
							3	23	66	96	107	110	451	54
8	conventional	7/23	5	.955	.960	95	1	28	104	111	164	167	155	54
							2	6	23	39	45	46	230	69
							3	7	34	59	67	69	341	71
9	carbaryl bran	7/23	5	.955	.955	(66,40)	1	28	115	152	164	167	481	32
							2	12	62	106	121	123	514	43
							3	13	66	115	131	134	550	46

TABLE 27.2 (continued)

Scenario No.	Control Type	Control Date	Eggs/ Laying Female /Day	Survival Rates After Control Year 1	Year 2	Percent[c] Mortality from Control	Accumulated Forage Destroyed (kg) to the End of Month Year	June	July	Aug.	Sept.	Oct.	Total Eggs Laid (1000s)	% Forage Loss Saved
10	carbaryl bran	7/23[h]	5	.950	.950	(66,40)	1	28	113	145	154	156	374	37
							2	7	35	56	61	62	216	57
							3	4	20	32	36	37	125	68
11	carbaryl: Nosema 2:1	6/23	4.5[b]	.945	.950	(58,36)	1	21	83	125	136	137	420	44
							2	10	48	76	84	85	325	56
							3	8	37	59	66	67	252	64
12	carbaryl: Nosema 2:1	6/23	4.0[b]	.940	.945	(58,36)	1	21	81	117	125	126	312	49
							2	7	31	47	51	51	177	65
							3	4	18	27	29	30	100	74
13	carbaryl: Nosema 1:2	6/23	4.0[b]	.935	.945	(40,25)	1	23	92	127	133	134	313	46
							2	7	31	46	49	50	177	64
							3	4	18	26	28	28	100	73
14	carbaryl: Nosema 1:2	6/23	3.5[b]	.930	.945	(40,25)	1	23	90	120	125	125	233	49
							2	6	23	34	37	37	132	68
							3	3	13	19	21	21	74	77
15	Nosema	6/23	3.0[b]	.925	.945	(0,0)	1	28	116	148	152	153	235	38
							2	6	23	33	35	35	132	63
							3	3	13	19	20	20	74	74
16	Nosema	6/23[f]	3.0[b]	.915	.945	(0,0)	1	28	110	133	135	135	156	45
							2	3	13	18	19	19	65	70
							3	1	5	7	8	8	27	80

[a] Average maximum longevity of 100 days in effect, egg mortality of 2% per day from mean hatch date for 100 days, 60% total survival of eggs over winter assumed. Preoviposition period = 20 days.

[b] In the Nosema scenarios egg laying rates revert back to five/day in years after control.

[c] Mortality of hatched population, not eggs, on day of control.

[d] End of the year forage loss relative to the no control scenario.

[e] Only 80% survival per day from 7/23 to 8/7 to allow for residual effect.

[f] Variance of hatch date and instar distributions increased by 10 days.

[g] Numbers in parentheses are predicted percent mortality from eating carbaryl in the wheat bran for the early and late hatching groups, respectively.

[h] Also allows for 50% overwintering egg survival rate and daily survival rate of eggs over summer of .9775.

Scenario 6 demonstrates the impact of an insecticide with a residual effect. Daily survival rate is reduced from .955 to .80 for 15 days after control (see footnote e). Grasshoppers hatching after control are subject to this reduced survival rate and lay fewer total eggs. Forage losses in the next year are reduced by another third due to this residual effect.

Scenario 7 allows for an increase in the variance of instar distributions. With an increase in the dispersion of instars over the season, substantially poorer control is attained.

The increase in variance causes more egg-laying days to be lost due to maximum longevity. More of the egg-laying females are older and consequently more die. In order to compensate for this unwanted effect, eggs laid in year one, two, and three, and forage losses in the next year have been multiplied by 1.23, 1.52 and 1.87, respectively. This causes the population to grow at the same rate as the other conventional control scenarios.

Conventional control scenario 8 allows the survival rate of hatched grasshoppers to increase by 0.5% in years after control. Mortality and out-migration of predators and parasites could allow for increased survival of grasshoppers in years after control. With about 10% fewer grasshoppers dying daily in years two and three, the population grows faster and more forage is lost in comparison to Scenario 3.

Results of the conventional control scenarios indicate that approximately 80% of forage losses over three years can be saved by conventional control with 95% mortality. Poor control timing, a dispersed population, or enhanced survival rates of surviving grasshoppers reduce both the efficacy of control and the reduction of forage losses.

Results of the wheat bran controls are given in Scenarios 9-16. Scenario 9 provides results for a population treated with carbaryl bran bait. Mortality of hatched grasshoppers for the two groups was calculated according to the group's susceptibility to wheat bran (Onsager 1982).

Scenario 10 allows for a decreased survival rate of grasshoppers after the carbaryl treatment. Since bran baits do not affect predators, a decreased survival rate can be explained in part by increased predator to prey ratio due to bait applications.

Scenarios 11 - 16 allow for decreased fecundity (eggs/laying female/day) and survival rates due to *Nosema* on wheat bran. A "reasonable" and "optimistic" scenario is allowed for each treatment. *Nosema* treatments are applied on 6/23 because the simulated population is primarily third instars on this date, which is the desired stage for treatment. These scenarios are all somewhat optimistic in that they are well timed, bran consumption by these susceptible species is favorable and the disease becomes well established in the population and carries into the next year.

Forage loss saved in the control year under the *Nosema* treatments ranges from 38 - 49%. Although conventional Scenarios 2 and 3 provide better control, the late conventional treatments save only 34 and 25% of potential forage loss. A well-timed *Nosema* treatment may control better than a late conventional treatment.

Well-timed conventional treatments save 70% of potential forage loss by the end of the third year. The more optimistic *Nosema* treatments attain similar control. At the end of the second year, none of the *Nosema* treatments have attained control comparable to Scenarios 2 and 3. Even with the very optimistic *Nosema* scenarios assumed here, some reduction in survival rate probably must be sustained in years after control to give long-term control comparable to a well-timed conventional treatment.

Scenario 7 demonstrated that a greater dispersion of instars is detrimental to conventional control. Age and species dispersion may enhance communicability of *Nosema* leading to higher infection rates and mortality. *Nosema* could have a comparative advantage in this situation, but given the parameters used here, conventional control would appear to be more efficient.

SUGGESTIONS FOR FURTHER RESEARCH

Many of the initializing parameters might themselves be affected by weather, grasshopper density, and other factors. Basic and applied research is needed to obtain better estimates of population parameters. The daily survival rate might be affected by many other factors. As forage losses are sensitive to daily survival rate, this would be a good area for further research. Egg production per female also varies as a function of weather, diet, and other conditions. Development of a method for predicting overwintering survival rates would foster better multiple year simulations. In the present model, survival rates and egg laying rates are not dependent on weather, grasshopper density, or diet composition. Predator to prey dynamics are not explicitly included. Refinement of population dynamics would allow for better estimates of forage losses.

NOTES

1. Data on egg densities are from Lavigne and Pfadt (1966). Spring egg surveys from two locations over five years show an average of 34,238 viable egg pods per acre (.4047 hectare) with a range of 7,841 to 54,014. The three major species in the complex were *Aulocara elliotti*, (Thomas), *Cordillacris occipitalis*,

(Thomas), and *Ageneotettix deorum*, (Scudder). Egg pods of these grasshoppers contain eight, three and four eggs per pod, respectively (Onsager and Mulkern 1963). Allowing each species to make up a third of the population, we estimate about five eggs per pod and 39,000 to 270,000 eggs per acre. A population of 250,000 eggs per acre (617,750 per hectare) is used in the simulations resulting in a maximum density of adult grasshoppers of about 8/sq. yd.

2. Estimated natural mortality rates are provided. The average daily survival rates of rangeland grasshopper species based upon three years of data are reproduced in Table 27.3.

Another study found that for *M. sanguinipes*, total survival of nymphs to the adult stage ranged from 48 - 89% with a nymphal period of 42 - 55 days (Pfadt 1949, and Pfadt and Smith 1972). These data suggest survival rates in excess of 98% per day. Data published by Pickford (1960) for *M. sanguinipes* also suggest nymphal survival rates in excess of 97%.

TABLE 27.3
Natural mortality rates of grasshoppers (from Onsager and Hewitt 1982).

Species[a]	Average Nymphal Survival		Average Adult Survival	
	Rate per day (percent)	Range (percent)	Rate per day (percent)	Range (percent)
A. coloradus	93	89-95	94	92-95
A. deorum	94	91-97	94	91-96
A. elliotti	90	87-92	83	60-97
M. infantilis	94	89-97	94	93-96
M. packardii	92	87-95	--	---
M. sanguinipes	94	93-96	95	94-96

[a]*Amphitornus coloradus* (Thomas), *Ageneotettix deorum* (Scudder), *Aulocara elliotti* Thomas, *Melanoplus infantilis* Scudder, *Melanoplus packardii* Scudder, *Melanoplus sanguinipes* (Fabricius)

Data collected in eastern Wyoming by R.E. Pfadt were utilized to make estimates of survival rates. Observations (n=247) made over many locations, species, and years were used to determine a grand mean daily survival rate of 98%. No significant difference was found between mean survival rates of the various locations, species, and years. However, the highly unbalanced nature of the data set made hypothesis testing difficult. In the simulation, a daily survival rate of 95.5% was found to result in a fairly stable population over the years.

3.   However, no conclusive data are available on egg survival rates. An overwintering survival rate of 50% suggest daily overwinter survival rates in excess of 99%. It is expected that factors affecting mortality will be more active during summer than winter. Consequently, an egg survival rate of 98% per day is used in the simulation from hatch date to day of maximum longevity for each species group.

4.   In the journal, *Insecticide-Acaricide Tests* (Sorensen 1977-1981), many trials of the efficacy of insecticides on rangeland grasshoppers can be found (see Table 27.4). In some of the trials in Table 27.4, densities were sampled twice or more after treatment. This allows for the residual effects of insecticides on survival rates to be considered. However, since both hatch and mortality occur during a period, survival rates in the insecticide treated plots must be compared to the control plot.

Experiment No. 237, Vol. 6 is used to estimate the daily survival rates of grasshoppers. Daily survival rates two to 12 days posttreatment were 1.03, .97, .94 and .94 for three dosages of Orthene and for one dosage of malathion, respectively, compared to 1.03 for the control.

In No. 230, Vol. 4 daily survival rates from three to seven days posttreatment were 0, .78 and .75 for three rates of Orthene as compared to .98 for malathion. This shows that Orthene has a greater residual effect than malathion.

The method of estimating mortality of grasshoppers from the carbaryl component of wheat bran baits is outlined by Onsager et al. (1980). The method considers the percent of the population which will eat wheat bran, allows for molting non-feeding grasshoppers and considers the efficiency of wheat bran in terms of numbers dying from eating consecutive flakes. The estimated mortalities for each species complex is provided in Table 27.5.

5.   Estimates of reduction in the daily survival rate due to *Nosema locustae* infection after treatment are made with data from several *Nosema* field trials where usable results were obtained. Survival rates for a period between observations were determined by the formula:

$$(9) \quad R_x = e^{(\frac{\ln S}{D})}$$

Where: R_x = the daily survival rate after treatment

S = proportion surviving during the entire period
$0 \leq S \leq 1$
D = the number of days in the period.

Survival rates were compared between *Nosema* and control plots to determine the reduction in survival rate attributable to *Nosema*.

TABLE 27.4.
Efficacy of insecticides for grasshoppers (from Sorensen 1977-1981).

Source	Chemical	Location	% Decrease
Vol. 4 No. 230	Orthene	Bozeman, Mt	97.5-100
Vol. 4 No. 230	Malathion	Bozeman, MT	98.7
Vol. 2 No. 178	Dimethoate	Bozeman, MT	91.4-94.8
Vol. 3 No. 185	Malathion	Bozeman, MT	88.5
Vol. 5 No. 310	Malathion	Clay Ctr., NE	72
Vol. 5 No. 368	Malathion	Bozeman, MT	95
Vol. 5 No. 369	Carbaryl	Bozeman, MT	35.5
Vol. 6 No. 236	Carbaryl	Lovington, NM	85.5-96.3
Vol. 6 No. 236	Malathion	Lovington, NM	77.0-97.5
Vol. 6 No. 237	Orthene	Wheatland, WY	64-90
Vol. 6 No. 237	Malathion	Wheatland, WY	95
Vol. 6 No. 238	Orthene	Vernal, UT	92-94

From the Platte County, Wyoming, trial (Pfadt 1982), an average reduction in the daily survival rate attributable to *Nosema* for two species(*A. elliotti* and *M. occidentalis* Thomas) was calculated to be 2.4%.

TABLE 27.5

Description of some grasshopper populations in Wyoming with calculation of expected mortalities from carbaryl for three dosages of carbaryl bait.[a]

Population and Susceptibility	Source	% in Hatching Groups[b]		% That will eat Wheat Bran Baits	Predicted Mortality From Carbaryl		
					Carbaryl:*Nosema* 1:2	Carbaryl:*Nosema* 2:1	Carbaryl Full Dose
1 very susceptible	Pfadt (1977) Douglas	early	85	91	45	65	73
		late	15	80	39	57	64
2 non susceptible	Pfadt (1977) Average of four other populations	early	92	43	21	31	34
		late	8	26	13	18	21
3	Pfadt (1979) Sheridan North treatment	early	85	87	43	62	70
		late	15	53	26	38	43
4 moderately susceptible	Pfadt(1979) Sheridan South treatment	early	86	82	40	58	66
		late	14	50	25	36	40
5	Pfadt (1979) Acme *Nosema* treatment	early	93	84	41	60	67
		late	7	43	21	31	34
6	Pfadt(1982) Platte County	early	77	90	44	64	72
		overwintering	22	0[c]			

[a] The method of calculating mortalities is outlined in Onsager et al. (1980). E-values of .055, .111, and .167 were assumed for the three dosages with 15 bran flakes available per grasshopper.

[b] The early hatch group consists of species hatching from day 15 to 36 and the late group from 46 to 76 [from Capinera and Sechrist (1982)] where day 15 = May 15.

[c] Non-susceptible because overwintering species would not be hatched at control time.

For the 1979 trial at Sheridan, Wyoming (Pfadt 1979), the calculated seasonal reductions attributable to *Nosema* bran were 1, 2 and 3% for the Gomphocerinae, Locustinae, and Melanoplinae, respectively. For carbaryl-*Nosema* baits of 2:1 and 1:2, for the same respective subfamilies, reductions attributable to the *Nosema* component were 4, 0.2, and 1% and 1.4, -3.0, and 2%, respectively. From two to five weeks posttreatment, survival rates were reduced by 2, 5, and 9% for the straight *Nosema*, 1:2 and 2:1 bran baits, respectively. From five to eight weeks after treatment, respective survival figures were 1.3, -3 and -1% relative to the control. For the period from eight to 11 weeks the *Nosema* survival rate is reduced by 4% relative to the control.

For a field trial in Montana (Henry and Onsager 1982). reductions in daily survival rate attributable to *Nosema* were 4% for both a high and low dosage.

Henry (1971) reports that reductions in density attributable to *N. locustae* in a field trial were 33.5, 23, and 34.5% at four, five, and six weeks post application, respectively. Allowing two weeks for the *Nosema* to take effect, these figures translate to reductions in daily survival rates of 3, 1.2, and 1.3%, respectively.

Data on reduced survival rates of grasshoppers due to *Nosema* in years after control are limited. Data from Henry and Onsager (1982) indicate a reduction in survival of .2 and .6% relative to the control in the year after control for two concentrations of *Nosema*. Data from Sheridan (Pfadt 1979) indicate a reduction in early season (5/28-7/11) daily survival rate relative to the "control" for the year after application of about 1.5%. Ewen and Mukerji (1980) hatched eggs from treated populations and found that about 10% had trace levels of *Nosema* infection.

In the simulation, survival rates are decreased by 1, 2, and 3% in the carbaryl:*Nosema* 2:1, 1:2, and straight *Nosema* scenarios, respectively, with larger reductions in the "optimistic" scenarios. Survival rate in years after control is allowed to be a maximum of 1% less than the standard 95.5%.

6. Pickford (1960) found *M. sanguinipes* females to lay from 0.5 to 4.5 eggs per day. Pfadt (1949) found that *M. sanguinipes* laid from 0.25 to 6 eggs per day depending on the age of the female. Young adult females laid 3.6 eggs per day, while in the late summer to early fall only 1.2 eggs were laid per female per day. Data for *M. sanguinipes* drawn from several studies indicate an average of 200 eggs per female. With an average adult longevity after preoviposition of about 40-65 days, three to five eggs per female per day is estimated. Fecundity is affected by age, plant foods eaten, temperature, grasshopper density and a wide variety of other factors. Oviposition rates of over eight eggs per female per day have been attained and could conceivably

occur under good conditions in the field. In the simulation, five eggs/adult female during the oviposition period are used.

Data from Henry and Oma (1981) provide a strong indication of reduced fecundity from grasshoppers with *Nosema* infection. With a 100% infection rate, infected *M. sanguinipes* females laid 33 - 62% fewer eggs than control females. *Melanoplus differentialis* (Thomas) females laid 30 - 100% fewer eggs.

Henry and Oma (1974) found maximum infection rates in a *Nosema* field trial to be about 30%. Onsager et al. (1981) found infection rates at six weeks to be 12.7, 8.1, and 4.0% for full dose, two-thirds, and one-third doses of *Nosema*, respectively. Henry (1971) found average infection rates over six concentrations of spores to be 3, 15.5, 27.5, and 29.0% at three, four, five, and six weeks posttreatment, respectively. Henry and Onsager (1982) found *Nosema* infection rates resulting from treatment dosages of 2.1×10^9 and 2.1×10^8 spores per ha to be 24.5 and 12.2%, respectively. In combination with results of Henry and Oma (1981) where 100% of the grasshoppers were infected, these infection data indicate an expected decline in fecundity of anywhere from 10 - 40% in field trials.

Ewen and Mukerji (1980) reported that egg deposition for a *Nosema*-treated population was 81.2% lower compared to the control population. With reported mortality data, the treated population had a reduced survival rate of about 1% relative to the control. The simulation model estimated that the daily egg laying rate of the treated population reduced by 60% to achieve the 81.2% reduction in eggs laid after accounting for the reduced survival rate of the population.

In Sheridan, population data from a *Nosema*-treated population indicated that the *Nosema*-treated population either had a lower fecundity rate or a lower overwintering survival rate. Populations on 7/11 for all *Nosema* treatments were 8 - 15% of pretreatment levels while the carbaryl and check treatments were 31 - 75% of pretreatment levels. In the simulation, a maximum reduction in fecundity of 30% is allowed due to *Nosema*.

7.   Pfadt (1949) summarized experimental results on *M. sanguinipes* and reported preoviposition periods of seven to 25 days in a laboratory experiment, and 21 to 42 days in an outdoor, cool climate. A field study in Montana found an average preoviposition period of 20 days, ranging from 15 to 45 (Hastings and Pepper 1964). In the simulation, a period of 20 days is used.

8.   Lavigne and Pfadt (1966) reported an average overwintering survival rate of 60% as indicated by spring egg surveys. Parker (1930) found a maximum survival rate in the laboratory of 72%, but no predators were present. This maximum survival rate was reduced considerably by high temperatures and low humidity. Soil moisture and winter extremes may also

increase mortality. In the simulation, an overwintering survival rate of 60% is used.

9. Data on the composition of species complexes are drawn from Pfadt (1977). The moderately susceptible population is drawn from Pfadt (1979) for the Sheridan South treatment. The early hatching group makes up 92% of the total and consists of 16% *Aeropedellus clavatus* (Thomas), 26% *A. deorum*, 19% *M. infantilis*, 20% *M. confusus* (Scudder). and 17% *M. sanguinipes*. The late group is primarily *Metator pardalinus* (Saussure) and *Melanoplus dawsoni* (Scudder).

10. Parker's 1930 work provides some data on the temperatures experienced by grasshoppers. Climbing to escape heat occurred at an average soil temperature of 112.6°F and 107.4°F for *M. sanguinipes* and *Camnula pellucida* (Scudder), respectively. Soil temperatures at which other activities occurred often ranged from 70° - 100°F. Temperature records kept from June 25 to July 21 show that soil surface temperatures usually approached or exceeded 120°F at noon and rarely fell below 70°F during the day.

The actual temperature experience by grasshoppers is affected by solar radiation, wind, behavior of the grasshoppers and other factors. Pepper and Hastings (1952) note that direct solar radiation, reflected heat, heat radiation, and heat conduction all should raise body temperatures of grasshoppers. They note that "increases in body temperatures over that of the air, from 0.5-11°C could be accounted for by the action of direct solar radiation." In the simulation model, average air temperature is increased by 15°F to account for the temperature experience of the grasshoppers.

11. For *Aulocara elliotti* in 1968 and 1976, the variance of the instar distribution was estimated to range from five to nine days. In 1968, this variance appeared to increase with each progressing instar. However, the simulation model does not incorporate this effect. In the simulation, the estimated variance for each species group depends primarily on the mean hatch date for each species in the group.

12. Development of the preference parameter "x" to which proportions in the biomass are raised was developed from literature in a subjective manner (Pfadt and Lavigne 1982, Fry et al. 1978, Ueckert et al. 1971, Ueckert and Hansen 1971, Mulkern et al. 1969, Mulkern 1979, Ueckert 1968, Marks 1966, and Joern 1979).

13. Total population of 617,750 eggs is evaluated at mean hatch date. Of those, 531,265 are in the early group and 86,485 are in the late group with mean hatch dates of June 10 and June 30, respectively and with variances of hatch date of 13 and 13.5 days, respectively. About 68% of hatch of the group occurs in 13

days. Species composition of the total population equals 13% *A. clavatus*, 21% *A. deorum*, 17% *M. confusus*, 10% *M. infantilis*, and 15% *M. sanguinipes*, with small proportions in seven other species.

REFERENCES

Blickenstaff, C.C., F.E. Skoog, and R.J. Daum. 1974. Long term control of grasshoppers. J. Econ. Entomol. 67:268-274.

Capinera, J.L., and T.S. Sechrist. 1982. Grasshoppers (Acrididae) of Colorado. Identification, biology and management. Colorado State Univ. Exp. Sta. Bull. 584S.

Ewen, A.B., and M.K. Mukerji. 1980. Evaluation of *Nosema locustae* as a control agent of grasshopper populations in Saskatchewan. J. Invertebr. Pathol. 35:295-303.

Fry, B., A. Joern, and P.L. Parker. 1978. Grasshopper food web analysis: use of carbon isotope ratios to examine feeding relationships among terrestrial herbivores. Ecology 59:498-506.

Gilbert. B.J. 1973. Flow of herbage to herbivores. Unpublished M.S. Thesis, Colorado State Univ.

Gyllenberg, G.A. 1974. A simulation model for testing the dynamics of a grasshopper population. Ecology 55:645-650.

Hastings, E., and J.H. Pepper. 1964. Population studies on the big-headed grasshopper *Aulocara elliotti*. Ann. Entomol. Soc. Amer. 57:216-220.

Henry, J.E. 1971. Experimental application of *Nosema locustae* for control of grasshoppers. J. Invertebr. Pathol. 18:389-394.

Henry, J.E.. and E.A. Oma. 1974. Effect of prolonged storage of spores on field applications of *Nosema locustae* against grasshoppers. J. Invertebr. Pathol. 23:371-377.

Henry, J.E. and E.A. Oma. 1981. Pest control by *Nosema locustae*, a pathogen of grasshoppers and crickets, p. 573-586. In: H.D. Burges (ed.) Microbial control of pests and plant diseases 1970-1980. Academic Press, New York.

Henry, J.E., and J.A. Onsager. 1982. A large-scale test of control of grasshoppers on rangeland with *Nosema locustae*. J. Econ. Entomol. 75:31-35.

Hewitt, G.B. 1979. Hatching and development of rangeland grasshoppers in relation to forage growth, temperature and precipitation. Environ. Entomol. 8:24-29.

Hewitt, G.B. and J.A. Onsager. 1982. A method for forecasting potential losses from grasshopper feeding on northern mixed prairie forages. J. Range Manage. 35:53-57.

Joern, A. 1979. Feeding patterns in grasshoppers (Orthoptera: Acrididae): factors influencing diet specialization. Oecologia 38:325-347.

418

Lavigne, R.J., and R.E. Pfadt. 1966. Parasites and predators of Wyoming rangeland grasshoppers. Wyoming Agr. Exp. Sta. Sci. Monogr. 3.

Mann, R., R.E. Pfadt, and J.J. Jacobs. 1986. A simulation model of grasshopper population dynamics and results for some alternative control strategies. Wyoming Agr. Exp. Sta. Sci. Monogr. 51.

Marks, W.D. 1966. Food habits of southeastern Wyoming rangeland grasshoppers as determined by analysis of crop contents. Unpublished M.S. Thesis, Univ. Wyoming.

Mitchell, J. 1973. A model of food consumption by three grasshopper species as determined by differential feeding trials. Unpublished Ph.D. dissertation, Colorado State Univ.

Mulkern, G.B. 1979. Population fluctuations and competitive regulations of grasshopper species. Trans. Am. Entomol. Soc. 106:1-41.

Mulkern, G.B., K.P. Pruess, H. Knutson, A.F. Hagen, J.B. Campbell, and J.D. Lambley. 1969. Food habits and preferences of grassland grasshoppers of the North Central Great Plains. North Dakota Agr. Exp. Sta. Bull. 481.

Onsager, J.A. 1982. A method for estimating economic thresholds for control of rangeland grasshoppers with malathion and carbaryl. Unpublished manuscript.

Onsager, J.A., and G.B. Hewitt. 1982. Rangeland grasshoppers: average longevity and daily rate of mortality among six species in nature. Environ. Entomol. 11:127-133.

Onsager, J.A., and G.B. Mulkern. 1963. Identification of eggs and egg-pods of North Dakota grasshoppers (Orthoptera: Acrididae). North Dakota Agr. Exp. Sta. Bull. 446.

Onsager, J.A., J.E. Henry, and R.N. Foster. 1980. A model for predicting efficacy of carbaryl bait for control of rangeland grasshoppers. J. Econ. Entomol. 73:726-729.

Onsager, J.A., N.E. Rees, J.E. Henry, and R.N. Foster. 1981. Integration of bait formulations of *Nosema locustae* and carbaryl for control of rangeland grasshoppers. J. Econ. Entomol. 74:183-187.

Parker, J.R. 1930. Some effects of temperature and moisture upon *Melanoplus mexicanus* Saussure and *Camnula pellucida* Scudder (Orthoptera). Montana Agr. Exp. Sta. Bull. 223.

Pepper, J.H., and E. Hastings. 1952. The effects of solar radiation on grasshopper temperatures and activities. Ecology 33:96-103.

Pfadt, R.E. 1949. Food plants as factors in the ecology of the lesser migratory grasshopper *Melanoplus mexicanus* (Sauss.). Wyoming Agr. Exp. Sta. Bull. 290.

Pfadt, R.E. 1977. Some aspects of the ecology of grasshopper populations inhabiting the shortgrass plains. p. 73-79. In: H.M. Kulman and H.C. Chiang, (eds.) Insect Ecology, papers presented in the A.C. Hodson Lectures, Minnesota Agr. Exp. Sta. Tech. Bull. 310.

Pfadt, R.E. 1979. 1979 Grasshopper sampling data from the Sheridan, Wyoming, optimum grasshopper test management trial. Wyoming Agr. Exp. Sta.

Pfadt, R.E. 1982. 1981 Platte County Wyoming optimum grasshopper pest management trial. Wyoming Agr. Exp. Sta.

Pfadt, R.E., and D.S. Smith. 1972. Net reproductive rate and capacity for increase of the migratory grasshopper, *Melanoplus sanguinipes* (F.). Acrida 1:149-165.

Pfadt, R.E., and R.J. Lavigne. 1982. Food habits of grasshoppers inhabiting the Pawnee site. Wyoming Agr. Exp. Sta. Sci. Monogr. 42.

Pickford, R. 1960. Survival, fecundity and population growth of *Melanoplus bilituratus* (Wlk.) (Orthoptera: Acrididae) in relation to date of hatching. Can. Entomol. 92:1-10.

Putnam, L.G. 1963. The progress of nymphal development in pest grasshoppers of western Canada. Can. Entomol. 95:1210-1216.

Randell, R.L., and M.K. Mukerji. 1974. A technique for estimating hatching of natural egg populations of *Melanoplus sanguinipes*. Can. Entomol. 106:801-812.

Shotwell, R.L. 1965. Forecasting the hatching period of grasshoppers from weather data. ARS 33-102, USDA, Washington, D.C.

Sorensen, K.A. (ed.) 1977-1981. Insecticide and Acaricide Tests, Vol.2-6. Entomol. Soc. Am. College Park, Maryland.

Ueckert, D.N. 1968. Seasonal dry weight composition in grasshopper diets on Colorado herbland. Ann. Entomol. Soc. Am. 61:1539-1544.

Ueckert, D.N., and R.M. Hansen. 1971. Dietary overlap of grasshoppers on sandhill rangeland in northeastern Colorado. Oecologia 8:276-295.

Ueckert, D.N., R.M. Hansen, and C. Terwilliger, Jr. 1971. Influence of plant frequency and certain morphological variations on diets of rangeland grasshoppers. J. Range Manage. 25:61-65.

FUTURE DEVELOPMENTS

28. RANGELAND PEST MANAGEMENT: PROBLEMS AND PERSPECTIVES

John L. Capinera
Department of Entomology
Colorado State University, Fort Collins, Colorado 80523

There are many levels of IPM that might be attempted on rangeland. Integration may occur narrowly, as through selection of two or more compatible pest suppression tactics for control of a single pest organism. Integration may also occur broadly, as through selection of rangeland management practices which facilitate economic livestock production, maximum rangeland forage production, and minimal pest pressure. The latter form of pest management, *interdisciplinary integrated pest management*, is not readily achievable in most cases. Often we know very little regarding the various costs and benefits of individual pest management options, let alone combinations of options spanning several traditional disciplines (e.g., range science, entomology, weed science, economics). Some of the major problems and prospects confronting range managers who might attempt rangeland IPM are:

1. A number of grazing management systems exist, and there is not complete agreement on which are most favorable for the shortgrass region (see Chapter 2). Animal performance, financial return, and long-term range condition are some of the major factors to consider when selecting grazing management systems.

2. Grass forage quality, whether it is measured by primary nutrient levels (Chapter 3) or secondary chemistry (Chapter 4), will affect animal performance, and ultimately financial return. Some pests also may be influenced by host plant quality. Forage quality, particularly the role of secondary plant compounds, is inadequately appreciated.

3. Grasses are rather tolerant of grazing, and have a surprisingly high capacity to compensate for herbivory. The

quality of forage may be enhanced by grazing (Chapter 5). Quantitative assessment of herbivory, especially by small, mobile pests, is difficult (Chapter 6). As previously noted, qualitative changes often are unknown.

4. Pricklypear cactus and broom snakeweed are important weed species in the shortgrass prairie (Chapters 7 and 8). Snakeweed, but not pricklypear, increases with overgrazing; both are affected by weather. Removal of snakeweed, and perhaps pricklypear, can result in greater forage availability. The economic basis for controlling these weeds must be examined carefully, however, and seems more favorable in the case of snakeweed.

5. Mormon cricket has a history of spectacular outbreaks throughout the western United States (Chapter 9). Its actual economic impact is not especially great during periods of adequate precipitation. It is a highly visible insect, however, and has a somewhat undeserved reputation as a serious pest of rangeland grasses.

6. Despite extensive research on the biology of rangeland grasshoppers (Chapters 10 and 11), much remains unknown. Species-specific differences tend to be overlooked, and information about crop-infesting species sometimes is applied erroneously to rangeland species. Regional differences in biology exist, and the relative importance of biotic mortality agents remains obscure. Behavioral and ecological components of grasshopper biology should be further exploited to enhance development of new suppression measures.

7. Grasshoppers have long been serious pests of rangeland, but insecticides have proved useful for reducing economic loss in recent times (Chapter 12). Much can be done to optimize insecticide use. Of paramount importance are careful sampling of the population and timing of insecticide application to assure that grasshoppers have hatched, yet have not begun oviposition. This strategy is more feasible in some regions than others, and is dependent on the species complex present (Chapter 13). Grasshopper sampling procedures should be dynamic, varying with objectives of the project and insect density (Chapter 15).

8. Microbial pathogens offer considerable promise for grasshopper control. Viruses, bacteria, and fungi should be studied further, but some genetic modification of these

pathogens may be necessary before they can be practically utilized (Chapter 14).

9. Range caterpillar sometimes is the dominant insect pest in the southern shortgrass prairie. Its abundance seems to be limited by abiotic factors, but biotic mortality agents from the Mexican population should not be overlooked as potential regulatory agents (Chapter 16). An effective IPM program based on redefinition of the economic threshold, sampling, and low-cost chemical control was developed (Chapter 17). This program may serve as a model for other similar efforts in the future.

10. A number of belowground invertebrates may be pests on rangeland. The most apparent are harvester ants (Chapter 18), principally because they forage aboveground. However, a variety of micro- and macroarthropods (Chapter 19), and nematodes (Chapter 20), have important roles in the shortgrass ecosystem. It has long been suspected that macro- and microarthropods, and nematodes, may limit plant productivity, but some members of these groups are beneficial or are important in nutrient cycling.

11. Among vertebrates inhabiting rangelands, prairie dogs are usually considered to be detrimental while birds are beneficial. Re-examination of the role of prairie dogs (Chapter 21) suggests that they may not be economic to control, and that they have valuable roles in enhancement of wildlife diversity and for sport (hunting). Birds are clearly beneficial in that they consume enormous quantities of insects, particularly grasshoppers (Chapter 22). It is not apparent, however, how important they are at *regulating* grasshopper populations.

12. Modeling offers opportunities to overcome some of the problems inherent in rangeland research and management: extensive areas, long response times, ecological complexity, and low values per unit area. Modeling approaches are particularly useful in integrating different disciplines, in synthesizing data, and in providing a conceptual framework (Chapter 23). New modeling approaches may be necessary, however, in that there are substantive differences between rangeland and agroecosystems. Phenology, population, and economic models (Chapters 24, 25, 26, and 27) offer unique and valuable insight into rangeland pest management strategies. Critical timing of grasshopper control, for example, is greatly enhanced when grasshopper consumption

and development can be predicted accurately. Questions about the economic basis for pest suppression are often best answered through modeling because it is impossible to determine experimentally the outcome of the myriad combinations of pest, production, weather, and financial variables. Lastly, gaps in the extant data base become apparent when models are being constructed, so models serve to guide research efforts. Unfortunately, modeling exercises in rangeland IPM are still relatively rare, despite their potential to facilitate *interdisciplinary* IPM.

This book serves to document not only what is known about rangeland IPM, but perhaps more importantly, what is not known. It should be readily apparent to the reader that data on such fundamental issues as economic thresholds for nearly all pests are absent. The *perception* of problems may not be borne out by reality, and the traditional solutions to pest problems may not be economically feasible. All too often, important economic decisions are being made on a political or historical basis. This is increasingly difficult to justify in the context of reduced profits in livestock production, particularly when the multiple-use nature of America's rangelands is considered. Hopefully this book will serve as a catalyst for re-examination of traditional perspectives.

INDEX